DATE DUE

APR 1 9 2008			

Photosynthetic plant cells are compartmentalised into subcellular organelles such as chloroplasts, mitochondria, peroxisomes, the cytosol and the vacuole. Although this compartmentation serves to isolate particular functions to particular subcellular locations, the successful metabolic activity of the cell is actually dependent on a controlled and coordinated interaction between these organelles. In this book, leading scientists have contributed reviews of current research on the interaction of organelles in processes such as C3, C4, C3–C4, and CAM photosynthesis, photorespiration, substrate and protein transport, respiration, lipid metabolism and organelle biogenesis. The result is a comprehensive, state-of-the-art volume which provides a rich source of reference and information for plant biochemists and their students.

SOCIETY FOR EXPERIMENTAL BIOLOGY
SEMINAR SERIES : 50

PLANT ORGANELLES
COMPARTMENTATION OF METABOLISM IN
PHOTOSYNTHETIC CELLS

SOCIETY FOR EXPERIMENTAL BIOLOGY SEMINAR SERIES

A series of multi-author volumes developed from seminars held by the Society for Experimental Biology. Each volume serves not only as an introductory review of a specific topic, but also introduces the reader to experimental evidence to support the theories and principles discussed, and points the way to new research.

PLANT ORGANELLES
COMPARTMENTATION OF METABOLISM IN PHOTOSYNTHETIC CELLS

Edited by

Alyson K. Tobin

Department of Cell and Structural Biology, University of Manchester

CAMBRIDGE
UNIVERSITY PRESS

Published by the Press Syndicate of the University of Cambridge
The Pitt Building, Trumpington Street, Cambridge CB2 1RP
40 West 20th Street, New York, NY 10011–4211, USA
10 Stamford Road, Oakleigh, Victoria 3166, Australia

First published 1992

Printed in Great Britain by
Redwood Press Limited, Melksham, Wiltshire

A catalogue record for this book is available from the British Library

Library of Congress cataloguing in publication data

Plant organelles: compartmentation of metabolism in photosynthetic cells/edited
by Alyson K. Tobin.
    p.   cm.–(Society for Experimental Biology Seminar Series: 50)
    1. Plant organelles.   2. Plants–Metabolism.   3. Photosynthesis.
I. Tobin, Alyson K.   II. Series: Seminar series (Society for Experimental Biology
(Great Britain)) ; 50.
QK725.P572 1992
581.87′34–dc20   92–16112   CIP

ISBN 0 521 40171 2 hardback

wv

For Tim, Alex and George

Contents

x   *Contents*

Contributors

BARON, A.C.
Plant Metabolism Research Unit, Department of Cell and Structural Biology, Williamson Building, University of Manchester, Manchester M13 9PL, UK.

BASSHAM, D.C.
Department of Biological Sciences, University of Warwick, Coventry CV4 7AL, UK.

BROOKS, A.
Department of Animal and Plant Sciences, University of Sheffield, Sheffield S10 2TN, UK.

BROSNAN, J.M.
Biology Department, University of York, York YO1 5DD, UK.

BRYCE, J.H.
Department of Biological Sciences, Heriot-Watt University, Riccarton, Edinburgh EH14 4AS, UK.

DAVIES, J.M.
Biology Department, University of York, York YO1 5DD, UK.

EDWARDS, G.E.
Department of Botany, Washington State University, Pullman, Washington, USA.

FAWCETT, T.
Plant Molecular Biology Group, Department of Biological Sciences, University of Durham, Durham DH1 3LE, UK.

FLÜGGE, U.I.
Institut für Botanik und Pharmazeutische Biologie mit Botanischem Garten der Universität Würzburg, Mittlerer Dallenbergweg 64, 8700 Würzburg, Germany.

FRICAUD, A.C.
Department of Biochemistry, University of Sussex, Falmer, Brighton, Sussex, BN1 9QG, UK.

HELDT, H.W.
Institut für Biochemie der Pflanze, Universität Göttingen, Untere Karspüle 2, 3400 Göttingen, Germany.

HULFORD, A.
Department of Biological Sciences, University of Warwick, Coventry CV4 7AL, UK.

JOHANNES, E.
Biology Department, University of York, York YO1 5DD, UK.

KRALL, J.P.
Biology Department, University of Essex, P.O. Box 23, Wivenhoe Park, Colchester CO4 3SQ, UK.

LEEGOOD, R.C.
Department of Animal and Plant Sciences, University of Sheffield, Sheffield S10 2TN, UK.

MASTERSON, C.
Department of Biological and Nutritional Sciences, Agriculture Building, Kings Walk, University of Newcastle upon Tyne NE1 7RU, UK.

MEADOWS, J.W.
Department of Biological Sciences, University of Warwick, Coventry CV4 7AL, UK.

MEIDAN, E.
Botany Department, University of Adelaide, Adelaide, SA 5000, Australia.

MOORE, A.L.
Department of Biochemistry, University of Sussex, Falmer, Brighton, Sussex BN1 9QG, UK.

MOULD, R.M.
Department of Biological Sciences, University of Warwick, Coventry CV4 7AL, UK.

RAWSTHORNE, S.
Cambridge Laboratory, Institute for Plant Science Research, John Innes Centre, Norwich NR4 7UJ, UK.

REA, P.A.
Plant Science Institute, Department of Biology, University of Pennsylvania, Philadelphia, PA 19104, USA.

ROBINSON, C.
Department of Biological Sciences, University of Warwick, Coventry CV4 7AL, UK.

ROGERS, W.J.
Physiologie Végétale Moleculaire, Université de Paris Sud XI, 91405 Orsay Cédex, France.

SANDERS, D.
Biology Department, University of York, York YO1 5DD, UK.
SCHNABL, H.
Institute of Agricultural Botany, University of Bonn, Meckenheimer Allee 176, 5300 Bonn, Germany.
SHACKLETON, J.B.
Department of Biological Sciences, University of Warwick, Coventry, CV4 7AL, UK.
SIEDOW, J.N.
Department of Botany, Duke University, Durham, NC 27706, USA.
SIMON, W.
Plant Molecular Biology Group, Department of Biological Sciences, University of Durham, Durham DH1 3LE, UK.
SLABAS, A.R.
Plant Molecular Biology Group, Department of Biological Sciences, University of Durham, Durham DH1 3LE, UK.
SLABAS, D.
Plant Molecular Biology Group, Department of Biological Sciences, University of Durham, Durham DH1 3LE, UK.
SMITH, J.A.C.
Department of Plant Sciences, University of Oxford, South Parks Road, Oxford OX1 3RB, UK.
SWINHOE, R.
Plant Molecular Biology Group, Department of Biological Sciences, University of Durham, Durham DH1 3LE, UK.
THOMAS, D.R.
Department of Biological and Nutritional Sciences, Agriculture Building, Kings Walk, University of Newcastle upon Tyne, NE1 7RU, UK.
TOBIN, A.K.
Plant Metabolism Research Unit, Department of Cell and Structural Biology, Williamson Building, University of Manchester, Manchester M13 9PL, UK.
TURPIN, D.H.
Department of Botany, University of British Columbia, Vancouver, BC V6T 1W5, Canada.
VOJNIKOV, V.
Department of Biochemistry, University of Sussex, Falmer, Brighton, Sussex BN1 9QG, UK.
VON CAEMMERER, S.
Plant Environmental Biology Group, Research School of Biological Sciences, Australian National University, Canberra, Australia.

WALLSGROVE, R.M.
IACR, Rothamsted Experimental Station, Harpenden, Herts. AL5 2JQ, UK.
WALTERS, A.C.
Department of Biochemistry, University of Sussex, Falmer, Brighton, Sussex BN1 9QG, UK.
WHITEHOUSE, D.G.
Department of Biochemistry, Polytechnic of East London, London E15 4LZ, UK.
WISKICH, J.T.
Botany Department, University of Adelaide, Adelaide, SA 5000, Australia.
WOOD, C.
Department of Biological and Nutritional Sciences, Agriculture Building, Kings Walk, University of Newcastle upon Tyne, NE1 7RU, UK.

Preface

Plant cells, like those of all eukaryotes, are divided into subcellular compartments, or 'organelles'. This compartmentation serves a number of functions. It may concentrate a metabolite and provide a more favourable environment in which the enzymes may operate. It may separate intermediates from enzymes and, hence, avoid futile cycles or unwanted by-products. It may also enable pathways to interact more efficiently by bringing together a number of enzymes within the same organelle. Compartmentation, however, also poses a number of potential problems. The substrate required by a mitochondrial enzyme, for example, may be manufactured in the chloroplast. There is thus a requirement for controlled and coordinated fluxes of intermediates between organelles. Similarly, most proteins are nuclear-encoded and synthesised in the cytoplasm and those localised within organelles will have to cross at least one boundary membrane. This book addresses the way in which the metabolism of photosynthetic tissue is compartmentalised between organelles and how this compartmentation leads to controlled interaction between the organelles.

In 1991, a meeting was held on the topic of 'metabolic interactions of organelles in photosynthetic tissue' as part of the annual meeting of the Society for Experimental Biology, at the University of Birmingham (jointly organised by the Intracellular Co-ordination Group and the Plant Metabolism Group of the SEB). Invited speakers were asked to contribute a review of their subject area and these have been collected together to produce this volume. The first chapter, by Wiskich and Meidan, is a general overview of the field. There then follows a review of the fluxes of metabolites between plant cells, by Heldt and Flügge, with particular emphasis on the phosphate translocator, which was discovered and characterised in their laboratories. It was once thought that the chloroplasts could carry out photosynthesis in complete independence from the rest of the cell. This is not, however, the case as there is now ample evidence that the photosynthetic products, three carbon phosphoesters, are exported to the cytosol in exchange for an import of inorganic phosphate from the cytosol. The characterisation of the phosphate translocator, which is responsible for this exchange, has led us to a greater understanding of the interactions which occur between organelles during photosynthesis. This theme is expanded upon in subsequent chapters. David Turpin reviews current schemes for the integration between photosynthetic and respiratory carbon and nitrogen metabolism in a model photosynthetic system, the green alga *Selenastrum*. There then follows a series of reviews on metabolic interactions during different photosynthetic processes, such as photorespiration (Wallsgrove *et al.*), C4 photosynthesis (Edwards and Krall), C3–C4 (Rawsthorne *et al.*) and CAM photosynthesis (Smith and Bryce). The last of these involves considerable flux of metabolites into and out of the vacuole. Aspects of the control of ion transport across the vacuolar membrane are then discussed in detail by Sanders *et al.*

The respiratory activity of photosynthetic tissue has often proved difficult to quantify, particularly when plants are in the light. Moore *et al.* discuss the mechanisms which may operate to regulate mitochondrial respiration in photosynthetic tissue. The final sections of the book deal with lipid metabolism, in terms of biosynthesis and assembly of some of the enzymes involved (Slabas *et al.*) and with the role of carnitine in lipid biosynthesis and degradation within different organelles (Wood *et al.*) Protein import from the cytosol to the chloroplast is discussed by Meadows *et al.* Finally, there are considerations of the interactions which occur between organelles in specific cells of a leaf: the guard cells (Schnabl) and developing and differentiating leaf cells (Tobin and Rogers).

I thank all of the contributors for their efforts in producing both a paper for the meeting and a review article for this book. Particular thanks go to Joe Wiskich, Gerry Edwards and David Turpin, who travelled long distances to attend the meeting. I also thank the group convenors, Steve Boffey and Mike Emes, for their help with the organisation of the meeting.

Alyson K. Tobin
Manchester

Acknowledgements

Financial support towards the organisation of this meeting was provided by ICI Seeds, Schering Agrochemicals Ltd, and Unilever. I am most grateful to these sponsors, whose contribution enabled overseas speakers to attend the meeting.

J.T. WISKICH and E. MEIDAN

Metabolic interactions between organelles in photosynthetic tissue: a mitochondrial overview

The photosynthetic cell is particularly active when illuminated and inter-actions occur among chloroplasts, mitochondria, peroxisomes and the cytoplasm, which, although not an organelle, represents a separate cellular compartment. For one organelle to affect the metabolism of another there needs to be a transfer of metabolites so that the second organelle experiences a changed metabolite condition (e.g. substrate:product; NADH:NAD; ATP:ADP). This metabolite transfer may occur directly between organelles when their membranes are in contact or, indirectly, into the medium common to both, i.e. the cytoplasm. These metabolic interactions are dependent on metabolite transport across organelle membranes (Heldt & Flügge, 1987).

The phosphate translocator of chloroplasts can elevate the cytoplasmic concentrations of triose-P, ATP and reduced NAD. These compounds influence mitochondrial respiration by acting as substrates or as feedback inhibitors, producing adenylate or respiratory control. Metabolic control results from complex, multi-component interactions and is best understood with metabolic control analysis (Kacser, 1987).

Much has been written about the persistence of mitochondrial activity (tricarboxylic acid cycle turnover, electron flow and oxidative phosphorylation) in illuminated leaves. The comments have ranged from a complete shutdown of mitochondrial activity imposed by the high cytosolic ATP/ADP ratios in the light (Heber, 1974); maintenance of TCA cycle activity (Azcón-Bieto & Osmond, 1983); increased rates of oxygen uptake (Azcón-Bieto & Day, 1983; Stitt et al., 1985 – although not necessarily linked to ATP synthesis) to a dependence of photosynthetic activity on mitochondrial oxidative phosphorylation (Krömer et al., 1988). The current view is that mitochondrial activity, including oxidative phosphorylation, does persist in leaf cells during photosynthesis. From a biochemical standpoint illumination induces a period of hectic activity – after all, the leaf cell is receiving 'free' energy – and it would be strange if its enzymic machinery were not fully utilised under these circumstances.

Society for Experimental Biology Seminar Series 50: *Plant organelles*, ed. A. K. Tobin.
© Cambridge University Press 1992, pp. 1–19.

We intend to concentrate on the general mitochondrial aspects of organelle interaction in this 'overview', knowing that some material will receive detailed treatment elsewhere.

Photosynthetic carbon metabolism

C3 photosynthesis

The transfer of photosynthetic carbon to the cytoplasm is usually discussed in reference to partitioning between sucrose synthesis in the cytoplasm and starch synthesis in the chloroplast (Kelly *et al.*, 1989). It appears that the initial fate of the fixed carbon is export from the chloroplast in exchange for P_i (Heldt & Flügge, 1987) and polymerisation to sucrose in the cytoplasm. If the rate of sucrose synthesis decreases the rate of triose-P export also decreases, leading to starch synthesis within the chloroplast (Stitt & Quick, 1989). Both of these effects are mediated via a decreased concentration of P_i (Fig. 1). Cytoplasmic triose-P is a glycolytic intermediate and can increase the supply of TCA cycle substrates to mitochondria. During C3 photosynthesis photorespiratory substrates are also made available to mitochondria and these can account

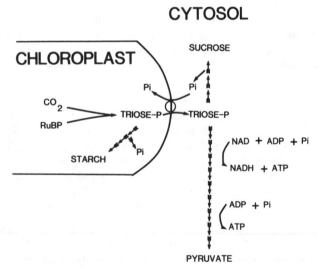

Fig. 1. The export of triose-P from chloroplasts. In the cytosol the triose-P may either be polymerised to sucrose or oxidised to pyruvate via glycolysis. The sequence of short arrows represents a series of reactions. RuBP, ribulose bisphosphate.

for some 15–20% of the initial post-illumination rates of respiration. The remaining respiration is related to the increased carbohydrate status of the leaf cells (Azcón-Bieto & Osmond, 1983). This correlation is not perfect because the rates of leaf respiration decrease with the increasing period of darkness but not necessarily in parallel with the decrease in carbohydrate content (Stitt *et al.*, 1985). This presumably reflects time-dependent changes in the flux control coefficients (Kacser, 1987). The rate of post-illumination respiration exceeds the rate of dark respiration and for some leaves this has been attributed to the non-phosphorylating, alternative oxidase (Azcón-Bieto & Day, 1983). These authors concluded that for wheat leaves the level of respiratory substrate determined both the rate of respiration and the extent to which the alternative oxidase was engaged. The increased level of mitochondrial substrates increases the level of reduced ubiquinone and hence flux through the alternative pathway (Dry *et al.*, 1989).

C4 photosynthesis

There is a special role postulated for mitochondria in the bundle sheath cells of the NAD-malic enzyme ('NAD-ME') and of the phosphoenolpyruvate carboxykinase ('PCK') types of C4 plants (Hatch, 1987). In the 'NAD-ME' type aspartate is the major C4 acid transported from the mesophyll to the bundle sheath cells and is decarboxylated by NAD-malic enzyme – exclusively a mitochondrial enzyme in higher plants. The decarboxylation process involves aspartate transamination to produce oxaloacetate, and its reduction to malate which is oxidatively decarboxylated to pyruvate and CO_2 (Fig. 2A). The role of mitochondria is to provide the enzymic system for decarboxylation (Gardeström & Edwards, 1985; Hatch, 1987). The nature of the transport processes across the inner mitochondrial membrane needs to be resolved. If this were the classic malate/aspartate shuttle the organic acid moving outwards would be malate, not pyruvate (Dry *et al.*, 1987). However, the classic malate/aspartate shuttle transports reducing power, whereas this shuttle functions as a CO_2-transporting system.

The biochemistry of the 'PCK' type of C4 photosynthesis has been the hardest to understand. It appears that cytosolic PEP carboxykinase and mitochondrial NAD-malic enzyme contribute about equally to the decarboxylating process in the bundle sheath (Hatch, 1987; Burnell & Hatch, 1988). The mitochondrial oxidation provides the ATP for the cytosolic PEP-carboxykinase activity (Hatch *et al.*, 1988), with some of the PEP so formed being transported directly to the mesophyll cells (Fig. 2B). The mitochondrial transporters described here are more straightforward. A

A

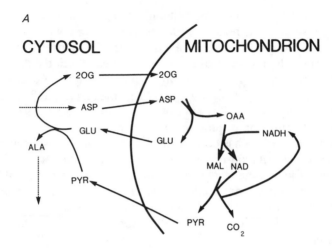

B

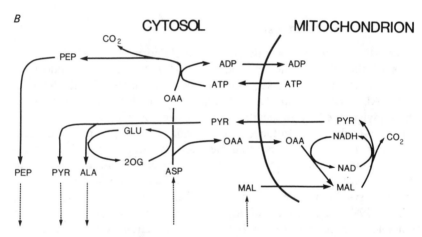

Fig. 2. The transport processes of bundle-sheath mitochondria in some C4 plants. *A*, The NAD-ME type of C4 photosynthesis. Some malate may be transported from the mesophyll but this would represent less than 10% of the total carbon flux. *B*, The PCK-type of C4 photosynthesis. Broken arrows represent movement to and from the mesophyll cells. 2OG, 2 oxoglutarate; ASP, aspartate; GLU, glutamate; MAL, malate; PYR, pyruvate; ALA, alanine.

very active and specific oxaloacetate transporter is known to occur in plant mitochondria (Day & Wiskich, 1982). The other transporters are also present in plant mitochondria (Day & Wiskich, 1984; Heldt & Flügge, 1987). In the PCK-type the mitochondria serve to provide energy for C4 photosynthesis as well as the enzymic system for C4-acid decarboxylation.

Nitrogen metabolism

The localisation of nitrate reductase in green cells has not been established with certainty, but most investigations would suggest that it is a cytosolic enzyme (Solomonson & Barber, 1990). Its reductant is NADH which could be generated in the cytoplasm by dehydrogenation of glycolytic substrate or by transfer of reducing equivalents from mitochondria or chloroplasts, via a malate/oxaloacetate shuttle. The subsequent reduction of nitrite to ammonia and its assimilation is considered to be chloroplastic – nitrite reduction being linked to ferredoxin via nitrite reductase and ammonia incorporation via glutamine synthetase and glutamate synthase (Fig. 3). This results in the synthesis of glutamate requiring a continual production of 2-oxoglutarate, which presumably comes from the mitochondrial TCA cycle. If this is true then anaplerotic reactions must be occurring at the same time to maintain the carbon pool within the mitochondrial matrix needed for continual operation of the TCA cycle.

Recent studies with the green alga *Selenastrum* sp. have highlighted the important role mitochondrial respiration can have in producing carbon skeletons and reducing power for nitrogen assimilation. Weger *et al.* (1988) and Weger & Turpin (1989) have shown that mitochondrial respiration provides both to the chloroplast to maintain nitrate/nitrite

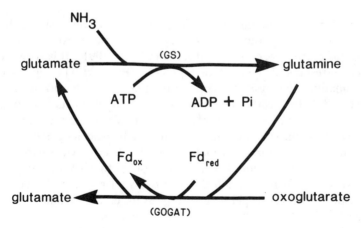

Fig. 3. The path of ammonia assimilation in the chloroplast. GS, glutamine synthetase; GOGAT, glutamate synthase.

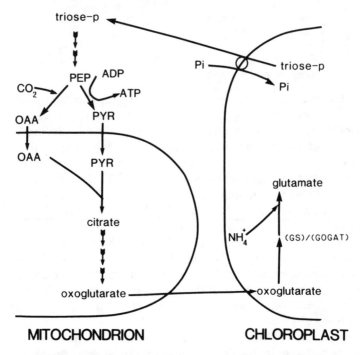

Fig. 4. The path of carbon during net ammonia assimilation in the green alga *Selenastrum minitum* (adapted from Turpin *et al.*, 1990). The sequence of short arrows represents a series of reactions. PYR, pyruvate; GS, glutamine synthetase; GOGAT, glutamate synthase.

reduction and ammonia assimilation during photosynthesis. Furthermore, a non-photosynthetic CO_2 requirement for ammonia assimilation has recently been reported (Amory *et al.*, 1991). It is suggested that this represents the anaplerotic process (Fig. 4). The significance of terminating glycolysis by converting phosphoenolpyruvate to oxaloacetate (via PEP carboxylase) and transporting it, or its reduction product, malate, into mitochondria has been discussed (Wiskich & Dry, 1985). Plant mitochondria possess malate dehydrogenase and NAD-malic enzyme and the import of malate, via the dicarboxylate carrier, would allow intramitochondrial generation of both OAA and pyruvate. However, plant mitochondria also possess an active and specific OAA transporter which, in parallel with pyruvate uptake, would also serve to maintain TCA cycle turnover. Both of these processes would serve as anaplerotic reactions helping to maintain the pool of organic acids within the mitochondria. As shown in Fig. 4, the process involves export of triose-P from the chloroplast and its glycolytic and TCA cycle conversion to oxoglutarate which is returned to the chloroplast.

Photorespiration

Photorespiration results from the oxygenase activity of ribulose bisphosphate carboxylase which hydrolyses ribulose bisphosphate to phosphoglycerate and phosphoglycolate. It is the subsequent processing of the phosphoglycolate that constitutes the metabolic pathway of photorespiration and involves three organelles (Fig. 5). The products of this process are phosphoglycerate, ammonia, CO_2 and reducing power in the form of NADH. There seems to be little controversy about the sequence of metabolic events but note that there are two transamination reactions in the peroxisome. One of these is thought to be satisfied by the return of serine from the mitochondria, and the other by reassimilation of the ammonia liberated in the mitochondria. This reassimilation process, or the photorespiratory N cycle (Givan *et al.*, 1988) occurs in the chloroplast (Fig. 3) and requires delivery to the chloroplasts of ammonia from the mitochondria and oxoglutarate from the peroxisomes. The ability of these compounds to enhance light-dependent O_2 evolution in isolated chloroplasts, as would be expected if glutamine synthetase and glutamate synthase activities were increased, has been demonstrated (Woo & Osmond, 1982; Dry & Wiskich, 1983).

The other transport process that has caused much debate is the origin of the reducing power for the peroxisomal reduction of hydroxypyruvate to glycerate. It is generally thought that the peroxisome imports malate, oxidising it to OAA and reducing NAD via malate dehydrogenase. In turn, the OAA would need to be reduced to malate elsewhere. Mitochondria oxidise glycine, producing NADH in the required stoichiometry, and are an obvious site for OAA reduction. Furthermore, they have their specific OAA transporters making the whole a closed, self-contained cycle.

However, there are problems with this proposal. The estimated rates of photorespiration are quite variable but the best estimates suggest they are many times faster than dark respiration (Dry & Wiskich, 1987) and it is difficult to see mitochondrial activity being maintained under such conditions. We know that the addition of OAA to isolated mitochondria inhibits oxygen uptake (Day & Wiskich, 1981) because it maintains the NAD pool in the oxidised state (Day *et al.*, 1985). Thus, the mitochondria would be maintained in an oxidised, de-energised state.

An alternative source of reducing power could be the chloroplast which could export it via a malate/OAA shuttle (Ebbighausen *et al.*, 1987). The

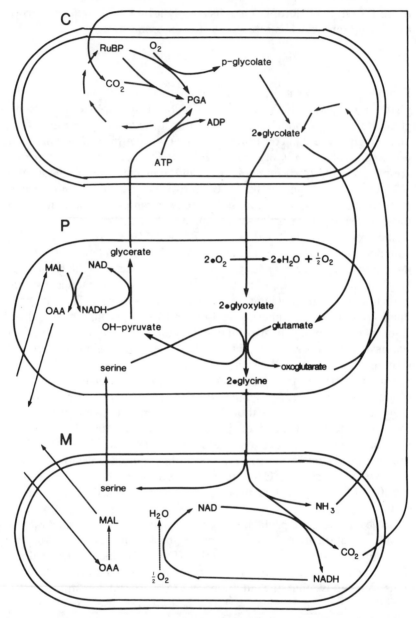

Fig. 5. The path of carbon flow in photorespiration. Broken arrows represent alternative methods of oxidising NADH. C, chloroplast; P, peroxisome; M, mitochondria; RuBP, ribulosebisphosphate; PGA, phosphoglycerate.

possibility of this being coupled to the transfer of the amino groups in the photorespiratory N cycle, between the chloroplast and peroxisome has been identified (Dry & Wiskich, 1983). Recent evidence that mitochondrial oxidative phosphorylation participates in photosynthetic metabolism, if substantiated, will further implicate the chloroplast as the source of peroxisomal reducing power.

Mitochondrial glycine oxidation

Glycine is metabolised in mitochondria by a combination of the glycine decarboxylase complex (Walker & Oliver, 1986) and serine hydroxymethyl transferase. The overall reaction is:

$$2 \text{ Gly} + NAD^+ \rightarrow \text{Ser} + CO_2 + NH_3 + NADH + H^+$$

and for simplicity we will refer to the whole process as glycine decarboxylase. The CO_2 and ammonia return to the chloroplast and the serine returns to the peroxisome. It is the re-oxidation of the NADH which remains unresolved. There is no doubt that it can be re-oxidised via the electron transport chain (Woo & Osmond, 1976; Moore *et al.*, 1977) or via malate dehydrogenase reducing OAA (Day & Wiskich, 1981). The rates of oxygen uptake with glycine (Wiskich & Dry, 1985) are sufficient to account for the estimated rates of tissue photorespiration (Dry & Wiskich, 1987). Furthermore, it was reported (Day & Wiskich, 1981) that the presence of TCA cycle intermediates did not affect the rate of glycine oxidation. Walker *et al.* (1982) suggested differential use of two NADH dehydrogenases by glycine and malate to explain the apparent independence of glycine oxidation from TCA cycle activity. Dry *et al.* (1983) referred to it as a preferential oxidation of glycine, as did Bergman & Ericson (1983), when they observed that it was the oxidation of TCA cycle acids that was inhibited in the presence of glycine and not vice versa.

These observations gave impetus to the idea that the NADH from glycine oxidation had special access to the electron transport chain and could be used to reduce oxygen. However, the mechanism by which this could be achieved was not at all clear, nor was the fate of the ATP which would have been generated in those mitochondria with little alternative oxidase activity.

These studies had shown us that glycine could stimulate ADP-limited (state 4) malate oxidation, as had been reported earlier (Walker *et al.*, 1982) and could be explained by an increased level of NAD reduction (Fig. 6*A*). The problem with malate oxidation via malate dehydrogenase is that the equilibrium favours oxidised NAD and any accumulation of

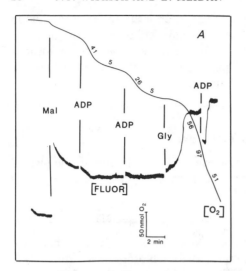 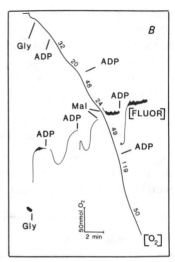

Fig. 6. Simultaneous measurement of oxygen uptake and NADH fluorescence of pea leaf mitochondria. Conditions as in Wiskich *et al.*, 1990. FLUOR, fluorescence; Mal, malate; Gly, glycine. Numbers along the trace represent oxygen uptake in nmol min^{-1} mg^{-1} protein. An upward deflection of the fluorescence trace represents increased NAD reduction. *A*, Malate added before glycine; *B*, glycine added before malate.

OAA will severely inhibit malate oxidation (this can happen even in the presence of glutamate if the aminotransferase does not remove OAA rapidly enough). Electron flux through the electron transport chain responds to the redox poise of the NAD pool (see p. 14) and the increased reduction leads to increased reduction of the ubiquinone pool, increased membrane potential (Moore *et al.*, 1991) and oxygen uptake. However, what was more difficult to explain was the stimulation of glycine state 4 oxidation on the addition of malate (Dry & Wiskich, 1985). In that paper we concluded that differential access to NAD and to electron transport chains was the only possible explanation. Any other explanation ran into thermodynamic or kinetic impossibilities – something not always appreciated and producing some conceptually novel, but scientifically unsound, hypotheses. The addition of malate during glycine oxidation in state 4 had a negligible effect on the level of NAD reduction, as judged by fluorescence (Day *et al.*, 1985; Fig. 6*B*) but did increase the level of reduction of the ubiquinone pool (Wiskich, 1991). More significantly, it increased the rate of glycine metabolism as judged by the rate of release of NH_3 (Wiskich *et al.*, 1990). In fact, all of the oxygen uptake

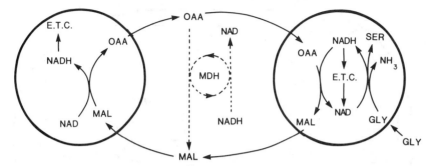

Fig. 7. A scheme for synergistic malate and glycine oxidation by two populations of mitochondria. MAL, malate; GLY, glycine; SER, serine; E.T.C., electron transport chain; MDH, malate dehydrogenase. Broken arrows represent the operation of an OAA-scavenging system added to the external medium.

under these conditions was equivalent to the rate of glycine metabolism. The answer was obvious: the addition of malate was enhancing glycine oxidation by setting up a malate/OAA shuttle. It was simplest to envisage two populations of mitochondria, both with TCA cycle enzymes but only one with glycine decarboxylase, and with malate and OAA diffusing between the two (Fig. 7). Thus, glycine would stimulate malate state 4 oxidation for the reasons given above, and malate would stimulate glycine state 4 oxidation by increasing the number of functional mitochondria.

This proposal has been tested in a number of ways, all of which showed it to be incorrect. First, phthalonate is a potent inhibitor of mitochondrial OAA transport (Day & Wiskich, 1981, 1982) but it did not prevent malate-induced stimulation of oxygen uptake and NH_3 release from glycine under state 4 conditions (Wiskich *et al.*, 1990). Secondly, the presence of an OAA-scavenging system in the medium (as shown in Fig. 7) also failed to reduce the rate of NH_3 release (Wiskich *et al.*, 1990). Thirdly, the distribution of gold-labelled antibody to glycine decarboxylase did not differ markedly from that of an antibody to complex I (NADH-ubiquinone reductase) in a preparation of isolated pea mitochondria (Tobin *et al.*, 1990). Thus, we are forced to the conclusion that the malate/OAA shuttle (Fig. 7) occurs not between mitochondria but *within* mitochondria (Fig. 8). This means we have a physical separation of glycine decarboxylase with its associated malate dehydrogenase from other malate dehydrogenases and that the two sets of malate dehydrogenases are operating in opposite directions. Such a scheme is not impossible: it simply means that the two microenvironments are dif-

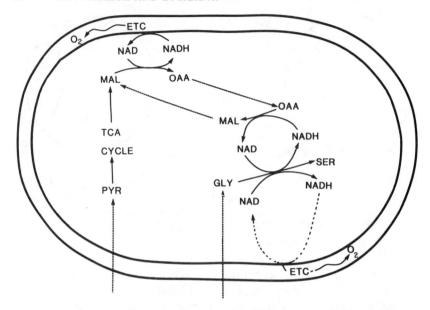

Fig. 8. A model for synergistic malate and glycine oxidation within a single mitochondrion. PYR, pyruvate; other abbreviations as in Fig. 7. Broken lines represent diffusional movement.

ferent – that with glycine decarboxylase will have a high level of NAD reduction and that without glycine decarboxylase a low level of NAD reduction, the latter being maintained by oxidation via the respiratory chain.

Two questions remain with respect to the scheme shown in Fig. 8. Why does the presence of malate increase the rate of glycine decarboxylase turnover and the rate of oxygen uptake under *state 4 conditions*? Common to both explanations is the intramitochondrial NADH/NAD ratio, which according to fluorescence measurements does not change significantly. However, the distribution of NADH within the mitochondrion may have changed together with the ratio of bound to soluble NADH, both of which fluoresce.

Glycine oxidation under state 4 conditions has a certain distribution of control coefficients (Kacser & Burns, 1979), with glycine decarboxylase activity being restricted by NADH binding to it (Pascal *et al.*, 1990) or by the availability of NAD. The addition of malate generates OAA which diffuses to the glycine decarboxylase complex where it is reduced via malate dehydrogenase, lowering the NADH/NAD ratio and thereby activating glycine metabolism. At the other site, presumably adjacent to the inner membrane, but now with a continual supply of malate and continual removal of OAA, the NADH/NAD ratio is increased. This new condition would enhance the activity of nearby NADH dehydrogenases, increasing the rate of electron flow. The overall level of NADH fluorescence need not have changed significantly, but the turnover of enzymic activity has increased dramatically. In other words, the control coefficients have been redistributed and their relativities changed. It has been known for some time that OAA is a powerful activator of glycine metabolism under state 4 conditions (Day & Wiskich, 1981). The stimulation of glycine state 4 oxidation by malate is such that state 3 rates of oxidation are attained (Wiskich *et al.*, 1990), i.e. very rapid rates of glycine metabolism, sufficient to maintain photorespiratory rates *in vivo* can be achieved without net phosphorylation of ADP to ATP. How this can be accomplished is discussed in the next section and does not preclude glycine oxidation being coupled to phosphorylation (Gardeström & Wigge, 1988). Finally, it should be noted that this scheme permits simultaneous oxidation of glycine and of TCA cycle substrates and simultaneous operation of the electron transport chain and the malate/OAA shuttle.

Control of mitochondrial electron transport

The protonmotive force (Δp), being the intermediate between electron flow and oxidative phosphorylation, can impose a restriction on the rate of oxygen uptake. Dry & Wiskich (1982) showed that high ATP/ADP ratios (>50) were needed to reduce electron flow significantly. The effect was also dependent on the adenylate concentration and they concluded that control was exerted not only by the ATP/ADP ratio but also by the absolute concentration of ADP.

If determined enzymically after acid extraction of the tissue or organelles, ATP and ADP analyses usually produce ratios of less than 10 (Stitt *et al.*, 1982; Gardeström & Wigge, 1988). Non-invasive analysis using ^{31}P-NMR produces significantly higher values, mainly owing to the low measurements of ADP, presumably made undetectable by being bound to proteins (Hooks *et al.*, 1989).

Nevertheless, it appears that there may be a transient increase in the cytosolic ATP/ADP ratio of leaf cells on illumination (Gardeström & Malmberg, 1990) but the steady-state level may be lower than that in the dark (Stitt *et al.*, 1982). This decrease would tend to either de-energise the mitochondria or increase their rate of oxidative phosphorylation, i.e. the rate of oxygen consumption would increase, which considering the current evidence may be significant. In fact, Gardeström & Wigge (1988) claim that photorespiratory glycine oxidation is coupled to ATP synthesis resulting in higher cytosolic ATP/ADP ratios.

Further support for active oxidative phosphorylation during photosynthesis is provided by the oligomycin sensitivity of CO_2 fixation in leaf protoplasts (Krömer et al., 1988). The function of this mitochondrially generated ATP is not known but it does not appear to be directly involved in CO_2 fixation, because the chloroplasts of disrupted protoplasts can maintain the control rates of photosynthesis without showing any sensitivity to oligomycin.

The low in vivo level of nucleotide diphosphates (<16 μM) detected with ^{31}P-NMR compared with acid extracts (Hooks et al., 1989) supports the conclusion that the absolute concentration of ADP available for uptake regulates mitochondrial respiration together with the ATP/ADP ratio (Dry & Wiskich, 1982). An adenine nucleotide transporter (Heldt & Flügge, 1987) in the inner membrane exchanges cytosolic ADP for matrix ATP, but cytosolic ATP can act as a competitive inhibitor. Thus, low ADP concentrations could severely limit the operation of the transporter, allowing it to possess a significant flux control coefficient. We have performed further analyses on the results reported by Dry & Wiskich (1982) and conclude that under our conditions of assay the absolute concentration of ADP is some 5–6 times more important in determining rates of respiration than is the ATP/ADP ratio (A.T. James, I.B. Dry and J.T. Wiskich, unpublished data).

Adenylate control is therefore transmitted to the electron transport chain via Δp. The driving force for electron flow is the potential difference between NADH/NAD and H_2O/O_2. In most cases we can assume the latter to be constant, making the available NADH/NAD ratio the significant factor; this in turn is dependent on the nature of the substrate and the substrate concentration. So an increase of substrate from the chloroplast, either via triose-P or via photorespiration, will tend to increase the reduction state of mitochondrial NAD. This will increase the rate of electron flow and of Δp. This increased Δp will tend to inhibit the rate of electron flow and at the same time increase the rate of H^+ diffusion into the mitochondria (i.e. increase H^+ conductance). In the steady state this rate of inward H^+ diffusion is equal to the rate of outward H^+ translocation, which is linked to electron flow. So in the coupled, steady state an increased Δp, if induced by an increased NADH/NAD ratio, will increase the rate of oxygen uptake. This is complicated by the observation (Whitehouse et al., 1989) that the relationship between inward H^+ diffusion (i.e. membrane H^+ conductance) and the magnitude of Δp is not linear. At membrane potentials of 180 mV or above, H^+ conductance increases rapidly for relatively small increases in membrane potential, i.e.

increases in Δp produce a disproportionate stimulation in the rate of oxygen uptake.

It has been shown that for the extreme case of ADP-limited (state 4) respiration malate can increase the rate of glycine oxidation (as assessed by oxygen uptake and ammonia release) to that of, or greater than, the ADP-stimulated rate (Fig. 5B; Dry & Wiskich, 1985; Wiskich *et al.*, 1990; Moore *et al.*, 1991). In parallel with this the steady-state level of reduced ubiquinone (Wiskich, 1991) and of membrane potential (Moore *et al.*, 1991) has also been observed. The increased membrane potential indicates that oxygen uptake is via the cytochrome pathway and this was confirmed by Moore *et al.* (1991) who observed the same effects in the presence of inhibitors of the alternative oxidase.

In this section we have tried to establish that

1. The NADH produced from photorespiratory glycine metabolism can be oxidised via the mitochondrial electron transport chain.
2. The rate of mitochondrial glyine oxidation is sufficient to account for the rate of photorespiratory flux.
3. The scheme presented in Fig. 7 allows for the TCA cycle to turn over simultaneously with glycine oxidation.
4. Rapid rates of glycine oxidation need not involve phosphorylation of ADP (although this is not precluded) and need not involve alternative oxidase activity.
5. The electron transport chain can be regarded in electrical terms but with a non-linear relationship between potential and current.
6. Some of the reducing power may be exported to the peroxisome via a malate/OAA shuttle.
7. In all probability, activity of leaf mitochondria is much greater in the light than in the dark.

Acknowledgements

We wish to thank the many people who have been associated with this work: I.B. Dry, D.A. Day, A.L. Moore, A.K. Tobin, A.T. James, J.H. Bryce, L. Mischis, A. Padovan and C. Robinson. Financial support was provided by the Australian Research Council and British Council, Academic Links and Interchange Scheme.

References

Amory, A.M., Vanlerberghe, G.C. & Turpin, D.H. (1991). Demonstration of both a photosynthetic and a non-photosynthetic CO_2 requirement for NH_4^+ assimilation in the green alga *Selenastrum minutum. Plant Physiology* **95**, 192–6.

Azcón-Bieto, J. & Day, D.A. (1983). Effect of photosynthesis and carbohydrate status on respiratory rates and the involvement of the alternative pathway in leaf respiration. *Plant Physiology* **72**, 598–603.

Azcón-Bieto, J. & Osmond, C.B. (1983). Relationship between photosynthesis and respiration. The effect of carbohydrate status on the rate of CO_2 production by respiration in darkened and illuminated wheat leaves. *Plant Physiology* **71**, 574–81.

Bergman, A. & Ericson, I. (1983). Effects of pH, NADH, succinate and malate on the oxidation of glycine in spinach leaf mitochondria. *Physiologia Plantarum* **59**, 421–7.

Burnell, J.N. & Hatch, M.D. (1988). Photosynthesis in phosphoenolpyruvate carboxylase-type C_4 plants: pathways of C_4 acid decarboxylation in bundle sheath cells of *Urochloa panicoides. Archives of Biochemistry and Biophysics* **260**, 187–99.

Day, D.A., Neuburger, M. & Douce, R. (1985). Interactions between glycine decarboxylase, the tricarboxylic acid cycle and the respiratory chain in pea leaf mitochondria. *Australian Journal of Plant Physiology* **12**, 119–30.

Day, D.A. & Wiskich, J.T. (1981). Glycine metabolism and oxaloacetate transport by pea leaf mitochondria. *Plant Physiology* **68**, 425–9.

Day, D.A. & Wiskich, J.T. (1982). Effect of phthalonic acid on respiration and metabolite transport in higher plant mitochondria. *Archives of Biochemistry and Biophysics* **211**, 100–7.

Day, D.A. & Wiskich, J.T. (1984). Transport processes of isolated plant mitochondria. *Physiologie Végétale* **22**, 241–61.

Dry, I.B., Day, D.A. & Wiskich, J.T. (1983). Preferential oxidation of glycine by the respiratory chain of pea leaf mitochondria. *FEBS Letters* **158**, 154–8.

Dry, I.B., Dimitriadis, E., Ward, D.A. & Wiskich, J.T. (1987). The photorespiratory hydrogen shuttle. Synthesis of phthalonic acid and its use in the characterization of the malate/aspartate shuttle in pea (*Pisum sativum*) leaf mitochondria. *Biochemical Journal* **245**, 669–75.

Dry, I.B., Moore, A.L., Day, D.A. & Wiskich, J.T. (1989). Regulation of alternative pathway activity in plant mitochondria: nonlinear relationships between electron flow and the redox poise of the quinone pool. *Archives of Biochemistry and Biophysics* **273**, 148–57.

Dry, I.B. & Wiskich, J.T. (1982). The role of the external adenosine triphosphate/adenosine diphosphate ratio in the control of plant mitochondrial respiration. *Archives of Biochemistry and Biophysics* **217**, 72–9.

Dry, I.B. & Wiskich, J.T. (1983). Characterization of dicarboxyate stimulation of ammonia, glutamine and 2-oxoglutarate-dependent O_2 evolution in isolated pea chloroplasts. *Plant Physiology* **72**, 291–6.

Dry, I.B. & Wiskich, J.T. (1985). Characteristics of glycine and malate oxidation by pea leaf mitochondria: evidence of differential access to NAD and respiratory chains. *Australian Journal of Plant Physiology* **12**, 329–39.

Dry, I.B. & Wiskich, J.T. (1987). Interactions of mitochondria with other metabolic processes – an overview. In *Plant Mitochondria: Structural, Functional and Physiological Aspects*, ed. A.L. Moore & R.B. Beechey, pp. 151–60. New York: Plenum Press.

Ebbighausen, H., Hatch, M.D., Lilley, R. McC., Krömer, S., Stitt, M. & Heldt, H.W. (1987). On the function of malate-oxaloacetate shuttles in a plant cell. In *Plant Mitochondria: Structural, Functional and Physiological Aspects*, ed. A.L. Moore & R.B. Beechey, pp. 171–80. New York: Plenum Press.

Gardeström, P. & Edwards, G.E. (1985). Leaf mitochondria (C_3 + C_4 + CAM). In *Higher Plant Cell Respiration*. Encyclopedia of Plant Physiology, New Series, Vol. 18, ed. R. Douce & D.A. Day, pp. 314–36. Berlin: Springer-Verlag.

Gardeström, P. & Malmberg, G. (1990). Subcellular ATP/ADP in barley protoplasts during photosynthesis induction. *Physiologia Plantarum* **79**, No. 2, Part 2, A60.

Gardeström, P. & Wigge, B. (1988). Influence of photorespiration on ATP/ADP ratios in the chloroplasts, mitochondria, and cytosol, studied by rapid fractionation of barley (*Hordeum vulgare*) protoplasts. *Plant Physiology* **88**, 69–76.

Givan, C.V., Joy, K.W. & Kleczkowski, L.A. (1988). A decade of photorespiratory nitrogen cycling. *Trends in Biochemical Sciences* **13**, 433–7.

Hatch, M.D. (1987). C_4 photosynthesis: a unique blend of modified biochemistry, anatomy and ultrastructure. *Biochimica et Biophysica Acta* **895**, 81–106.

Hatch, M.D., Agostino, A. & Burnell, J.N. (1988). Photosynthesis in phosphoenolpyruvate carboxykinase-type C_4 plants: activity and role of mitochondria in bundle sheath cells. *Archives of Biochemistry and Biophysics* **261**, 357–67.

Heber, U. (1974). Metabolite exchange between chloroplasts and cytoplasm. *Annual Review of Plant Physiology* **25**, 393–421.

Heldt, H.W. & Flügge, U.I. (1987). Subcellular transport of metabolites in plant cells. In *The Biochemistry of Plants*, Vol. 12, ed. D.D. Davies, pp. 49–85. Orlando, FL: Academic Press.

Hooks, M.A., Clark, R.A., Nieman, R.H. & Roberts, J.K.M. (1989).

Compartmentation of nucleotides in corn root tips studied by ^{31}P-NMR and HPLC. *Plant Physiology* **89**, 963–9.

Kacser, H. (1987). Control of metabolism. In *The Biochemistry of Plants*, Vol. 11, ed. D.D. Davies, pp. 39–67. Orlando, FL: Academic Press.

Kacser, H. & Burns, J. A. (1979). Molecular democracy: who shares the controls. *Biochemical Society Transactions* **7**, 1149–60.

Kelly, G.J., Holtum, J.A.M. & Latzko, E. (1989). Photosynthesis. Carbon metabolism: new regulators of CO_2 fixation, the new importance of pyrophosphate, and the old problem of oxygen involvement revisited. *Progress in Botany* **50**, 74–101.

Krömer, S., Stitt, M. & Heldt, H.W. (1988). Mitochondrial oxidative phosphorylation participating in photosynthetic metabolism of a leaf cell. *FEBS Letters* **226**, 352–6.

Moore, A.L., Dry, I.B. & Wiskich, J.T. (1991). Regulation of electron transport in plant mitochondria under state 4 conditions. *Plant Physiology* **95**, 34–40.

Moore, A.L., Jackson, C., Dench, J., Morris, P. & Hall, D.O. (1977). Intramitochondrial localization of glycine decarboxylase in spinach leaves. *Biochemical and Biophysical Research Communications* **78**, 483–91.

Pascal, N., Dumas, R. & Douce, R. (1990). Comparison of the kinetic behaviour toward pyridine nucleotides of NAD^+-linked dehydrogenases from plant mitochondria. *Plant Physiology* **94**, 189–93.

Solomonson, L.P. & Barber, M.J. (1990). Assimilatory nitrate reductase: functional properties and regulation. *Annual Review of Plant Physiology and Plant Molecular Biology* **41**, 225–53.

Stitt, M., Lilley, R. McC. & Heldt, H.W. (1982). Adenine nucleotide levels in the cytosol, chloroplasts, and mitochondria of wheat leaf protoplasts. *Plant Physiology* **70**, 971–7.

Stitt, M. & Quick, W.P. (1989). Photosynthetic carbon partitioning: its regulation and possibilities for manipulation. *Physiologia Plantarum* **77**, 633–41.

Stitt, M., Wirtz, W., Gerhardt, R., Heldt, H.W., Spencer, C., Walker, D.A. & Foyer, C. (1985). A comparative study of metabolite levels in plant leaf material in the dark. *Physiologia Plantarum* **166**, 354–64.

Tobin, A.K., Thorpe, J.R., Day, D.A., Wiskich, J.T. & Moore, A.L. (1990). Immunogold localisation of glycine decarboxylase in isolated pea leaf mitochondria. *Physiologia Plantarum* **79**, No 2, Part 2, A72.

Turpin, D.H., Botha, F.G., Smith, R.G., Feil, R., Horsey, A.K. & Vanlerberghe, G.C. (1990). Regulation of carbon partitioning to respiration during dark ammonium assimilation by the green alga *Selenastrum minutum*. *Plant Physiology* **93**, 166–75.

Walker, J.L. & Oliver, D.J. (1986). Glycine decarboxylase multienzyme complex, purification and partial characterization from pea leaf mitochondria. *Journal of Biological Chemistry* **261**, 2214–21.

Walker, G.H., Oliver, D.J. & Sarojini, G. (1982). Simultaneous oxidation of glycine and malate by pea leaf mitochondria. *Plant Physiology* **70**, 1465–9.

Weger, H.G., Birch, D.G., Elrifi, I.R. & Turpin, D.H. (1988). Ammonia assimilation requires mitochondrial respiration in the light. A study with the green alga *Selenastrum minutum*. *Plant Physiology* **86**, 688–92.

Weger, H.G. & Turpin, D.H. (1989). Mitochondrial respiration can support NO_3^- and NO_2^- reduction during photosynthesis. Interactions between photosynthesis, respiration, and N assimilation in the N-limited green alga *Selenastrum minutum*. *Plant Physiology* **89**, 409–15.

Whitehouse, D.G., Fricaud, A.C. & Moore, A.L. (1989). The role of nonohmicity in the regulation of electron transport in plant mitochondria. *Plant Physiology* **91**, 487–91.

Wiskich, J.T. (1991). Compartmentation and metabolic control in plant mitochondria. In *Molecular Approaches to Compartmentation and Metabolic Regulation*, ed. A.H.C. Huang & L. Taiz, pp. 30–6. Beltsville, MD: American Society of Plant Physiologists.

Wiskich, J.T., Bryce, J.H., Day, D.A. & Dry, I.B. (1990). Evidence for metabolic domains within the matrix compartment of pea leaf mitochondria. *Plant Physiology* **93**, 611–16.

Wiskich, J.T. & Dry, I.B. (1985). The tricarboxylic acid cycle in plant mitochondria: its operation and regulation. In *Higher Plant Cell Respiration*. Encyclopedia of Plant Physiology, New Series, Vol. 18, ed. R. Douce & D.A. Day, pp. 281–313. Berlin: Springer-Verlag.

Woo, K.C. & Osmond, C.B. (1976). Glycine decarboxylation in mitochondria isolated from spinach leaves. *Australian Journal of Plant Physiology* **3**, 771–85.

Woo, K.C. & Osmond, C.B. (1982). Stimulation of ammonia and 2-oxoglutarate-dependent O_2 evolution in isolated chloroplasts by dicarboxylates and the role of the chloroplast in photorespiratory nitrogen recycling. *Plant Physiology* **69**, 591–6.

HANS WALTER HELDT and ULF INGO
FLÜGGE

Metabolite transport in plant cells

In a green plant, cell metabolism is highly compartmentalised. Photosynthesis and photorespiration involve the participation of three different organelles: chloroplasts, mitochondria and peroxisomes. Mitochondria and chloroplasts are surrounded by two membranes; the outer membrane is freely permeable to small molecules like metabolites (Pfaff et al., 1968; Heldt & Sauer, 1971), owing to the presence of pores formed by porins. In chloroplasts these pores allow the passage of substances up to a molecular weight of 10 000 (Flügge & Benz, 1984), whereas an exclusion limit of 4000–6000 was found in mitochondria from animal tissues (Zalman et al., 1980). Thus in mitochondria and chloroplasts the inner boundary membranes represent the border between metabolic compartments and are the site of metabolite translocators. Peroxisomes are surrounded by a single membrane. Recent studies have suggested that the boundary membrane of animal peroxisomes also contains porins, allowing the passage of metabolites of molecular mass up to 800 Da (van Veldhoven et al., 1987). This raises the question, to what extent and by what means peroxisomal metabolism is compartmentalised. This chapter presents a summary of our current knowledge of the processes by which metabolites are transferred between different subcellular compartments and between cells in the course of photosynthetic metabolism.

Phosphate translocator in chloroplasts of C3 plants

In C3 plants the phosphate–triose phosphate–3-phosphoglycerate translocator, in short called phosphate translocator (Heldt & Rapley, 1970) accepts at its binding site either P_i or a phosphate molecule attached to the end of a three-carbon chain, such as 3-phosphoglycerate or dihydroxyacetone phosphate. Three-carbon compounds in which the phosphate is attached to carbon atom C-2, such as 2-phosphoglycerate or phosphoenolpyruvate, show little interaction with the translocator. There is a minor interaction of the translocator with erythrose 4-phosphate; pentose

Society for Experimental Biology Seminar Series 50: *Plant organelles*, ed. A. K. Tobin.
© Cambridge University Press 1992, pp. 21–47.

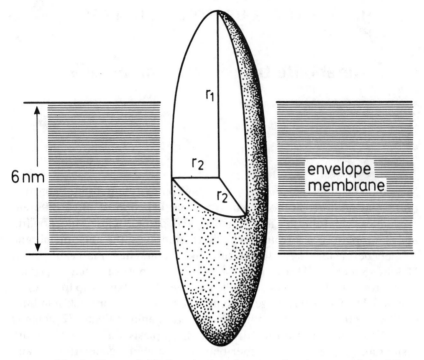

Fig. 1. Shape of the dimeric phosphate translocator, as evaluated from hydrodynamic studies (Flügge, 1985).

and hexose phosphates are not transported at all (Fliege *et al.*, 1978; Flügge & Heldt, 1991). Transport by the phosphate translocator occurs by strict counterexchange: for each molecule transported inwards, one molecule is transported outwards.

The phosphate translocator protein is the largest protein fraction of the chloroplast envelope, amounting to about 15% of the total envelope protein in spinach (Flügge & Heldt, 1981). It has been isolated after solubilisation in detergents and reconstituted into artificial membranes (Flügge & Heldt, 1981). Hydrodynamic studies of the isolated trans-locator–detergent micelle revealed that, in its functional state, the trans-locator exists as a dimer consisting of two identical subunits (Flügge, 1985). Rotational diffusion studies of the eosin-5-isothiocyanate-labelled translocator revealed that the cross-sectional area of the protein embed-ded in the membrane is about 9 nm^2 (Wagner *et al.*, 1989). This agrees well with hydrodynamic studies suggesting that the portion of the dimeric translocator protruding from both sides of the membrane is large enough to render the translocator accessible to its substrates (Fig. 1).

The phosphate translocator protein is coded for in the nucleus (Flügge, 1982) and is synthesised in the cytosol as a higher molecular mass precursor protein with an amino-terminal extension, the transit peptide (Flügge & Wessel, 1984). Such nuclear-coded proteins are subsequently transported into the chloroplasts where they are processed to their mature sizes by specific proteases (see Meadows *et al.*, this volume). Recently, the cDNA sequences of the chloroplast phosphate translocator precursor proteins from both spinach (Flügge *et al.*, 1989) and pea chloroplasts (Willey *et al.*, 1991) have been elucidated. The transit peptides of the precursor proteins were found to contain 80 (spinach) and 72 (pea) amino acid residues. The two transit peptides have only about 45% homology but they both contain a positively charged amphiphilic α-helix. Such amphiphilic structures are essential features of mitochondrial transit peptides but are completely absent in stromal or thylakoid proteins (von Heijne *et al.*, 1989). Experimental studies revealed that this structural element might have the general ability to direct precursor proteins to the chloroplast inner envelope membrane (Dreses-Werringloer *et al.*, 1991).

The mature phosphate translocator proteins from spinach and pea chloroplasts contain 324 and 330 amino acid residues, respectively, of which 87% are identical and 5% isofunctional substitutions. These amino acid sequences do not show any homology with known mitochondrial or bacterial transport proteins, nor with any other known nucleotide or amino acid sequences. Apparently, the plastid phosphate translocator represents a class of transport proteins distinct from mitochondrial and bacterial translocators. The molecular masses of the mature phosphate translocators from pea and spinach, calculated from the deduced amino acid sequences (35 957 and 35 603 Da, respectively) are significantly higher than those determined by SDS polyacrylamide gel electrophoresis (*c.* 29 000–30 000). The apparent discrepancy is probably attributable to an unusually high binding of detergent resulting from the hydrophobic nature of the translocator proteins. Indeed, hydrophobicity distribution analyses indicate that each monomer of the phosphate translocator protein contains at least six extended regions of strong hydrophobicity. Each of these putative transmembrane segments consists of about 20–23 amino acid residues which appear to be able to form α-helical structures, large enough to span the inner envelope membrane, thus anchoring the protein in the membrane (Fig. 2). Some of these α-helices show an alteration between pronounced hydrophilic as well as hydrophobic regions (Willey *et al.*, 1991). It is proposed that, in the functional dimeric phosphate translocator protein, the 12 transmembrane helices are arranged in such a way that the hydrophilic regions are directed towards the inside of the protein thus forming a hydrophilic pore through which the substrates

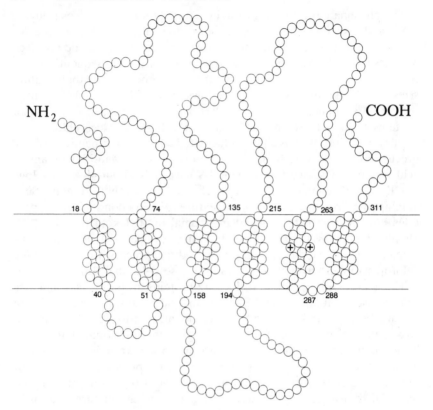

Fig. 2. Topology of the phosphate translocator protein in the chloroplast inner envelope membrane. The positive charges symbolise the location of lysine-273 and arginine-274 in the transmembrane helix V.

could be transported across the membrane. The hydrophobic regions of the α-helices could serve to anchor the protein in the membrane. Based on computer-aided modelling of the individual helices and on additional information an alignment of the 12 α-helices in the dimeric translocator protein was proposed and is depicted in Fig. 3 (Wallmeier et al., 1992). According to this picture the functional translocator exhibits an approx-imately C_2-symmetry with a C_2-symmetry axis perpendicular to the mem-brane plane (Fig. 3). The resulting structure has a total length of 5.0 nm and a width of 2.6 nm which is in agreement with values obtained by hydrodynamic and rotational diffusion studies of the isolated translocator protein (Flügge, 1985; Wagner et al., 1989).

Earlier studies showed that the chloroplast phosphate translocator

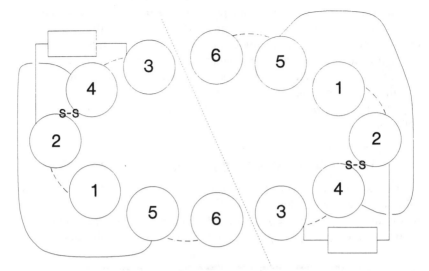

Fig. 3. Hypothetical arrangement of the transmembrane helices of the dimeric phosphate translocator protein forming a hydrophilic translocation channel (Wallmeier *et al.*, 1992).

transports twice negatively charged anions (A^{2-}: Fliege *et al.*, 1978), and that a lysine and an arginine residue are involved in the binding (Flügge & Heldt, 1986). The fact that only one substrate binding site per dimeric phosphate translocator molecule was found indicates that the substrate binding site is exposed to only one side of the membrane at a time, thereby catalysing the observed counterexchange of substrates in a gated pore mode (Gross *et al.*, 1990). The only charged amino acid residues contained in the part of the protein embedded in the membrane and which point to the centre of the translocation channel appear to be a lysine and an arginine residue (Lys-273 and Arg-274 of the spinach translocator protein, helix V: see Fig. 2). It is feasible that these two cationic residues are involved in the binding of A^{2-} transported by the phosphate translocator.

Very recently, Blobel's group has cloned and sequenced a putative chloroplast import receptor (Schnell *et al.*, 1990) which had previously been identified by anti-idiotypic antibodies raised against a chloroplast transit peptide (Pain *et al.*, 1988). Surprisingly, Schnell *et al.* (1990) found that the primary structure of this protein, named p36, turned out to be identical with that described previously for the chloroplast phosphate translocator (Flügge *et al.*, 1989; Willey *et al.*, 1991). This has led to the suggestion that p36 is not responsible for chloroplast phosphate transport activity (Schnell *et al.*, 1990). This proposal has been recently rejected by Flügge *et al.* (1991) by presenting clear evidence that the major chloroplast envelope polypeptide mentioned above is indeed the phosphate translocator and not the protein import receptor. The observation that antibodies directed against p36 react with components involved in precursor binding might be explained best by assuming that both the import receptor and the chloroplast phosphate translocator share some common epitopes which are recognised by the antibody.

Carbon transport during photosynthesis of a C3 plant

Figure 4 summarises major transport processes involved in carbon metabolism of a C3 plant. Whereas the CO_2 required for carbon assimilation is taken up by unspecific permeation (Werdan *et al.*, 1972), the photosynthetic products triose-P and PGA are released via the phosphate trans-

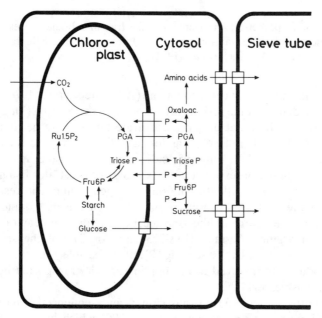

Fig. 4. Transport processes involved in photosynthetic metabolism of a C3 plant. Abbreviations for this and Figs 5–9: PGA, 3-phosphoglycerate; Fru6P, fructose 6-phosphate; Fru1.6P$_2$, fructose 1,6-bisphosphate; Glc6P, glucose 6-phosphate; triose P, triose phosphate; Ru1.5P$_2$, ribulose 1,5-bisphosphate; Ru5P, ribulose 5-phosphate; PEP, phosphoenolpyruvate; Pyr, pyruvate; Mal, L-malate; OAA, oxaloacetate; OH Pyr, hydroxypyruvate.

locator in exchange with P_i. Photosynthetic products can be stored temporarily within the chloroplasts as starch. Starch mobilisation occurring simultaneously via hydrolytic and phosphorylitic degradation (Peavey *et al.*, 1977; Stitt & Heldt, 1981) results in the formation of hexose phosphates which, after conversion to triose phosphates, can be exported via the phosphate translocator, or in the formation of glucose and maltose, which are released by a glucose translocator (Schäfer *et al.*, 1977; Herold *et al.*, 1981).

The triose-P exported from the chloroplasts is precursor for the synthesis of sucrose, and PGA for synthesis of amino acids. Sucrose and amino acids (in the case of spinach mainly glutamate, glutamine and aspartate: Riens *et al.*, 1991) are the main products translocated via the sieve tubes to other parts of the plant. There have been indications that the transfer of substances from the cytosol to the sieve tubes involves a crossing through the apoplastic space (for review see Turgeon, 1989). This concept of apoplastic phloem loading has been tested recently by the expression of yeast cell wall invertase in leaves of transgenic tobacco, *Arabidopsis* and potato. The introduction of sucrose hydrolysis catalysing invertase into the apoplast resulted in an accumulation of carbohydrates including sucrose in the source leaves (von Schaewen *et al.*, 1990) and an inhibition of the phloem loading of sucrose (Heineke *et al.*, 1992). These results clearly demonstrate that the apoplastic space is indeed involved in the phloem loading process.

As the apoplastic phloem loading involves the crossing of two membranes, one might expect that a set of translocators is required for each membrane. In plasma membrane vesicles from sugar beet leaves a sucrose translocator (Lemoine & Delrot, 1989; Bush, 1990) and several amino acid translocators (Li & Bush, 1990) have been characterised. All transport mediated by these translocators was shown to be driven by proton symport. At present it is not known in which of the two membranes of the apoplastic phloem loading pathway the above-mentioned translocators are located. In spinach leaves a comparison of the concentrations in the cytosol, as determined by non-aqueous subcellular fractionation of frozen leaves (Gerhardt & Heldt, 1984) and in the phloem sap, as collected from the stylet of aphids which had been severed by laser beam (Barlow & McCully, 1972), revealed characteristic differences in the phloem loading of sucrose and amino acids (Riens *et al.*, 1991). Whereas the transfer of sucrose from the cytosol to the sieve tube was shown to involve a concentration increase by a factor of about 10, amino acids were not found to be concentrated to any significant extent. Thus the ratio of sucrose to total amino acids changed from 0.8 in the cytosol to about 5 in the phloem sap. Similar results have been obtained with barley leaves (Winter *et al.*, 1992). It has been postulated from these results that the translocation of amino acids via the sieve tubes requires the mass flow of sucrose driven by active sucrose transport.

Metabolite transport in C4 plants

In C4 plants the accumulation of CO_2 for photosynthesis in the bundle sheath chloroplasts requires an extensive traffic of metabolites between cells and subcellular compartments (Hatch, 1987; see also Edwards & Krall, this volume). Figure 5 summarises transport processes occurring during photosynthesis in maize. The mesophyll and bundle sheath cells are connected by plasmodesmata allowing the passage of molecules up to a molecular mass of 850–900 Da (Weiner *et al.*, 1988). Metabolite flow between the mesophyll and bundle sheath cells is driven by diffusion gradients whose existence has been shown by the use of techniques of intercellular metabolite analysis in maize leaves (Leegood, 1985; Stitt & Heldt, 1985; Weiner & Heldt, 1992).

The oxaloacetate is formed in the cytosol of the mesophyll cells from the capture of the atmospheric CO_2 by phosphoenolpyruvate carboxylase. It is transported into the chloroplast by a specific translocator, which

MESOPHYLL CELL   BUNDLE SHEATH CELL

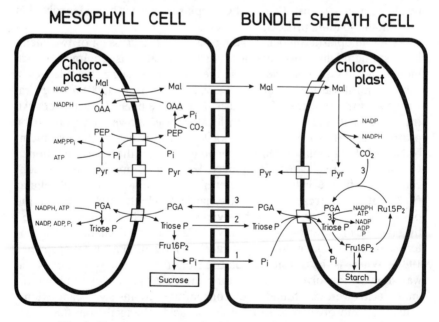

Fig. 5. Transport processes in C4 metabolism of a maize leaf.

has a very high affinity for oxaloacetate with very low competition by malate (Hatch *et al.*, 1984). In the stroma oxaloacetate is reduced to malate. It is not known how the subsequent release of malate is related to oxaloacetate uptake, since in maize mesophyll chloroplasts transport of malate is relatively little affected by oxaloacetate (Day & Hatch, 1981*a*). It seems likely that transport of oxaloacetate and malate involves different binding sites. The uptake of malate into bundle sheath chloroplasts has not yet been characterised. The pyruvate formed in the bundle sheath chloroplasts by malic enzyme activity has to be transferred to the mesophyll chloroplasts for the regeneration of phosphoenolpyruvate. In both mesophyll (Huber & Edwards, 1977*a*; Flügge *et al.*, 1985; Ohnishi & Kanai, 1987) and bundle sheath (Ohnishi & Kanai, 1987) chloroplasts a rapid and specific transport of pyruvate has been shown, although the properties of pyruvate transport in the two types of chloroplasts appear to be different. In chloroplasts from mesophyll cells, but not from bundle sheath cells, pyruvate is transported into the chloroplasts against a concentration gradient. In mesophyll chloroplasts from the C4 NADP-malic enzyme type plants *Zea mays* and *Sorghum bicolor*, this active transport was found to be driven by a H^+ gradient (Flügge *et al.*, 1985; Ohnishi & Kanai, 1990) whereas in the C4 NAD-malic enzyme plant *Panicum miliaceum* and in the C4 PEP carboxykinase type plants *Urochloa panicoides* and *Panicum maximum* active pyruvate transport in mesophyll chloroplasts was found to be driven by a Na^+ gradient, but not by a H^+ gradient (Ohnishi *et al.*, 1990).

For the conversion of pyruvate into phosphoenolpyruvate occurring in the mesophyll chloroplasts, P_i has to be taken up, and the phosphoenolpyruvate formed has to be released to the cytosol in order to be carboxylated. This transport is also catalysed by a phosphate translocator. In contrast to chloroplasts of the C3 plants spinach and pea, mesophyll chloroplasts from C4 plants also transport phosphoenolpyruvate and 2-phosphoglycerate. Measurements of the effects of inhibitors on the transport of the various substances and of the competition between substrates and back-exchange clearly showed that in C4 mesophyll chloroplasts PEP, triose-P, 3-phosphoglycerate and P_i are transported by a single translocator (Huber & Edwards, 1977*b*; Day & Hatch, 1981*b*; Rumpho & Edwards, 1985; Gross *et al.*, 1990). Since in maize bundle sheath chloroplasts the capacity of non-cyclic electron transport is very low (Nakano & Edwards, 1987), the redox equivalents required for CO_2 fixation have to be delivered to the bundle sheath chloroplasts from the mesophyll chloroplasts, most probably by a PGA/triose-P shuttle. Such a redox transfer implies that in the bundle sheath and mesophyll chloroplasts the PGA/triose-P exchanges proceed in opposite directions. A comparison of the properties of the phosphate translocator in mesophyll and bundle sheath chloroplasts from the C4 plant *Panicum miliaceum* revealed that although the bundle sheath chloroplasts also transported PEP, the kinetic properties of the translocators from the two types of chloroplasts were different with respect to their affinity for PEP, triose-P and PGA (Ohnishi *et al.*, 1989), indicating that the chloroplast phosphate translocator can be different in different cells of a leaf of a single species.

Phosphate translocator in root plastids

Transport studies with plastids from pea roots showed an uptake of P_i which was inhibited by triose-P or PGA. These results indicated that the root plastids possessed a phosphate translocator similar to that in C3 chloroplasts (Emes & Traska, 1987). Subsequent studies revealed that these plastids also transport glucose 6-phosphate, but not glucose 1-phosphate (Borchert *et al.*, 1989). The measurement of the concentration dependence on the uptake of the various substances, the competitive inhibition by one another and back-exchange, provided evidence that pea root plastids contain a phosphate translocator which is similar to chloroplast translocators in transporting P_i, triose-P, and PGA and (as in C4 plants) also PEP, but which differs in that it also transports glucose 6-phosphate (Borchert *et al.*, 1989; Heldt *et al.*, 1991). A major function of plastids contained in pea root tissue is the reduction of nitrite to ammonia for which redox equivalents are provided by the oxidative pentose phosphate pathway (Bowsher *et al.*, 1989). In these plastids, the phosphate translocator serves the purpose of supplying glucose 6-phosphate from the cytosol as a substrate for the oxidative pentose phosphate pathway (Fig. 6). Because of the lack of fructose 1,6-bisphosphatase activity in the plastid stroma (Entwistle & ap Rees, 1988), the triose-P formed by the oxidative pentose phosphate pathway in the plastids cannot be converted there into hexose phosphates. For this reason the oxidative pentose phosphate pathway in the root plastids involves a glucose 6-P–triose-P exchange (L. Harborth, S. Borchert and H. W. Heldt, unpublished data).

Transfer of redox equivalents and of ATP between subcellular compartments

In a leaf cell performing photosynthesis, hydroxypyruvate reduction occurring as part of the photorespiratory pathway in the perosixomes and nitrate reduction proceeding as a partial step of nitrate assimilation in the

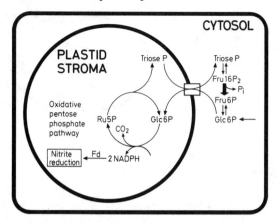

Fig. 6. Transport involved in metabolism of a pea root plastid.

cytosol both require a supply of redox equivalents, which might be served either from photosynthetic electron transport in the chloroplasts or from mitochondrial substrate oxidation. A prerequisite for a redox transfer between the compartments is the existence of redox gradients. For an estimation of redox gradients the redox states of pyridine nucleotides in the subcellular compartments of a leaf need to be known. Using a non-aqueous fractionation procedure the NADPH/NADP ratio in the stroma of illuminated spinach leaves was determined as 0.5 (Heineke *et al.*, 1991). It has not yet been possible to determine the mitochondrial NADH/NAD ratio by the same method, but measurements with isolated mitochondria indicated that in the steady-state of leaf metabolism the NADH/NAD ratio in the mitochondria may be in the order of 10^{-1} (Rustin *et al.*, 1987; Krömer & Heldt, 1991*a*). The cytosolic NADH/ NAD ratio of illuminated spinach leaves was estimated from cytosolic metabolite ratios assayed by non-aqueous fractionation of leaves to be about 10^{-3} (Heineke *et al.*, 1991). There are good reasons for assuming that the NADH/NAD ratio in the peroxisomes might not be very different from the ratio in the cytosol. These results lead to the conclusion that there are redox gradients in a leaf cell allowing the transfer of redox equivalents from either the chloroplasts or the mitochondria to the cytosol and the peroxisomes.

Transfer of redox equivalents from the chloroplasts to the cytosol can occur by malate/oxaloacetate shuttle (Fig. 7) or by triose-P/PGA shuttle (Fig. 8). Spinach chloroplasts have a high capacity for redox transfer by malate/OAA shuttle, the properties of oxaloacetate and malate transport being very similar to those of C4 mesophyll chloroplasts (see above).

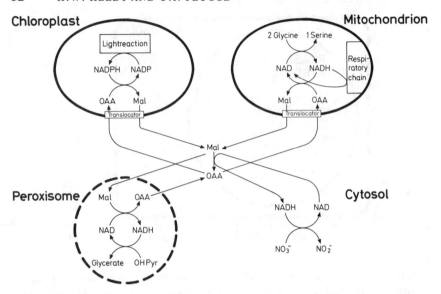

Fig. 7. Transfer of redox equivalents between subcellular compartments by malate–oxaloacetate shuttles.

Redox transfer by the shuttle is controlled at the site of the stromal NADP-malate dehydrogenase in such a way that this enzyme is active only when the stromal NADPH/NADP ratio is very high (Scheibe, 1987). In this way, the large redox gradient between the stromal NADPH/NADP and the cytosolic NADH/NAD is maintained and only excessive redox equivalents generated by photosynthesis are released to the cytosol. With the triose-P/PGA shuttle, the provision of redox equivalents in the form of NADH is accompanied by the export of ATP. This shuttle, catalysed by the phosphate translocator, also does not appear to be limited in its rate by the transport. Experimental results suggest that, in the triose-P/PGA shuttle, the redox gradient between the NADPH/NADP ratio in the stroma and the NADH/NAD ratio in the cytosol is maintained primarily by a limitation of the conversion of triose-P to PGA in the cytosol (Heineke et al., 1991).

In a leaf cell the mitochondria also have the capacity for a very rapid redox transfer from the matrix to the cytosol by malate/OAA shuttle (Ebbighausen et al., 1985). The transport of malate and OAA facilitating this shuttle is very different from the known dicarboxylate transport of animal mitochondria. In plant mitochondria both substances can be transported independently of one another or of other anions by electrogenic uniport (Zoglowek et al., 1988). The redox gradient between the com-

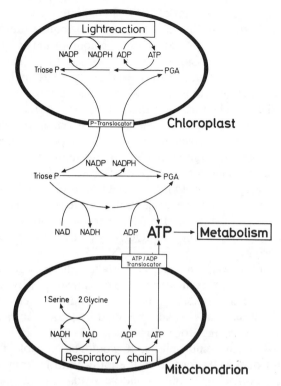

Fig. 8. Transfer of ATP and of redox equivalents between subcellular compartments.

partments is probably maintained by regulation of malate transport (Zoglowek *et al.*, 1988; Krömer & Heldt, 1991*a*).

The transfer of redox equivalents into the peroxisomes for reduction of hydroxypyruvate also occurs by malate/OAA shuttle (Yu & Huang, 1986; Heldt *et al.*, 1990). This shuttle does not involve a translocator, as will be discussed in detail below. Although there is no specific transport involved, a direct reduction by cytosolic NADH of hydroxypyruvate contained in the peroxisomes does not seem possible, as the access of NADH to the peroxisomal hydroxypyruvate reductase is restricted, owing to a diffusion limitation at the low NADH concentrations (10^{-6} M) occurring in the cytosol (see above).

In isolated chloroplasts, as well as in isolated mitochondria, the capacities of redox transfer by malate/OAA shuttles largely exceed the rates required for hydroxypyruvate reduction and nitrite reduction at maximum rates of leaf photosynthesis (Heldt *et al.*, 1990). In the photorespiratory cycle, the amount of NADH generated from oxidation of glycine to serine is equal to the amount of NADH required for reduction of hydroxypyruvate in the peroxisomes. The equilibrium of hydroxypyruvate reduction is far to the side of glycerate ($K_{eq} = 10^{-5}$). Therefore, because of the high activities of mitochondrial and peroxisomal malate/OAA shuttles, this reaction would act as a drain, sequestering the redox equivalents from the mitochondria. For this reason the NADH generated from glycine oxidation would not be available for mitochondrial electron transport, unless the redox equivalents required in the peroxisomes were to be provided from another source, namely via the malate/OAA shuttle from the chloroplastic electron transport. Because of the accepted stoichiometry of non-cyclic photosynthetic electron transport (NADPH/ATP = 0.75) and the stoichiometry of the Calvin cycle (NADPH/ATP = 0.66) there seems to be a surplus of NADPH in the chloroplasts. This surplus could be transferred via malate/OAA shuttles for use in the cytosol and peroxisomes.

The redox gradient between the mitochondria and the cytosol makes it seem impossible that redox equivalents can be transferred by malate/OAA shuttle from the chloroplasts to the mitochondria directly. But in C3 plants photosynthesis is always accompanied by photorespiration. If the demand of the peroxisomes for redox equivalents to reduce hydroxypyruvate is served by the chloroplasts, it would be possible that an equivalent amount of NADH formed in the mitochondria from glycine oxidation is made available for mitochondrial oxidation. In this way, though indirectly, redox transfer by malate/OAA shuttle from the chloroplasts could provide redox equivalents for mitochondrial ATP synthesis to proceed at more than 50% of its maximal capacity (Krömer & Heldt, 1991b). Such a cooperation of chloroplast and mitochondrial metabolism appears to have the advantage of preventing the overreduction of photosynthetic electron transport carriers and of improving the quantum yield of ATP formation. Whereas in cyclic photophosphorylation, according to the accepted stoichiometry, 2 quanta of light yield 0.66 mol ATP, for non-cyclic photophosphorylation connected with the oxidation of the formed redox equivalents by mitochondrial electron transport, from 4 quanta of light about 4 mol ATP can be formed, representing a three-fold higher yield. Thus, in balance, redox equivalents formed by photosynthetic electron transport can ultimately provide the fuel for mitochondrial ATP synthesis. In this way the ATP demand in the cytosol during photosynthesis can be met very efficiently.

It is a matter of debate, to what extent in a photosynthesising leaf cell the ATP demand of the cytosol is met by photophosphorylation via the triose-P/PGA shuttle (Yin et al., 1990) and by oxidative phosphorylation as discussed above. Both pathways have a high capacity for supply of ATP to the cytosol and it may depend on the metabolic conditions which of these capacities is primarily utilised. In barley protoplasts and in whole barley leaves the contribution of mitochondrial oxidative phosphorylation to photosynthetic metabolism has been demonstrated by the use of oligomycin. Low concentrations of this substance, which strongly inhibit mitochondrial ATP synthesis but do not affect chloroplast photophosphorylation, were shown to inhibit photosynthesis of barley protoplasts by 40–60%. This inhibition was reversed and the full rate of photosynthesis regained when the protoplasts were ruptured so as to leave the chloroplasts intact (Krömer *et al.*, 1988). Similarly, oligomycin inhibited the photosynthesis of whole barley leaves (Krömer & Heldt, 1991*b*). Metabolite analysis revealed that in the cytosol oligomycin caused an accumulation of glucose 6-phosphate connected with shortage of ATP, and in the stroma an increase of the reductive state of the NADPH system. These results, together with the experiments of Gardeström & Wigge (1988), showing that in protoplasts mitochondrial glycine oxidation is coupled to oxidative phosphorylation in the light, clearly demonstrate that during photosynthesis of a plant leaf cell, mitochondrial oxidative phosphorylation contributes to the ATP supply of the cell and prevents overreduction of the chloroplast redox carriers, by oxidising surplus reductive equivalents generated by photosynthetic electron transport.

The mitochondrial ADP/ATP translocator involved in oxidative phosphorylation is similar in animals and plants. First evidence for the existence of an ADP/ATP translocator came from experiments with rat liver mitochondria, showing that atractyloside inhibited the phosphorylation of added ADP but not of internal ADP (Heldt *et al.*, 1965). Mitochondrial ADP/ATP transport is an electrogenic process, resulting in an asymmetric distribution of ADP and ATP on the two sides of the inner mitochondrial membrane. With isolated rat liver mitochondria (Heldt *et al.*, 1972) and whole liver tissue (Soboll *et al.*, 1978) and with wheat leaf protoplasts (Stitt *et al.*, 1982) the ATP/ADP ratio outside the mitochondria was found to be much higher than in the matrix. Thus energy generated by mitochondrial electron transport in the form of an electrochemical proton gradient is required not only for the actual synthesis of ATP at the matrix side of the mitochondrial inner membrane, but also for transferring the newly formed ATP to the higher phosphorylation potential of the cytosolic ATP/ADP system (Klingenberg & Heldt, 1982). The membrane protein involved in ADP/ATP translocation has been characterised as a dimer consisting of two identical subunits of 31 kDa, catalysing a gated-pore mechanism (Klingenberg, 1981). The assay of the amino acid sequences revealed a high degree of homology between the ADP/ATP translocator proteins from beef heart (Aquila *et al.*, 1982) and maize (Baker & Leaver, 1985).

Chloroplasts also contain an ATP/ADP translocator catalysing a counterexchange, although its activity is rather low (Heldt, 1969). In chloroplasts from mature spinach leaves the activity of the ATP/ADP translocator was found to be more than one order of magnitude lower than the activity of the phosphate translocator. In pea, chloroplasts from young leaves showed a higher translocator activity than chloroplasts from mature leaves (Robinson & Wiskich, 1977). Apparently the activity of the ATP/ADP translocator is dependent on the developmental state of the plastid. The chloroplast ATP/ADP translocator is not suited for an ATP export from the chloroplasts, as the transport is highly specific for external ATP. Inward transport of ADP is 10 times slower than that of ATP. Moreover, in leaves, not only in the dark but also in the steady state of photosynthesis in the light, the ATP/ADP ratio is higher in the cytosol than in the chloroplast stroma (Santarius & Heber, 1965; Keys & Whittingham, 1969; Sellami, 1976; Stitt *et al.*, 1982; Heineke *et al.*, 1991). Because of the gradient in ATP/ADP ratios the ATP/ADP translocator should facilitate an import of ATP into chloroplasts. It may not only supply the chloroplasts in the dark with ATP, but also enable to a limited extent in the light a transfer of ATP generated by mitochondrial oxidative phosphorylation into the chloroplasts to minimise the energetically more costly cyclic photophosphorylation (see above). A role of the ATP/ADP translocator in the protein import into chloroplasts has been demonstrated (Flügge & Hinz, 1986).

The chloroplast ATP/ADP translocator differs in its specificity markedly from the mitochondrial ADP/ATP translocator and is, in contrast to the mitochondrial translocator, not inhibited by atractyloside (Heldt, 1969). Bongkrekic acid, a strong inhibitor of mitochondrial ADP/ATP transport, showed no effect on ATP/ADP transport in spinach chloroplasts (Flügge & Hinz, 1986), but an inhibitory effect of this substance has been reported on ATP transport in pea chloroplasts (Woldegiorgis *et al.*, 1983). Akazawa *et al.* (1991) recently reported that amyloplasts from sycamore cell cultures transport ATP, ADP, AMP and ADPglucose in a competitive mode: transport of ATP and of ADPglucose was inhibited by carboxyatractyloside, whereas transport of ADP and AMP was nearly insensitive to this inhibitor. Moreover, it has been shown that inner envelope membranes of sycamore amyloplasts may contain a 32 kDa protein which cross-reacts with an antiserum directed against the ADP/ATP translocator from mitochondria in *Neurospora crassa* (Ngernprasirtsiri *et al.*, 1989). It would be very surprising indeed if plastids contained a mitochondrial translocator. Further experiments are required in order definitely to exclude the possibility that previous experimental results have been caused by mitochondrial contamination of plastid preparations.

Transfer of metabolites between subcellular compartments involved in the photorespiratory pathway

The photorespiratory cycle is a pathway of extreme compartmentation (Tolbert, 1980*b*; see also Wallsgrove *et al.*, this volume). The recycling of the CO_2 fixation by-product phosphoglycolate requires altogether 20 passages of intermediates across organelle membranes. Figure 9 summarises these events. Two molecules of glycolate, resulting from the hydrolysis of phosphoglycolate in the stroma, are exported from the chloroplasts in exchange for one molecule of glycerate, which is subsequently phosphorylated via glycerate kinase yielding 3-phosphoglycerate. This exchange is catalysed by a translocator, having similar affinities for glycolate, glyoxylate, D-glycerate and D-lactate (Howitz & McCarty, 1985). By the same translocator, glycolate and glycerate can be either transported together by counterexchange, or individually by H^+ symport (or OH^- antiport). This flexibility makes it possible that the amount of glycerate returning to the chloroplasts is only half of the amount of glycolate released. The glycolate has to enter the peroxisomal compartment to be oxidised to glyoxylate, followed by transamination to glycine by reaction with equal amounts of serine and glutamate. Glycine leaves the peroxisomal compartment and is taken up into the mitochondrial matrix where oxidation of two molecules of glycine results in the formation of one molecule of serine, accompanied by reduction of NAD and release of CO_2 and NH_3. The mechanism of glycine uptake into the mitochondria, and of the release of serine, remains to be elucidated. Because of the very high fluxes required in photorespiratory metabolism one would expect that these fluxes are translocator-mediated, and there are indeed experimental results supporting the existence of a specific transport of glycine and serine in plant mitochondria (Walker *et al.*, 1982; Yu *et al.*, 1983).

Serine enters the peroxisome for transamination to hydroxypyruvate and reduction to glycerate, which is then transferred to the chloroplast stroma (see above). The redox equivalents (NADH) required for hydroxypyruvate reduction in the peroxisomes can be supplied by malate/OAA shuttle from the mitochondria, as described above. Ammonia released in the course of glycine oxidation reaches the chloroplast stroma, probably by diffusion, where it is refixed by the glutamine synthetase–glutamate

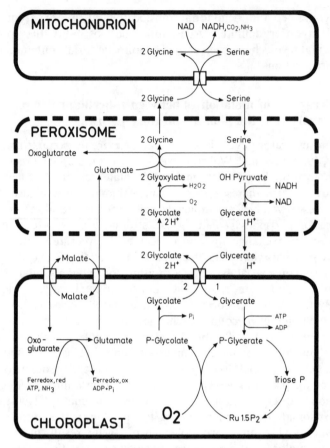

Fig. 9. Transport processes involved in photorespiration.

synthase pathway. For this, 2-oxoglutarate, derived from transamination in the peroxisomes, has to enter the chloroplast stroma and the glutamate formed from NH_3 refixation is released again. The uptake of 2-oxoglutarate into and the release of glutamate from the chloroplasts is facilitated by two different translocators, each catalysing a counterexchange with malate (Woo *et al.*, 1987).

Photorespiratory metabolism requires that four different metabolites (glycolate, glutamate, serine and malate) enter the peroxisomes and another four (glycine, 2-oxoglutarate, glycerate and OAA) are released from them. The question had been repeatedly asked as to whether specific translocators are required for this movement of metabolites. There were in the past, however, no conclusive experimental results which could provide an answer to this problem.

Enzymes contained in intact peroxisomes show latency (Tolbert, 1980*a*). Latency of enzymes is also found in chloroplasts and mitochondria, where it is caused by the impermeability of the inner boundary membranes and is abolished when this membrane is ruptured by osmotic shock or detergent treatment. Therefore in mitochondria and chloroplasts the latency of different enzymes is equally affected by a rupture of the boundary membrane. In intact peroxisomes, however, different enzymes show different latency, and upon ageing different enzymes are differently affected (Heupel *et al.*, 1991). The incubation of intact peroxisomes for 10 min at room temperature in distilled water, a treatment resulting in complete rupture of chloroplasts and mitochondria and hence full release of latency, had in peroxisomes only little effect on the latency of enzymes. Electron microscopic studies revealed that peroxisomes exposed to 'osmotic shock' altered their shape and decreased their size but retained a compact matrix devoid of a continuous boundary membrane (Heupel *et al.*, 1991).

Isolated peroxisomes are able to synthesise glycerate at a high rate when serine, glycolate, malate and NAD are added together as substrates (Yu & Huang, 1986). When this reaction is started by adding peroxisomes to a medium containing these four metabolites, no lag in glycerate formation is observed, even if the reaction is carried out at minute peroxisomal concentrations (Heupel *et al.*, 1991). The reaction sequence of glycerate synthesis involves the formation of glyoxylate, hydroxypyruvate and NADH as intermediates. The absence of any lag phase in glycerate synthesis clearly demonstrates that the intermediates of the reaction sequence do not leak out into the medium. This indicates that peroxisomal metabolism is highly compartmentalised. When peroxisomes are disrupted by a detergent (e.g. Triton X-100), which abolishes the latency of enzymes but does not affect the activities of the enzymes *per se*, a very long lag phase is observed and the final rate of glycerate synthesis is much lower than with intact peroxisomes. With peroxisomes, which had been subjected to 'osmotic shock' no such lag period was observed (Heupel *et al.*, 1991) indicating that these 'shocked' peroxisomes had retained their metabolic compartmentation, although an intact boundary membrane was no longer present (see above).

These results clearly demonstrate that the confinement of intermediates in the peroxisomal compartment is not the result of selective permeability of a boundary membrane, as in chloroplasts or mitochondria, but a function of the peroxisomal matrix *per se*. Apparently a structural arrangement of enzymes in the peroxisomal membrane leads to metabolite channelling, preventing the release of intermediates such as H_2O_2, glyoxylate and hydroxypyruvate, but enabling ready access for reactants such as glycolate, malate and serine from outside.

If compartmentation of peroxisomal metabolism were the sole consequence of permeability properties of the peroxisomal boundary membrane, one might have expected at least three different translocators of very high specificity, e.g. one translocator transporting glycolate but not glyoxylate, another transporting glycerate but not hydroxypyruvate. Whereas there is no evidence available for the existence of any such highly selective translocators, experimental evidence with animal peroxisomes suggests that the peroxisomal boundary membrane may contain unspecific protein channels, allowing the passage of metabolites of molecular mass up to 800 Da (van Veldhoven et al., 1987). In view of the similarities between peroxisomes in animal and plant tissues it seems likely that plant peroxisomes also contain pores, allowing the passage of the intermediates of the photorespiratory pathway into and out of the peroxisomes.

Acknowledgements

The authors are grateful to Mr Bernd Raufeisen for drawing the figures. Own work mentioned here has been supported by the Deutsche Forschungsgemeinschaft.

References

Akazawa, T., Pozueta-Romero, J. & Ardila, F. (1991). Adenylate translocators and starch biosynthesis in the plant cell. In *Molecular Approaches to Compartmentation and Metabolic Regulation*, ed. A. H.C. Huang & L. Taitz, pp. 74–85. Belstville, MD: American Society of Plant Physiologists.

Aquila, H., Misra, D., Eulitz, M. & Klingenberg, M. (1982). Complete amino acid sequence of the ADP/ATP carrier from beef heart mitochondria. *Hoppe-Seylers Zeitschrift der Physiologischen Chemie* **363**, 345–9.

Baker, A. & Leaver, C.J. (1985). Isolation and sequence analysis of a cDNA encoding the ATP/ADP translocator of *Zea mays* L. *Nucleic Acids Research* **13**, 5857–67.

Barlow, C.A. & McCully, M.E. (1972). The ruby laser instrument for cutting the stylets of feeding aphids. *Canadian Journal of Zoology* **50**, 1497–8.

Borchert, S., Große, H. & Heldt, H.W. (1989). Specific transport of phosphate, glucose 6-phosphate, dihydroxyacetone phosphate and 3-phosphoglycerate into amyloplasts from pea roots. *FEBS Letters* **253**, 183–6.

Bowsher, C.G., Hucklesby, D.B. & Emes, M.J. (1989). Nitrite reduction and carbohydrate metabolism in plastids purified from roots of *Pisum sativum* L. *Planta* **177**, 359–66.

Bush, D.R. (1990). Electrogenicity, pH dependence, and stoichiometry of the proton-sucrose symport. *Plant Physiology* **93**, 1590–6.

Day, D.A. & Hatch, M.D. (1981*a*). Dicarboxylate transport in maize mesophyll chloroplasts. *Archives of Biochemistry and Biophysics* **211**, 738–42.

Day, D.A. & Hatch, M.D. (1981*b*). Transport of 3-phosphoglyceric acid, phosphoenolpyruvate, and inorganic phosphate in maize mesophyll chloroplasts, and the effect of 3-phosphoglyceric acid on malate and phosphoenolpyruvate production. *Archives of Biochemistry and Biophysics* **211**, 743–9.

Dreses-Werringloer, U., Fischer, K., Wachter, E., Link, T.A. & Flügge, U.-I. (1991). cDNA sequence and deduced amino acid sequence of the precursor of the 37 kD inner envelope membrane polypeptide from spinach chloroplasts: its transit peptide contains an amphiphilic α-helix as the only detectable structural element. *European Journal of Biochemistry* **195**, 361–8.

Ebbighausen, H., Chen, J. & Heldt, H.W. (1985). Oxaloacetate translocator in plant mitochondria. *Biochemica et Biophysica Acta* **810**, 184–99.

Emes, M.J. & Traska, A. (1987). Uptake of inorganic phosphate by plastids purified from the roots of *Pisum sativum* L. *Journal of Experimental Botany* **38**, 1781–8.

Entwistle, G. & ap Rees, T. (1988). Enzymic capacities of amyloplasts from wheat (*Triticum aestivum*) endosperm. *Biochemical Journal* **255**, 391–6.

Fliege, E., Flügge, U.-I., Werdan, K. & Heldt, H.W. (1978). Specific transport of inorganic phosphate, 3-phosphoglycerate and triosephosphates across the inner membrane of the envelope in spinach chloroplasts. *Biochimica et Biophysica Acta* **502**, 232–47.

Flügge, U.-I. (1982). Biogenesis of the chloroplast phosphate translocator. *FEBS Letters* **140**, 273–6.

Flügge, U.-I. (1985). Hydrodynamic properties of the Triton X-100 solubilized chloroplast phosphate translocator. *Biochimica et Biophysica Acta* **815**, 299–305.

Flügge, U.I. & Benz, R. (1984). Pore forming activity in the outer membrane of the chloroplast envelope. *FEBS Letters* **169**, 85–9.

Flügge, U.-I., Fischer, K., Gross, A., Sebald, W., Lottspeich, F. & Eckerskorn, C. (1989). The triose phosphate 3-phosphoglycerate-phosphate translocator from spinach chloroplasts: nucleotide sequence of a full-length cDNA clone and import of the *in vitro* synthesized precursor protein into chloroplasts. *EMBO Journal* **8**, 39–46.

Flügge, U.-I. & Heldt, H.W. (1981). The phosphate translocator of the chloroplast envelope. Isolation of the carrier protein and reconstitution of transport. *Biochimica et Biophysica Acta* **638**, 296–304.

Flügge, U.-I. & Heldt, H.W. (1986). Chloroplast phosphate–triose phosphate–phosphoglycerate translocator: its identification, isolation and reconstitution. *Methods of Enzymology* **125**, 716–30.

Flügge, U.I. & Heldt, H.W. (1991). Metabolite translocators of the chloroplast envelope. *Annual Review of Plant Physiology and Plant Molecular Biology* **42**, 129–44.

Flügge, U.-I. & Hinz, G. (1986). Energy dependence of protein translocation into chloroplasts. *European Journal of Biochemistry* **160**, 563–70.

Flügge, U.-I., Stitt, M. & Heldt, H.W. (1985). Light driven uptake of pyruvate into mesophyll chloroplasts from maize. *FEBS Letters* **183**, 335–9.

Flügge, U.-I., Weber, A., Fischer, K., Lottspeich, F., Eckershorn, Ch., Waegemann, K. & Soll, J. (1991). The major chloroplast envelope polypeptide is the phosphate translocator and not the protein import receptor. *Nature* **353**, 364–7.

Flügge, U.-I. & Wessel, D. (1984). Cell-free synthesis of putative precursors for envelope membrane polypeptides of spinach chloroplasts. *FEBS Letters* **168**, 355–9.

Gardeström, P. & Wigge, B. (1988). Influence of photorespiration on ATP/ADP in the chloroplasts, mitochondria, and cytosol, studied by rapid fractionation of barley (*Hordeum vulgare*) protoplasts. *Plant Physiology* **88**, 69–76.

Gerhardt, R. & Heldt, H.W. (1984). Measurement of subcellular metabolite levels in leaves by fractionation of freeze-stopped material in nonaqueous media. *Plant Physiology* **75**, 542–7.

Gross, A., Brückner, G., Heldt, H.W. & Flügge, U.-I. (1990). Comparison of the kinetic properties, inhibition and labelling of the phosphate translocators from maize and spinach mesophyll chloroplasts. *Planta* **180**, 262–71.

Hatch, M.D. (1987). C_4 photosynthesis: a unique blend of modified biochemistry, anatomy and ultrastructure. *Biochimica et Biophysica Acta* **895**, 81–106.

Hatch, M.D., Dröscher, L., Flügge, U.I. & Heldt, H.W. (1984). A specific translocator for oxaloacetate transport in chloroplasts. *FEBS Letters* **178**, 15–19.

Heineke, D., Riens, B., Große, H., Hoferichter, P., Peter, U., Flügge, U.I. & Heldt, H.W. (1991). Redox transfer across the inner chloroplast envelope membrane. *Plant Physiology* **95**, 1131–7.

Heineke, D., Sonnewald, U., Büssis, D., Günter, G., Leidreiter, K., Wilke, I., Raschke, K., Willmitzer, L. & Heldt, H. W. (1992).

Apoplastic expression of yeast-derived invertase in potato: effects on photosynthesis, leaf solute composition, water relations and tuber composition. *Plant Physiology* (in press).

Heldt, H.W. (1969). Adenine nucleotide translocation in spinach chloroplasts. *FEBS Letters* **5**, 11–14.

Heldt, H.W., Flügge, U.I. & Borchert, S. (1991). Diversity of specificity and function of phosphate translocators in various plastids. *Plant Physiology* **95**, 341–3.

Heldt, H.W., Heineke, D., Heupel, R. Krömer, S. & Riens, B. (1990). Transfer of redox equivalents between subcellular compartments of a leaf cell. *Proceedings of the VIIIth International Conference of Photosynthesis*, Vol. IV, ed. M. Baltscheffsky, pp. 15.1–7. Dordrecht: Academic Publications.

Heldt, H.W., Jacobs, H. & Klingenberg, M. (1965). Endogenous ADP of mitochondria, an early acceptor of oxidative phosphorylation as disclosed by kinetic studies with C^{14} labelled ADP and ATP and with atractyloside. *Biochemical and Biophysical Research Communications* **18**, 174–9.

Heldt, H.W., Klingenberg, M. & Milovancev, M. (1972). Difference between the ATP/ADP ratio in the mitochondrial matrix and in the extramitochondrial space. *European Journal of Biochemistry* **30**, 434–40.

Heldt, H.W. & Rapley, L. (1970). Specific transport of inorganic phosphate, 3-phosphoglycerate and dihydroxyacetone phosphate and of dicarboxylates across the inner membrane of spinach chloroplasts. *FEBS Letters* **10**, 143–8.

Heldt, H.W. & Sauer, F. (1971). The inner membrane of the chloroplast envelope as the site of specific metabolite transport. *Biochimica et Biophysica Acta* **234**, 83–91.

Herold, A., Leegood, R.C., McNeil, P.H. & Simon, P.R. (1981). Accumulation of maltose during photosynthesis in protoplasts isolated from spinach leaves treated with mannose. *Plant Physiology* **67**, 85–8.

Heupel, R., Markgraf, T., Robinson, D.G. & Heldt, H.W. (1991). Compartmentation studies of spinach leaf peroxisomes. Evidence for channeling of photorespiratory metabolites in peroxisomes devoid of intact boundary membrane. *Plant Physiology* **96**, 971–9.

Howitz, K.T. & McCarty, R.E. (1985). Substrate specificity of the pea chloroplast glycolate transporter. *Biochemistry* **24**, 3645–50.

Huber, S.C. & Edwards, G.E. (1977a). Transport in C_4 mesophyll chloroplasts. Characterization of the pyruvate carrier. *Biochimica et Biophysica Acta* **462**, 583–602.

Huber, S.C. & Edwards, G.E. (1977b). Transport in C_4 mesophyll chloroplasts. Evidence for an exchange of inorganic phosphate and phosphoenolpyruvate. *Biochimica et Biophysica Acta* **462**, 603–12.

Keys, A.J. & Whittingham, C.P. (1969). Nucleotide metabolism in chloroplast and nonchloroplast components of tobacco leaves. *Progress in Photosynthesis Research* **1**, 352–8.

Klingenberg, M. (1981). Membrane protein oligomeric structure and transport function. *Nature* **290**, 449–54.

Klingenberg, M. & Heldt, H.W. (1982). The ADP–ATP translocation in mitochondria and its role in intracellular compartmentation. In *Metabolic Compartmentation*, ed. H. Sies, pp. 101–22. New York: Academic Press.

Krömer, S. & Heldt, H.W. (1991*a*). Respiration of pea leaf mitochondria and redox transfer between the mitochondrial and extramitochondrial compartment. *Biochimica et Biophysica Acta* **1057**, 42–50.

Krömer, S. & Heldt, H.W. (1991*b*). On the role of mitochondrial oxidative phosphorylation in photosynthesis metabolism as studied by the effect of oligomycin on photosynthesis in protoplasts and leaves of barley. *Plant Physiology* **95**, 1270–6.

Krömer, S., Stitt, M. & Heldt, H.W. (1988). Mitochondrial oxidative phosphorylation participating in photosynthesis metabolism of a leaf cell. *FEBS Letters* **226**, 352–6.

Leegood, R.C. (1985). The intercellular compartmentation of metabolites in leaves of *Zea mays*. *Planta* **164**, 163–71.

Lemoine, R. & Delrot, S. (1989). Proton-motive-force-driven sucrose uptake in sugar beet plasma membrane vesicles. *FEBS Letters* **249**, 129–33.

Li, Z.C. & Bush, D.R. (1990). pH dependent amino acid transport into plasma membrane vesicles isolated from sugar beet leaves. Evidence for carrier-mediated electrogenic flux through multiple transport systems. *Plant Physiology* **94**, 268–77.

Nakano, Y. & Edwards, G.E. (1987). Hill reaction, hydrogen peroxide scavenging, and ascorbate peroxidase activity of mesophyll and bundle sheath chloroplasts of NADP-malic enzyme type C_4 species. *Plant Physiology* **85**, 294–8.

Ngernprasirtsiri, J., Takabe, T. & Akazawa, T. (1989). Immunochemical analysis shows that an ATP/ADP translocator is associated with the inner envelope membrane of amyloplasts from *Acer pseudoplatanus* L. *Plant Physiology* **89**, 1024–7.

Ohnishi, J., Flügge, U.-I. & Heldt, H.W. (1989). Phosphate translocator of mesophyll and bundle sheath chloroplasts of a C_4 plant, *Panicum miliaceum* L. Identification and kinetic characterization. *Plant Physiology* **91**, 1507–11.

Ohnishi, J., Flügge, U.-I., Heldt, H.W. & Kanai, R. (1990). Involvement of Na^+ in active uptake of pyruvate in mesophyll chloroplasts of some C_4 plants: Na^+/pyruvate transport. *Plant Physiology* **94**, 950–9.

Ohnishi, J.I. & Kanai, R. (1987). Pyruvate uptake by mesophyll and bundle sheath chloroplasts of a C_4 plant *Panicum miliaceum*. *Plant and Cell Physiology* **28**, 1–10.

Ohnishi, J.I. & Kanai, R. (1990). Pyruvate uptake induced by a pH jump in mesophyll chloroplasts of maize and sorghum, NADP-malic enzyme type C₄ species. *FEBS Letters* **269**, 122–4.

Pain, D., Kanwar, Y.S. & Blobel, G. (1988). Identification of a receptor for protein import into chloroplasts and its localization to envelope contact zones. *Nature* **331**, 232–7.

Peavey, D.G., Steup, M. & Gibbs, M. (1977). Characterization of starch breakdown in isolated spinach chloroplasts. *Plant Physiology* **60**, 305–8.

Pfaff, E., Klingenberg, M., Ritt, E. & Vogell, W. (1968). Korrelation des unspezifisch permeablen mitochondrialen Raumes mit dem 'Intermembran-Raum'. *European Journal of Biochemistry* **5**, 222–32.

Riens, B., Lohaus, G., Heineke, D. & Heldt, H.W. (1991). Amino acid and sucrose content in the cytosolic chloroplastic and vacuolar compartments and in the phloem sap of spinach leaves. *Plant Physiology* **97**, 227–33.

Robinson, S.P. & Wiskich, J.T. (1977). Pyrophosphate inhibition of carbon dioxide fixation in isolated pea chloroplasts by uptake in exchange for endogenous adenine nucleotides. *Plant Physiology* **59**, 422–7.

Rumpho, M.E. & Edwards, G.E. (1985). Characterization of 4,4'-diisothiocyano-2,2'-disulfonic acid stilbene inhibition of 3-phosphoglycerate-dependent O₂ evolution in isolated chloroplasts. Evidence for a common binding site on the C₄ phosphate translocator for 3-phosphoglycerate, phosphoenolpyruvate, and inorganic phosphate. *Plant Physiology* **78**, 537–44.

Rustin, P., Neuburger, M., Douce, R. & Lance, C. (1987). In *Plant Mitochondria*, ed. A.L. Moore & R.B. Beechey, pp. 89–92. New York: Plenum Press.

Santarius, K.A. & Heber, U. (1965). Changes in the intracellular levels of ATP, ADP, P and regulatory function of the adenylate system in leaf cells during photosynthesis. *Biochimica et Biophysica Acta* **102**, 39–54.

Schäfer, G., Heber, U. & Heldt, H.W. (1977). Glucose transport into spinach chloroplasts. *Plant Physiology* **60**, 286–9.

Scheibe, R. (1987). NADP-malate dehydrogenase in C₃ plants: Regulation and role of a light activated enzyme. *Physiologia Plantarum* **71**, 393–400.

Schnell, D.J., Blobel, G. & Pain, D. (1990). The chloroplast import receptor is an integral membrane protein of chloroplast envelope contact sites. *Journal of Cell Biology* **111**, 1825–38.

Sellami, A. (1976). Evolution des adenosine phosphates et de la charge énergetique dans les compartiments chloroplastique et nonchloroplastique des feuilles de blé. *Biochimica et Biophysica Acta* **423**, 524–39.

Soboll, S., Scholz, R. & Heldt, H.W. (1978). Subcellular metabolite concentrations. Dependence of mitochondrial and cytosolic ATP systems on the metabolic state of perfused rat liver. *European Journal of Biochemistry* **87**, 377–90.

Stitt, M. & Heldt, H.W. (1981). Physiological rates of starch breakdown in isolated intact spinach chloroplasts. *Plant Physiology* **68**, 755–61.

Stitt, M. & Heldt, H.W. (1985). Generation and maintenance of concentration gradients between the mesophyll and bundle sheath in maize leaves. *Biochimica et Biophysica Acta* **808**, 400–14.

Stitt, M., Lilley, R. McC. & Heldt, H.W. (1982). Adenine nucleotide levels in the cytosol, chloroplasts and mitochondria of wheat leaf protoplasts. *Plant Physiology* **70**, 971–7.

Tolbert, N.E. (1980a). Microbodies – peroxisomes and glyoxisomes. In *The Biochemistry of Plants*, Vol. 1, ed. P.K. Stumpf & E.E. Conn, pp. 359–89. New York: Academic Press.

Tolbert, N.E. (1980b). Photorespiration. In *The Biochemistry of Plants*, Vol. 2, ed. P.K. Stumpf & E.E. Conn, pp. 488–525. New York: Academic Press.

Turgeon, R. (1989). The sink–source transition in leaves. *Annual Review of Plant Physiology and Plant Molecular Biology* **40**, 119–38.

van Schaewen, A., Stitt, M., Schmidt, R., Sonnewald, U. & Willmitzer, L. (1990). Expression of yeast-derived invertase in the cell wall of tobacco and *Arabidopsis* plants leads to inhibition of sucrose export, accumulation of carbohydrate and inhibition of photosynthesis, and strongly influences the growth and habitus of transgenic tobacco plants. *EMBO Journal* **9**, 3033–44.

van Veldhoven, P.P., Just, W.W. & Mannaerts, G.P. (1987). Permeability of the peroxisomal membrane to cofactors of β-oxidation. *Journal of Biological Chemistry* **262**, 4310–18.

von Heijne, G., Steppuhn, J. & Herrmann, R.G. (1989). Domain structure of mitochondrial and chloroplast targeting peptides. *European Journal of Biochemistry* **180**, 535–45.

Wagner, R., Apley, E.C., Gross, A. & Flügge, U.I. (1989). The rotational diffusion of chloroplast phosphate translocator and of lipid molecules in bilayer membranes. *European Journal of Biochemistry* **182**, 165–73.

Walker, G.H., Sarojini, G. & Oliver, D.J. (1982). Identification of a glycine transporter from pea leaf mitochondria. *Biochemical and Biophysical Research Communications* **107**, 856–61.

Wallmeier, H., Weber, A., Gross, A. & Flügge, U.-I. (1992). Insights into the structure of the chloroplast phosphate translocator protein. In *Transport and Receptor Proteins of Plant Membranes*, ed. D.T. Clarkson and D.T. Cooke. New York: Plenum Publishing Corp. (in press).

Weiner, H., Burnell, J.N., Woodrow, J.E., Heldt, H.W. & Hatch, M.D. (1988). Metabolite diffusion into bundle sheath cells from C_4 plants. Relation to C_4 photosynthesis and plasmodesmatal function. *Plant Physiology* **88**, 815–22.

Weiner, H. & Heldt, H.W. (1992). Inter- and intracellular distribution of amino acids and other metabolites in maize (*Zea mays* L.) leaves. *Planta* (in press).

Werdan, K., Heldt, H.W. & Geller, G. (1972). Accumulation of bicarbonate in intact chloroplasts following a pH gradient. *Biochimica et Biophysica Acta* **283**, 430–41.

Willey, D.I., Fischer, K., Wachter, E., Link, T.A. & Flügge, U.-I. (1991). Molecular cloning and structural analysis of the phosphate translocator from pea chloroplasts and its comparison to the spinach phosphate translocator. *Planta* **183**, 451–61.

Winter, H., Lohaus, G. & Heldt, H.W. (1992). Phloem transport of amino acids in relation to their cytosolic levels in barley leaves. *Plant Physiology* (in press).

Woldegiorgis, G., Voss, S., Shrago, E., Werner-Washburne, M. & Keegstra, K. (1983). An adenine nucleotide–phosphoenolpyruvate counter-transport system in C_3 and C_4 plant chloroplasts. *Biochemical and Biophysical Research Communications* **116**, 945–51.

Woo, K.C., Flügge, U.I. & Heldt, H.W. (1987). A two translocator model for the transport of 2-oxoglutarate and glutamate in chloroplasts during ammonia assimilation in the light. *Plant Physiology* **84**, 624–32.

Yin, Z.-H., Dietz, K.-J. & Heber, U. (1990). Light dependent pH changes in leaves of C_3 plants. Effect of inhibitors of photosynthesis and of the developmental state of the photosynthetic apparatus on cytosolic. *Planta* **182**, 262–9.

Yu, C., Claybrook, D.L. & Huang, A.H.C. (1983). Transport of glycine, serine and proline in spinach leaf mitochondria. *Archives of Biochemistry and Biophysics* **227**, 180–7.

Yu, C. & Huang, A.H.C. (1986). Conversion of serine to glycerate in intact spinach leaf peroxisomes: role of malate dehydrogenase. *Archives of Biochemistry and Biophysics* **245**, 125–33.

Zalman, L.S., Nikaido, H. & Kagawa, Y. (1980). Mitochondrial outer membrane contains a protein producing nonspecific diffusion channels. *Journal of Biological Chemistry* **255**, 1771–4.

Zoglowek, C., Krömer, S. & Heldt, H.W. (1988). Oxaloacetate and malate transport of plant mitochondria. *Plant Physiology* **87**, 109–15.

DAVID H. TURPIN

Metabolic interactions during photosynthetic and respiratory nitrogen assimilation in a green alga

This chapter addresses some of the interactions that occur between primary nitrogen assimilation and photosynthetic and respiratory metabolism. Much of the pioneering work in this area was carried out by Syrett (Syrett, 1953, 1956a,b, 1981) and Bassham, Kanazawa and co-workers (Bassham et al., 1981; Kanazawa et al., 1970, 1972, 1983). It is not coincidental that this work has often employed algal cells. Single-celled algae have C/N ratios between 7 and 12 compared with >20 in higher plants. Therefore, the relative importance of nitrogen in carbon metabolism of single-celled algae is much greater than in higher plants and the interactions between N assimilation and carbon metabolism are much more apparent.

For several years our group has studied the interactions between photosynthesis, respiration and N assimilation in the green alga *Selenastrum minutum*. Our approach has been to grow this alga in chemostat cultures under N-limited conditions. The addition of a source of inorganic nitrogen (NH_4^+, NO_2^- or NO_3^-) to these cells activates the assimilation of N into amino acids allowing study of the corresponding changes in metabolism. This chapter reviews some of the progress made using this system.

Nitrogen assimilation by N-limited and N-sufficient algal cells

The primary assimilation of inorganic N into amino acids and protein requires ATP, reducing power and carbon skeletons in the form of keto-acids. In photosynthetic tissues, ATP and reducing power are supplied by either photosynthetic or respiratory processes; however, most of the carbon skeletons used in amino acid synthesis are intermediates of respiratory metabolism (Fig. 1). As a result, an increase in the rate of primary N assimilation requires an increase in the flow of carbon through respiratory pathways leading to dramatic interactions between photosyn-

Society for Experimental Biology Seminar Series 50: *Plant organelles*, ed. A. K. Tobin.
© Cambridge University Press 1992, pp. 49–78.

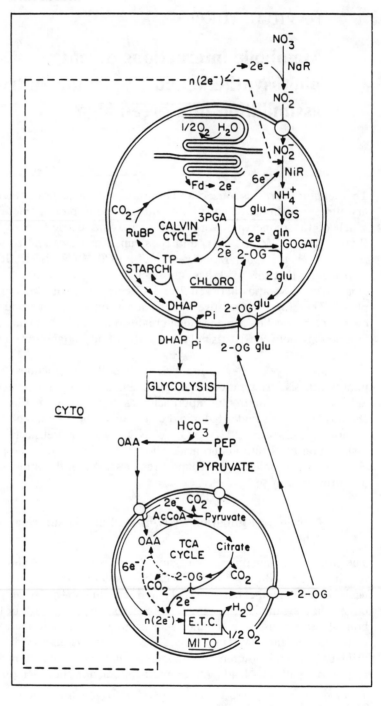

thetic and respiratory metabolism (Bassham *et al.*, 1981; Syrett, 1981; Turpin, 1991).

Carbon skeletons for amino acid synthesis are ultimately the products of photosynthesis. That N assimilation depends on photosynthetic CO_2 fixation has been demonstrated most clearly in N-sufficient algal cells. These algae do not accumulate carbohydrate reserves (e.g. starch), therefore the supply of carbon to amino acid synthesis is dependent upon recent photosynthate (Syrett, 1981). Withholding either light or CO_2 inhibits N assimilation (Grant, 1968; Grant & Turner, 1969; Thacker & Syrett, 1972; Boussiba *et al.*, 1984; Romero *et al.*, 1985; Guerrero & Lara, 1987; Lara *et al.*, 1987; Larsson & Larsson, 1987; Syrett, 1988; Rigano *et al.*, 1990; Ullrich *et al.*, 1990; Amory *et al.*, 1991). Inhibitors of photosynthetic carbon fixation (DCMU or DL-glyceraldehyde) mimic this effect (Thacker & Syrett, 1972; Lara *et al.*, 1987; Amory *et al.*, 1991). The requirement for CO_2 and light diminishes if the cells are provided with a metabolisable source of organic carbon (e.g. glucose, acetate) (Eisele & Ullrich, 1977; Syrett, 1981; Schlee *et al.*, 1985).

In contrast, algal cells grown under N limitation store large quantities of starch. Under these conditions N assimilation (NO_3^-, NO_2^- and NH_4^+) becomes independent of recent photosynthate and occurs in both light and dark (Syrett, 1981; Amory *et al.*, 1991). N-limited algal cells, therefore, afford an opportunity to study the interaction of N assimilation and respiration in the dark without the complicating effects of photosynthesis. At the same time this tissue may also be used to study 'photosynthetic' N assimilation. The fact that large portions of the world's oceans and lakes have inorganic N concentrations in the sub-micromolar range make N-limited algal cells not only a useful physiological tool but a relevant one. The balance of this chapter will focus on our studies with N-limited cells of the green alga *Selenastrum minutum*.

Effects of N assimilation on respiratory gas exchange

Figure 1 illustrates that primary N assimilation, unlike photorespiratory N assimilation, requires the mobilisation of new carbon to provide the

Fig. 1. Interactions between photosynthesis, respiration and N assimilation. This figure provides a simplified representation of N assimilation into glutamate. 2-OG, 2-oxoglutarate; E.T.C., mitochondrial electron transport chain; Fd, ferredoxin; glu, glutamate; OAA, oxaloacetate; PEP, phospho*enol*pyruvate; PGA, 3-phosphoglycerate; RuBP, ribulose bisphosphate; TP, triose phosphate; NaR, nitrate reductase; NiR, nitrite reductase; MITO, mitochondrion; CYTO, cytoplasm; CHLORO, chloroplast; DHAP, dihydroxyacetone phosphate.

carbon skeletons for amino acid synthesis. Hence the onset of N assimilation must activate respiratory carbon flow (Syrett, 1953).

In the dark:

During dark N assimilation by *S. minutum* the activation of respiratory carbon flow is manifested as an increase in respiratory CO_2 release (Table 1). The source of N being assimilated determines the degree of respiratory CO_2 release. NH_4^+ assimilation resulted in a 2.5-fold increase in CO_2 release while NO_3^- assimilation yielded a 3.5-fold increase. The effect of NO_2^- was intermediate. As the rates of dark NO_3^- and NO_2^- assimilation are approximately one-third that of NH_4^+ (Table 1) we can calculate that the CO_2 respired per N assimilated is approximately five-fold greater for NO_3^- than NH_4^+ (Table 1). This reflects the high reductant requirement for NO_3^- reduction (Table 2).

The different reductant requirements among N sources should affect the mechanism by which respiratory NAD(P)H is oxidised. Figure 2 provides a diagrammatic representation of the fate of respiratory electrons during NH_4^+ and NO_3^- assimilation. During NH_4^+ assimilation, respiratory O_2 consumption increases in conjunction with CO_2 evolution

Table 1. *Nitrogen assimilation and respiratory gas exchange by N-limited* Selenastrum minutum *in the dark and light*

N source	N assimilation	Gross CO_2 release	Gross O_2 uptake	Incremental CO_2/N
		μmol mg^{-1} Chl h^{-1}		
Dark				
None	0	75	98	—
NH_4^+	169	187	223	0.66
NO_2^-	55	234	132	2.89
NO_3^-	53	262	113	3.53
Light (250 μE m^{-2} s^{-1})				
None	0	37	94	—
NH_4^+	167	97	221	0.35
NO_2^-	160	165	134	0.80
NO_3^-	155	188	94	0.97

From Weger & Turpin, 1989.

Table 2. *The energetics of CO_2 and N assimilation*

Process	Reactions[a]	ATP/e^- requirement
1. Photosynthetic CO_2 fixation	$CO_2 + RuBP + 3ATP + 4e^- \rightarrow \frac{1}{3} Triose\text{-}P + RuBP$	0.75
2. N transport (N = NH_4^+, NO_2^- or NO_3^-)	$N_{out} + ATP \rightarrow N_{in}$	
3. Nitrate reduction (NR)	$NO_3^- + 2e^- \rightarrow NO_2^-$	
4. Nitrite reduction	$NO_2^- + 6e^- \rightarrow NH_4^+$	
5. Ammonium assimilation (GS/GOGAT)	$NH_4^+ + Glu + 2\text{-}OG + ATP + 2e^- \rightarrow 2Glu$	
6. Amino acid interconversion (transaminase)	$Glu + keto\text{-}acid \rightarrow amino\ acid + 2\text{-}OG$	
7. Protein synthesis	$Amino\ acid + 4ATP + polypeptide_{(n)} \rightarrow polypeptide_{(n+1)}$	
8. Nitrate assimilation into protein (2+3+4+5+7)	$NO_3^- + keto\text{-}acid + 6ATP + 10e^- + polypeptide_{(n)} \rightarrow polypeptide_{(n+1)}$	0.6
9. Nitrite assimilation into protein (2+4+5+7)	$NO_2^- + keto\text{-}acid + 6ATP + 8e^- + polypeptide \rightarrow polypeptide_{(n+1)}$	0.75
10. NH_4^+ assimilation into protein (2+5+7)	$NH_4^+ + keto\text{-}acid + 6ATP + 2e^- + polypeptide \rightarrow polypeptide_{(n+1)}$	3.0

[a]Table 1 outlines the ATP and reductant requirements of the various processes associated with photosynthetic carbon reduction and the assimilation of N into protein. The equations are not balanced.

A NH$_4^+$ ASSIMILATION

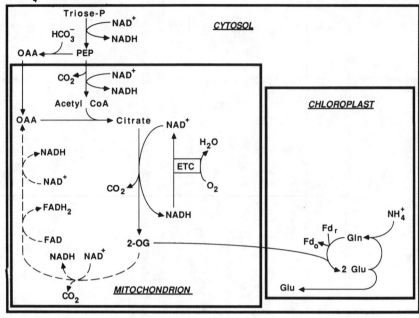

B NO$_3^-$/NO$_2^-$ ASSIMILATION

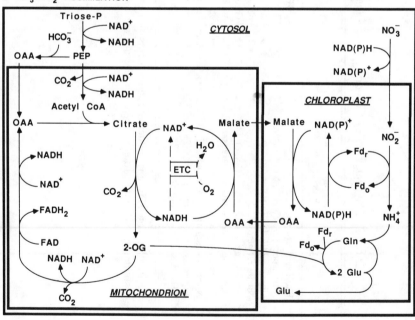

(Table 1). This implies that increased respiratory carbon flow to amino acid synthesis is coupled to the mitochondrial electron transport chain during NH_4^+ assimilation and O_2 serves as the respiratory electron acceptor (Weger *et al.*, 1988, 1990*a*; Weger & Turpin, 1989). On the other hand, NO_3^- assimilation causes only a minor increase in respiratory O_2 uptake despite the dramatic increase in respiratory CO_2 release (Table 1). In this instance NO_3^- serves as the sink for respiratory electrons. Experiments carried out anaerobically clearly demonstrated that NH_4^+ assimilation and its associated CO_2 release was inhibited more than NO_3^- assimilation (Weger & Turpin, 1989).

The coupling of respiratory carbon flow to mitochondrial electron transport during NH_4^+ assimilation could be facilitated by either the cytochrome or alternate oxidase. It has been hypothesised that the alternate oxidase may play a role in supporting respiratory carbon flow for biosynthesis. We were unable to test this hypothesis with *S. minutum* as it lacks any capacity for alternate pathway activity, implying that electron transport is through cytochrome oxidase (Weger *et al.*, 1990*b*). In contrast, *Chlamydomonas reinhardtii* exhibits a high capacity for alternate oxidase activity (Weger *et al.*, 1990*a*). By employing inhibitors of the cytochrome and alternate oxidase (CN^- and SHAM respectively) and capitalising on the observation that the cytochrome and alternate oxidases discriminate differently against $^{18}O_2$, (Guy *et al.*, 1989*a*) we showed that the alternate oxidase was not engaged during respiratory carbon flow to amino acid synthesis (Weger *et al.*, 1990*a*).

In the light:

The primary assimilation of N into amino acids requires carbon flow through respiratory pathways, regardless of whether the cells are in the

---

Fig. 2. Proposed pathways of TCA cycle electron and carbon flow during transient NH_4^+ assimilation (*A*) and transient NO_3^-/NO_2^- assimilation (*B*) by N-limited *S. minutum* in the light and dark. Cytosolic triose phosphate would be provided by the chloroplast, either from the Calvin cycle or from starch breakdown. The actual mechanism of reductant shuttling during inorganic N assimilation is unknown, and has been represented in (*B*) as malate–oxaloacetate exchange across the mitochondrial and chloroplast envelopes. During photosynthesis the light reactions would also provide photogenerated reductant (NADPH and Fd_r), and therefore contribute to NO_3^-/NO_2^- reduction in the light (not illustrated). 2-OG, 2-oxoglutarate; OAA, oxaloacetate; Fd_r, reduced ferredoxin; Fd_o, oxidised ferredoxin (modified from Weger & Turpin, 1989).

dark or photosynthesising in the light. To overcome the difficulty of measuring respiratory CO_2 release and O_2 consumption while the cells are carrying out photosynthetic CO_2 fixation and O_2 evolution, the O_2 and CO_2 in the medium were replaced with $^{18}O_2$ and $^{13}CO_2$. A membrane inlet mass spectrometer allowed observation of photosynthesis as the disappearance of $^{13}CO_3$ and appearance of $^{16}O_2$ and of respiration as $^{18}O_2$ uptake and $^{12}CO_2$ release (Peltier & Thiebault, 1985; Weger et al., 1988; Weger & Turpin, 1989). The pattern of respiratory gas exchange found during photosynthesis was very similar to that in the dark. NH_4^+ assimilation increased respiratory CO_2 release and O_2 consumption indicating respiratory electron transport during photosynthesis (Table 1). NO_3^- assimilation resulted in a much greater stimulation in CO_2 release but again little change in respiratory O_2 consumption occurred. The implications of these results were that not only was respiratory carbon flow required to provide carbon for amino acid synthesis during photosynthesis, but that during NO_3^- assimilation respiratory carbon flow was also responsible for providing some reductant for NO_3^- reduction (Weger & Turpin, 1989).

The requirements for anaplerotic carbon fixation

If tricarboxylic acid (TCA) cycle intermediates are consumed by N assimilation, they must be replaced by reactions that carboxylate phosphoenolpyruvate (PEP) and produce oxaloacetic acid (OAA) (Fig. 1). The major enzyme catalysing this function in S. minutum is PEP carboxylase (PEPCase) (Schuller et al., 1990).

In the dark:

The enhancement of dark HCO_3^- fixation observed during N assimilation demonstrates the relationship between inorganic N assimilation and anaplerotic HCO_3^- fixation (Syrett, 1956b). In S. minutum, dark carbon fixation increases 40-fold upon addition of NH_4^+ to N-limited cells (Fig. 3A). To determine the stoichiometry between NH_4^+ assimilation and dark anaplerotic HCO_3^- fixation we established the rate of NH_4^+ assimilation as an independent variable (Vanlerberghe et al., 1990b). The correlation between dark carbon fixation and the rate of NH_4^+ assimilation was extremely high ($r^2 = 0.99$) and exhibited a slope of 0.3 mol C fixed per mol N assimilated (Fig. 3B). A calculation based on the amino acid composition of total cellular protein indicated that anaplerotic HCO_3^- fixation would be required in the assimilation of 34% of the N incorporated into S. minutum protein (Vanlerberghe et al., 1990b). If we

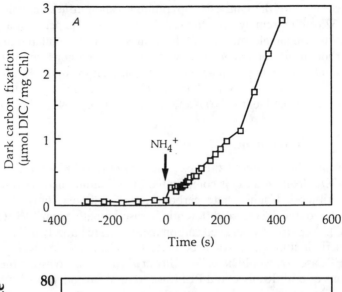

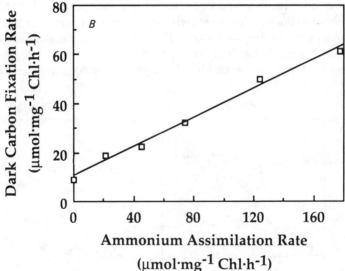

Fig. 3. *A*, Effect of NH_4^+ assimilation on dark carbon fixation by N-limited *S. minutum*. Cells at a density of 2 μg Chl ml$_{-1}$ were supplied with $Na_2{}^{13}CO_3$ (specific radioactivity of 6.1 μCi μmol^{-1} DIC). Total DIC was 4.3 mM. NH_4Cl (2 mM) was added at time zero. Data are means of three replicate experiments (from Schuller *et al.*, 1990). *B*, Steady-state *in vivo* PEPCase activity as a function of NH_4^+ assimilation rate. This line has a slope of 0.3 mol C · mol N and a r^2 of 0.991 (from Vanlerberghe *et al.*, 1990*b*).

accounted for the synthesis of pyrimidines and purines this value declined to 32.6%, extremely close to the 30% determined experimentally. Hence, short-term anaplerotic HCO_3^- fixation is precisely regulated to match the carbon needs of amino acid synthesis. The demonstration that NH_4^+ assimilation required anaplerotic carbon fixation by PEPCase was provided by the inhibition of dark NH_4^+ assimilation through stringent removal of ambient CO_2/HCO_3^- (Amory et al., 1991).

In the light:

During photosynthesis the assessment of in vivo PEPCase activity by the HCO_3^- fixation assay is complicated by the simultaneous fixation of CO_2 by ribulose bisphosphate carboxylase/oxygenase (Rubisco). However, Rubisco and PEPCase differentially discriminate against ^{13}C (relative to ^{12}C). At pH 7, if whole cell carbon fixation were entirely by Rubisco, total discrimination (D_T) would be close to 37‰ if fixation were entirely by PEPCase, D_T would be 0.5‰ (Guy et al., 1989b). To assess the contributions of both Rubisco and PEPCase to total carbon fixation by N-limited S. minutum during photosynthesis, we examined changes in the isotopic composition of substrate CO_2/HCO_3^- under control conditions and during NH_4^+ assimilation (Guy et al., 1989b).

Figure 4 presents a family of model-generated curves representing

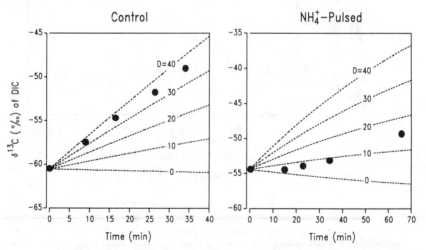

Fig. 4. Model generated scenarios for the change in $\delta^{13}C$ of dissolved inorganic carbon over time for a range of discrimination values plotted with experimental results for representative control and NH_4^+ pulsed experiments (from Guy et al., 1989b).

expected trends in the $\delta^{13}C$ value of dissolved inorganic carbon (DIC) over time for a range of discrimination factors. Superimposed on these curves are the actual measured $\delta^{13}C$ values of medium DIC. Under control conditions it is clear that D_T remained constant at about 35‰, confirming that Rubisco was responsible for almost all carbon fixation. In sharp contrast, change in the $\delta^{13}C$ of the DIC during NH_4^+ assimilation was much less pronounced and D_T, integrated over the first 35 min, remained below 10‰. This markedly lower discrimination factor indicated a large increase in carbon fixation by PEPCase during NH_4^+ assimilation in the light. These results provide strong evidence that the anaplerotic processes so easily measured in the dark also occur during photosynthesis.

The source of carbon for amino acid synthesis

In the dark:

During dark N assimilation, starch reserves are the major source of carbon for amino acid synthesis (Miyachi & Miyachi, 1985). The rate of starch degradation increases with the rate of NH_4^+ assimilation (Fig. 5).

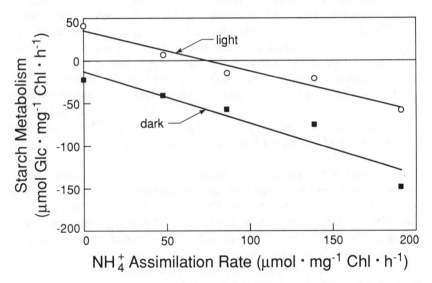

Fig. 5. The rate of net starch metabolism as a function of the rate of ammonium assimilation in the dark (■) and light (○). The slope of the dark line is 0.62 glucose equivalents per NH_4^+ assimilated. In the light the slope is 0.64 (K.D.M. Vlossak and D.H. Turpin, unpublished data).

During dark NH_4^+ assimilation starch is degraded at a rate of 0.62 hexose equivalents per NH_4^+ which equals c. 3.7 carbons per NH_4^+. The C/N ratio of S. minutum protein is 3.7, once again indicating a tight regulation of carbon metabolism in response to N assimilation. Assimilation of NO_3^- requires much higher rates of respiration than does NH_4^+ owing to the high reductant demands of NO_3^- assimilation. During NO_3^- assimilation more starch is mobilised per NO_3^- assimilated (K. D. M. Vlossak and D. H. Turpin, unpublished data). This excess carbon mobilised relative to the C/N content of protein is accounted for by the high rates of respiratory CO_2 release during NO_3^- assimilation (Table 1).

In the light:

During photosynthesis there are two potential sources of carbon for amino acid synthesis, starch and recent photosynthate. Recent photosynthate appears to provide most of the carbon for amino acid synthesis provided the carbon demands for N assimilation do not exceed the capacity of photosynthesis for carbon fixation. For example, in N-limited S. minutum the maximum rate of photosynthetic carbon fixation is between 150 and 250 μmol CO_2 mg^{-1} Chl h^{-1}. Given that the C/N ratio of protein is 3.7, carbon from photosynthesis would be able to sustain a rate of NH_4^+ assimilation of c. 50 μmol mg^{-1} Chl h^{-1}. Without N, illuminated cells accumulate starch. As the rate of NH_4^+ assimilation increases the rate of net starch accumulation decreases (Fig. 5). When N assimilation reaches c. 50 μmol mg^{-1} Chl h^{-1} all available new photosynthate is directed to N assimilation and there is no net change in starch content (Fig. 5). As the rate of N assimilation increases above this value, photosynthesis cannot meet the carbon demands of amino acid synthesis and starch degradation is observed in the light. The stoichiometry between starch mobilisation and the rate of NH_4^+ assimilation is similar to that observed in the dark (Fig. 5).

The onset of net starch degradation in the light is coincident with a dramatic decline in photosynthetic carbon fixation. The proximate cause of this decline is the limitation of Rubisco by ribulose 1,5-bisphosphate (RuBP) (Elrifi & Turpin, 1986; Elrifi et al., 1988). At present we do not know the regulatory mechanisms involved in suppressing photosynthetic carbon fixation at high rates of N assimilation but our view is that in the short term net starch degradation is incompatible with net photosynthetic CO_2 fixation. When the carbon demands for amino acid synthesis exceed the capacity of the Calvin cycle to provide the carbon, starch degradation is induced and as a consequence photosynthetic carbon reduction suppressed. Clearly this is a transient phenomenon, but the suppression in photosynthetic metabolism may last several hours or until either the available starch reserves are depleted or the added nitrogen is completely assimilated (Elrifi & Turpin, 1986).

Regulation of respiratory carbon flow to N assimilation

It is clear that the assimilation of N into amino acids causes a stimulation of respiratory metabolism in both the light and the dark. In this section I examine the regulatory properties of some of the key enzymes of carbohydrate metabolism in *S. minutum* and propose a model in which respiratory carbon flow is regulated by the carbon demands of amino acid biosynthesis.

The onset of dark NH_4^+ assimilation causes rapid changes in the levels of respiratory metabolites which have suggested that pyruvate kinase (PK), PEP carboxylase and phosphofructokinase (PFK) are important in regulating carbon flow to N assimilation (Bassham *et al.*, 1981; Elrifi & Turpin, 1987). We have studied the regulatory properties of these enzymes from *S. minutum*.

Pyruvate kinase

There is both a cytosolic (PK_c) and a plastid (PK_p) pyruvate kinase in *S. minutum* (Lin *et al.*, 1989*a*). The cytosolic form is highly regulated by metabolites important in N assimilation (Lin *et al.*, 1989*b*). The K_m values of PK_c for PEP and ADP are 90 and 50 μM respectively. It appears that this enzyme is limited, at least partially, by *in vivo* ADP supply (Turpin *et al.*, 1990, Vanlerberghe *et al.*, 1990*a*). The most potent inhibitors are glutamate ($I_{50} = 0.27$ mM) and P_i ($I_{50} = 1.8$ mM) while dihydroxyacetone phosphate (DHAP) is a strong activitor ($K_a = 27$ μM) (Fig. 6). DHAP is able to overcome the inhibition by P_i and glutamate.

PEP carboxylase

Two forms of PEPCase have been identified in *S. minutum*, $PEPC_1$, and $PEPC_2$ (Schuller *et al.*, 1990). $PEPC_1$ is strongly regulated by respiratory metabolites and the amino acids glutamate and glutamine (Fig. 7). The key inhibitors are glutamate ($I_{50} = 3.6$ mM), 2-oxoglutarate ($I_{50} = 2.1$ mM), aspartate ($I_{50} = 1.4$ mM) and malate ($I_{50} = 3.5$ mM). Glutamine and DHAP are activators with K_a values of 0.2 and 1.6 mM respectively. Both DHAP and Gln are capable of overcoming inhibition by Glu and 2-OG. $PEPC_2$ was also regulated by a similar suite of metabolites but to a much lesser extent. At present we do not know whether the two forms of

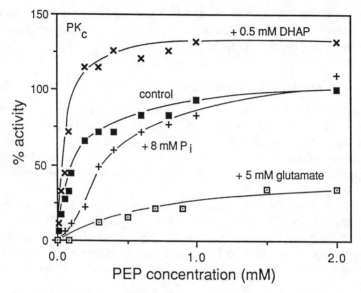

Fig. 6. Dependence of PK$_c$ activity on concentration of PEP. Enzyme activity was measured in the absence (■) or presence of 0.5 mM DHAP (×), 5 mM glutamate (□), or 8 mM P$_i$ (+); determined with saturating concentrations of ADP and cofactors (from Lin *et al.*, 1989*b*).

PEPCase are interconvertible forms of the same protein or distinct gene products.

Phosphofructokinase

S. minutum differs from most higher plants in having only one ATP-dependent phosphofructokinase (Botha & Turpin, 1990*a*). The kinetic and regulatory properties of *S. minutum* PFK are very similar to the plastid form in other organisms (Fig. 8) (Kelly & Latzko, 1977; Garland & Dennis, 1980; Kelly *et al.*, 1985; Botha & Small, 1987; Dennis & Greyson, 1987*a,b*; Botha *et al.*, 1988; Knowles *et al.*, 1990) suggesting that it is localised in the chloroplast. The algal PFK is strongly inhibited by PEP, 3-phosphoglycerate (PGA), 2-OG, and activated by low concentrations of P$_i$. In the absence of any effectors the enzyme exhibits a Hill coefficient of 2.0 (pH 7.2) with respect to F6P. The presence of 5 mM P$_i$ decreased the Hill coefficient to *c.* 1.1 while PEP, PGA and 2-OG (2 mM) all caused an increase in the Hill coefficient to between 3 and 4. Phosphate alleviated the inhibition by PEP, PGA and 2-OG (Fig. 8; Botha & Turpin, 1990*a*).

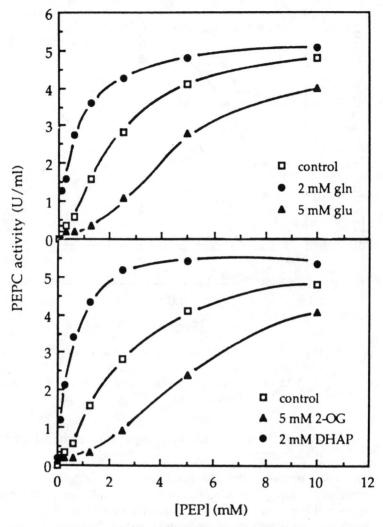

Fig. 7. Metabolite effects on the PEP saturation kinetics of PEP carbox-ylase (PEPC$_1$) from *S. minutum* (from Schuller *et al.*, 1990).

In addition to lacking a cytosolic-type PFK, *S. minutum* does not exhibit PP$_i$-dependent phosphofructokinase (PFP) activity nor does it produce any polypeptides which react with antibodies raised to the α or β subunits of potato PFP (Botha & Turpin, 1990*a*).

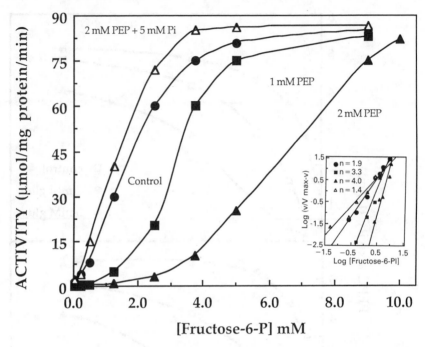

Fig. 8. F6P saturation curve of the *S. minutum* PFK in the presence of PEP, and P_i at pH 7.25. The Mg^{2+} concentration was 5 mM and that of ATP 0.5 mM. Inset shows a Hill plot of the data and reports Hill co-efficients (from Botha & Turpin, 1990*b*).

A model for activation of respiratory carbon flow to N assimilation

We have measured the changes in the concentration of respiratory metabolites during the initiation of NH_4^+ assimilation in the dark. The types and sequences of changes observed have allowed us to develop a model for the regulation of respiratory carbon flow to NH_4^+ assimilation (Turpin *et al.*, 1990; Fig. 9). Similar changes have been observed during NH_4^+ assimilation in the light (Smith *et al.*, 1989).

The onset of NH_4^+ assimilation via glutamine synthetase (GS) caused a rapid drop in glutamate and ATP and an increase in ADP, AMP and glutamine (Ohmori & Hattori, 1978; Turpin *et al.*, 1990). Subsequent metabolism of the glutamine by glutamate synthase (GOGAT) caused a decline of 2-OG.

The first major change observed in glycolytic metabolites was a 70% decline in PEP and a 2.5-fold increase in Pyr within 5 sec of NH_4^+

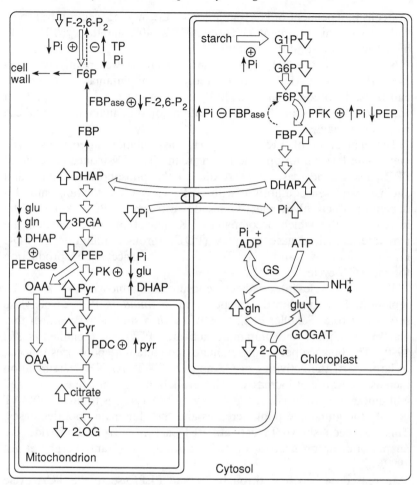

Fig. 9. A diagrammatic representation of key processes in the regulation of carbon partitioning between starch, respiration and NH_4^+ assimilation in the green alga *S. minutum*. Enzyme activation ⊕; Enzyme inhibition, ⊖; Increased metabolite levels in response to NH_4^+ assimilation, ⌃; Decreased metabolite levels in response to NH_4^+ assimilation, ⌄ (modified from Turpin *et al.*, 1990).

addition, which indicated a rapid activation of PK. The estimated ADP concentrations before NH_4^+ addition were below the $S_{0.5}$ for PK (Lin *et al.*, 1989a; Turpin *et al.*, 1990). Thus, PK may be adenylate-restricted *in vivo*. Upon NH_4^+ addition this restriction may be alleviated by ADP from increased GS activity. To test this hypothesis we increased ADP in control cells independent of N assimilation by uncoupling electron transport with CCCP. This treatment increased ADP by 40% and the Pyr/PEP ratio by 2.6-fold indicating PK activation.

Nevertheless, the NH_4^+-induced ADP increase was relatively short-lived. By the time ADP had recovered, levels of glutamate had declined and triose-P (predominantly DHAP) had nearly tripled. The long-term activation of PK during NH_4^+ assimilation could be maintained by these metabolite changes.

The activation of PK during NH_4^+ assimilation affects glycolytic metabolite levels both up- and downstream. The PK-induced decline of PEP causes a decrease in PGA due to the phosphoglucomutase and enolase catalysed equilibrium with PEP. Thus PK activation caused a depletion of glycolytic metabolites downstream of PGA kinase. The increase in Pyr which accompanies PK activation could activate the pyruvate dehydrogenase complex (PDC) through inhibition of PDC kinase (Schuller & Randall, 1989), thus increasing acetyl CoA production for the TCA cycle.

The TCA cycle intermediates consumed by biosynthesis must be replenished by anaplerotic reactions. In *S. minutum* PEP carboxylase serves this role (Schuller *et al.*, 1990). *In vitro* studies have shown that glutamate inhibits and glutamine activates PEPCase (Schuller *et al.*, 1990). The balance between glutamate and glutamine pool sizes reflects, in large part, the activities of GS and GOGAT. As NH_4^+ assimilation increases because of increased NH_4^+ availability, glutamate is consumed and glutamine produced. If the supply of 2-OG begins to limit GOGAT activity the glutamine pool increases and the glutamate pool decreases. This has been shown to activate PEPCase *in vivo* and provide the anaplerotic carbon needed for NH_4^+ assimilation (Vanlerberghe *et al.*, 1990*b*).

Increased carbon flow through PK and PEPCase to the TCA cycle requires activation of plastid PFK. PFK from *S. minutum* is strongly inhibited by PEP and PGA at *in vivo* concentrations of these metabolites (Botha & Turpin, 1990*a*). The PK-mediated decline in PEP and PGA may activate PFK, thus increasing the pool size of FBP. The increased levels of FBP would result in an increased rate of triose-P production in the plastid. In turn, this would cause an increase in the rate of triose-P export from the chloroplast via the P_i translocator with a concomitant increase in plastid P_i and a decrease in cytosolic P_i. Any increase in plastid P_i would further support PFK activation (Fig. 8). It would also serve to inhibit any plastid FBPase activity remaining in the dark thereby preventing futile cycling between FBP and F6P (Botha & Turpin, 1990*b*). An increase in plastid P_i could also serve to activate starch degradation via starch phosphorylase (Steup, 1988), providing G1P for glycolytic metabolism.

During the first few minutes of NH_4^+ assimilation most of the cytosolic triose-P would be expected to move down the lower half of cytosolic glycolysis for use in respiratory metabolism and amino acid biosynthesis. There are changes in F-2,6-P_2, a compound which plays a role in the partitioning of cytosolic triose-P between respiration and gluconeogenesis through its effect on cytosolic FBPase (Botha & Turpin, 1990*b*; Turpin *et al.*, 1990). Figure 9 provides an overview of the regulation of respiratory carbon flow to amino acid synthesis.

Partitioning of photosynthetic electron flow between CO_2 fixation and N assimilation

Photosynthetically generated reducing power and ATP can be used in many metabolic processes including CO_2 fixation and inorganic N assimilation. Table 2 illustrates the relative demands for ATP and reductant associated with the assimilation of CO_2 to carbohydrate and various forms of N into protein. The observation that high rates of N assimilation by N-limited *S. minutum* causes a suppression of photosynthetic CO_2 fixation (Elrifi & Turpin, 1986; Elrifi *et al.*, 1988) makes this a useful system for studying the partitioning of photosynthetic electrons between CO_2 fixation and N assimilation. Figure 10 illustrates the short-term effects of NH_4^+ and NO_3^- assimilation on photosynthetic gas exchange and chlorophyll *a* fluorescence quenching.

NH_4^+ assimilation

The dramatic decline of carbon fixation during photosynthetic NH_4^+ assimilation by N-limited *S. minutum* (Fig. 10*B*) results from a limitation of Rubisco by RuBP (Elrifi & Turpin, 1986; Elrifi *et al.*, 1988). Most of the remaining carbon fixation is by phospho*enol*pyruvate carboxylase serving in an anaplerotic function to replenish TCA intermediates consumed in amino acid synthesis (Guy *et al.*, 1989*b*). The reduced but significant photosynthetic O_2 evolution (Fig. 10*B*), suggests that photosynthetically generated reducing power is consumed in NH_4^+ assimilation, presumably by the GOGAT reaction.

The simultaneous effects of transient NH_4^+ assimilation on photochemical (Q_q) and non-photochemical (Q_{NP}) Chl *a* fluorescence quenching can be explained from the changes in photosynthetic gas exchange (Holmes *et al.*, 1989). The decline in Calvin cycle ATP consumption and possible increase in the transthylakoid pH gradient would

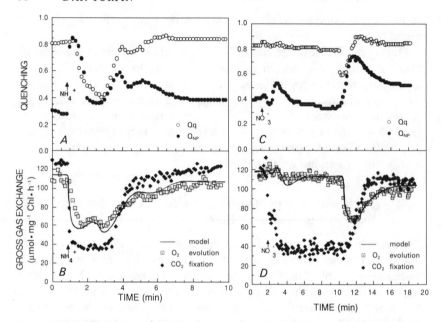

Fig. 10. Effects of N assimilation on fluorescence quenching and gross photosynthetic gas exchange. *A*, Changes in Q_q and Q_{NP} during transient NH_4^+ assimilation. *B*, Simultaneous changes in gross photosynthetic O_2 evolution ($\square$) and CO_2 fixation ($\blacklozenge$) measured by mass-spectrometry. Solid line represents the model predictions of photosynthetic O_2 evolution based on Equation 1 and the values of Q_q and Q_{NP} reported in *A*. *C*, Changes in Q_q and Q_{NP} during NO_3^--induced transients. *D*, Simultaneous changes in gross photosynthetic CO_2 fixation ($\blacklozenge$) and O_2 evolution ($\square$) measured by mass spectrometry. Solid line represents the rate of O_2 evolution calculated from Equation 1 and the values of Q_q and Q_{NP} in *C* (modified from Holmes *et al.*, 1989).

lead to the transient increase in Q_{NP}. This would decline once ATP consumption increased during protein synthesis and as a result of state transitions (see below). The Q_q decline has been suggested to result from the decreased NADPH consumption by glyceraldehyde 3-P dehydrogenase accompanying the decline in photosynthetic CO_2 fixation (Turpin & Weger, 1988).

NO_3^-/NO_2^- assimilation

The assimilation of NO_3^- by N-limited *S. minutum* also results in RuBP limitation of photosynthetic carbon fixation (Elrifi & Turpin, 1986; Elrifi *et al.*, 1988) (Fig. 10*D*). O_2 evolution, however, shows little change because NO_3^- and NO_2^- serve as physiological electron acceptors allowing maintenance of high rates of O_2 evolution (Fig. 10*D*). Because photosynthetic electron flow was maintained, photochemical quenching remained relatively unaffected (Fig. 10*C*). Following the completion of NO_3^- assimilation, but prior to the recovery of CO_2 fixation, there was no physiological acceptor for photogenerated electrons. Pools of reduced Q_A accumulated resulting in a decrease in photochemical quenching and water photolysis. Only when Calvin cycle induction has occurred, allowing the oxidation of Q_A, do photochemical quenching and O_2 evolution recover (Fig. 10*C*). This induction process is similar to that observed upon illumination (Walker *et al.*, 1983; Schreiber & Bilger, 1985; Sivak & Walker, 1985; Quick & Horton, 1986).

Use of fluorescence quenching to predict photosynthetic electron flow

Recently several groups have explored the use of room temperature Chl *a* fluorescence to predict photosynthetic electron flow (Weis & Berry, 1987; Genty *et al.*, 1989; Seaton & Walker, 1990). Weis & Berry (1987) have shown a linear correlation between the corrected apparent quantum yield of photosynthesis and non-photochemical quenching (Q_{NP}) of Chl *a* fluorescence. A similar empirical relationship could be defined between Chl *a* fluorescence and photosynthetic electron flow by holding *S. minutum* at a variety of light intensities and measuring gross O_2 evolution with a mass spectrometer (Eqn 1) (Holmes *et al.*, 1989):

$$J = I \cdot Q_q[0.4177 - 0.3282 \, Q_{NP}] \tag{1}$$

where J = photosynthetic electron flow (μmol O_2 mg^{-1} Chl h^{-1}), I = incident light intensity (μE m^{-2} s^{-1}), Q_q = photochemical quenching and Q_{NP} = non photochemical quenching. The constants are empirically derived (see Weis & Berry, 1987; Holmes *et al.*, 1989).

Ideally, a model predicting photosynthetic electron flow from fluorescence measurements should be applicable under transient as well as steady-state conditions and regardless of whether the electrons are used for CO_2 fixation or NO_3^-/NO_2^- reduction. We therefore examined the utility of the model during transient assimilation of different N sources in *S. minutum*.

During NH_4^+ assimilation the demand for photosynthetic electron flow declines dramatically. Excellent agreement was found between the observed rate of photosynthetic electron flow (Fig. 10*B*) and rates predicted from the measured values of fluorescence quenching (reported in Fig. 10A). Transient NO_3^- assimilation tests the utility of the model during electron flow to acceptors other than CO_2. Again, the predictions of photosynthetic electron flow under these conditions are in excellent agreement with observed rates (Fig. 10B). Further, the model predicts the transient decline in photosynthetic electron flow which occurs following the complete assimilation of NO_3^-, prior to induction of the Calvin cycle. These results show that room temperature fluorescence provides an accurate estimate of photosynthetic electron flow regardless of the electron acceptor, under both steady-state and transient conditions.

Metabolic control of the light state transition

Photosynthesising cells may modulate their capacity for ATP and NADPH production by controlling the relative activities of cyclic and non-cyclic electron flow by state transitions (Williams & Allen 1987). Recently, it has been proposed that state transitions act in altering ATP and NADPH production in response to changes in carbon (Horton & Lee, 1986; Horton, 1987, 1989; Horton et al., 1990) and N metabolism (Turpin & Bruce, 1990). Assimilation of different N sources results in large changes in cellular ATP/NADPH demands; when NH_4^+ is assimilated, the cellular ATP/NADPH requirements increase dramatically relative to those of CO_2 fixation (Table 2). Following 20 min of NH_4^+ assimilation the 77K fluorescence spectra indicated a decrease in PS II fluorescence (F686) and an increase in PS I fluorescence (F717) (Fig. 11). Once NH_4^+ assimilation was completed, the cells reverted to the control state. The increased respiratory metabolism which accompanies NH_4^+ assimilation and the decline in photosynthetic CO_2 fixation causes an increase in the NADPH/NADP$^+$ ratio (Mohanty et al., 1991). This correlates with a reduction in the intersystem photosynthetic electron transport chain (plastoquinone, PQ) (Mohanty et al., 1991) which drives the cells to a low photoactive state for PS II or state 2 (Bennoun, 1982; Horton & Lee, 1986; Bulte et al., 1990; Gans & Rebeille, 1990; Horton et al., 1990). Recently, it has been reported that a thylakoid-bound NAD(P)H-dependent PQ reductase activity was capable of reducing PQ in green algae in the dark. This activity may provide a mechanism whereby elevated levels of NADPH, brought about by increased respiratory carbon flow to NH_4^+ assimilation, may serve to reduce the PQ pool (Bennoun, 1982; Gfeller & Gibbs, 1985; Peltier et al., 1987; Bulte et al., 1990; Gans & Rebeille, 1990; Wilhelm & Duval, 1990). Together, these results imply that NH_4^+ assimilation decreases PS II activity and may increase the energy available to PS I which, in turn, serves to increase the ATP/

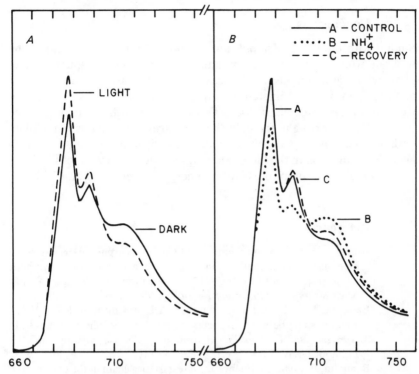

Fig. 11. 77K fluorescence emission spectra of *S. minutum*. *A*, Cells dark-adapted for 20 min are compared with cells light-adapted (78 μE m^{-2} s^{-1}) for 20 min at room temperature prior to freezing in liquid nitrogen. *B*, Fluorescence emission spectra of light-adapted cells before (A), during (B) and after (C) NH$_4^+$ assimilation. The excitation wavelength was 435 nm. The spectra have not been normalised (from Turpin & Bruce, 1990).

NADPH production ratio, bringing it in line with the metabolic requirements of NH$_4^+$ assimilation (Table 2).

NO$_3^-$ reduction and assimilation has a higher demand for NADPH than does NH$_4^+$. It is not surprising, therefore, that NO$_3^-$ assimilation did not cause an increase in the NADPH/NADP$^+$ ratio despite a large increase in respiratory carbon flow (Mohanty *et al.*, 1991). As would be expected, NO$_3^-$ assimilation did not cause reduction in PQ or a change in the distribution of excitation energy between PS I and PS II (Turpin & Bruce, 1990; Mohanty *et al.*, 1991).

Conclusions

Studies with the green alga *Selenastrum minutum* illustrate some of the ways in which primary N assimilation interacts with respiratory and photosynthetic metabolism. Changes in both the rate of N assimilation and the form of N assimilated modulate respiratory metabolism to provide carbon and reducing power. This occurs in both the light and the dark. When occurring in the light these changes in metabolism modulate both photosynthetic CO_2 fixation and O_2 evolution. Feedback from anabolic respiration in the light modulates the quantum yield of PS II and controls the distribution of excitation energy between PS I and PS II.

References

Amory, A.M., Vanlerberghe, G.C. & Turpin, D.H. (1991). Demonstration of both a photosynthetic and a nonphotosynthetic CO_2 requirement for NH_4^+ assimilation in the green alga *Selenastrum minutum*. *Plant Physiology* **95**, 192–6.

Bassham, J.A., Larsen, P.O., Lawyer, A.L. & Cornwell, K.C. (1981). Relationships between nitrogen metabolism and photosynthesis. In *Nitrogen and Carbon Metabolism*, ed. J.D. Bewley, pp. 135–63. London: Dr W. Junk.

Bennoun, P. (1982). Evidence for a respiratory chain in the chloroplast. *Proceedings of the National Academy of Sciences, USA* **79**, 4352–6.

Botha, F.C., Cawood, M.E. & Small, J.G.C. (1988). Kinetic properties of the ATP-dependent phosphofructokinase isoenzymes from cucumber seeds. *Plant and Cell Physiology* **29**, 415–21.

Botha, F.C. & Small, J.G.C. (1987). Comparison of the activities and some properties of pyrophosphate and ATP dependent fructose-6-phosphate 1-phosphotransferases of *Phaseolus vulgaris* seeds. *Plant Physiology* **83**, 722–77.

Botha, F.C. & Turpin, D.H. (1990*a*). Molecular, kinetic, and immunological properties of the 6-phosphofructokinase from the green alga *Selenastrum minutum*. *Plant Physiology* **93**, 871–9.

Botha, F.C. & Turpin, D.H. (1990*b*). Fructose 1,6-bisphosphatase in the green alga *Selenastrum minutum*. I. Evidence for the presence of isoenzymes. *Plant Physiology* **93**, 1460–5.

Boussiba, S., Resch, C.M. & Gibson, J.J. (1984). Ammonia uptake and retention in some cyanobacteria. *Archives of Microbiology* **138**, 287–92.

Bulte, L., Gans, P., Rebeille, R. & Wollman, F.A. (1990). ATP control of state transitions *in vivo* in *Chlamydomonas reinhardtii*. *Biochimica et Biophysica Acta* **1020**, 72–80.

Dennis, D.T. & Greyson, M.F. (1987*a*). Fructose-6-phosphotransferase isoenzymes from *Cucumis sativus*. *Plant and Cell Physiology* **29**, 195–9.

Dennis, D.T. & Greyson, M.F. (1987*b*). Fructose-6-phosphate metabolism in plants. *Physiologia Plantarum* **69**, 395–404.

Eisele, R. & Ullrich, W.R. (1977). Effect of glucose and CO_2 on nitrate uptake and coupled OH^- flux in *Ankistrodesmus braunii*. Plant Physiology **59**, 18–21.

Elrifi, I.R., Holmes, J.H., Weger, H.G., Mayo, W.P. & Turpin, D.H. (1988). RuBP limitation of photosynthetic carbon fixation during NH_3 assimilation. Interactions between photosynthesis, respiration, and ammonium assimilation in N-limited green algae. *Plant Physiology* **87**, 395–401.

Elrifi, I.R. & Turpin, D.H. (1986). Nitrate and ammonium induced photosynthetic suppression in N-limited *Selenastrum minutum*. *Plant Physiology* **81**, 273–9.

Elrifi, I.R. & Turpin, D.H. (1987). The path of carbon flow during NO_3-induced photosynthetic suppression in N-limited *Selenastrum minutum*. *Plant Physiology* **83**, 97–104.

Gans, P. & Rebeille, R. (1990). Control in the dark of the plastoquinone redox state by mitochondrial activity in *Chlamydomonas reinhardtii*. *Biochimica et Biophysica Acta* **1015**, 150–5.

Garland, W.J. & Dennis, D.T. (1980). Plastid and cytosolic phosphofructokinase from the developing endosperm of *Ricinus communis* II. Comparison of the kinetic and regulatory properties of isoenzymes. *Archives of Biochemistry and Biophysics* **204**, 310–17.

Genty, B., Briantais, J.M. & Baker, N.R. (1989). The relationship between the quantum yield of photosynthetic electron transport and quenching of chlorophyll fluorescence. *Biochimica et Biophysica Acta* **990**, 87–92.

Gfeller, R.P. & Gibbs, M. (1985). Fermentative metabolism of *Chlamydomonas reinhardtii*: Role of plastoquinone. *Plant Physiology* **77**, 509–11.

Grant, B.R. (1968). Effect of carbon dioxide concentration and buffer system on nitrate and nitrite assimilation in *Dunaliella tertiolecta*. *Journal of General Microbiology* **54**, 327–36.

Grant, B.R. & Turner, I.M. (1969). Light-stimulated nitrate and nitrite assimilation in several species of algae. *Comparative Biochemistry and Physiology* **29**, 995–1004.

Guerrero, M.G. & Lara, C. (1987). Assimilation of inorganic nitrogen. In *The Cyanobacteria*, ed. P. Fay & C. Van Bealen, pp. 163–86. Amsterdam: Elsevier Science Publishers.

Guy, R.D., Berry, J.A., Fogel, M.L. & Hoering, T.C. (1989*a*). Differential fractionation of oxygen isotopes by cyanide-resistant and cyanide-sensitive respiration in plants. *Planta* **177**, 483–91.

Guy, R.D., Vanlerberghe, G.C. & Turpin, D.H. (1989*b*). Significance of phospho*enol* pyruvate carboxylase during ammonium assimilation:

74   D.H. TURPIN

carbon isotope discrimination in photosynthesis and respiration by the N-limited green alga *Selenastrum minutum*. *Plant Physiology* **89**, 1150–7.

Holmes, J.H., Weger, H.G. & Turpin, D.H. (1989). Chlorophyll *a* fluorescence predicts total photosynthetic electron flow to CO_2 or NO_3^-/NO_2^- under transient conditions. *Plant Physiology* **91**, 331–7.

Horton, P. (1987). Interplay between environmental and metabolic factors in the regulation of electron transport in higher plants. In *Progress in Photosynthesis Research*, ed. J. Biggins, pp. 681–8. Dordrecht: Martinus Nijhoff.

Horton, P. (1989). Interaction between electron transport and carbon assimilation: Regulation of light harvesting and photochemistry. In *Photosynthesis*, ed. W.R. Briggs, pp. 393–406. New York: Alan R. Liss.

Horton, P. & Lee, P. (1986). Observation of enhancement and state transition in isolated intact chloroplasts. *Photosynthesis Research* **10**, 297–302.

Horton, P., Lee, P. & Fernyhough, P. (1990). Emerson enhancement, photosynthetic control and protein phosphorylation in isolated maize mesophyll chloroplasts: dependence upon carbon metabolism. *Biochimica et Biophysica Acta* **1017**, 160–6.

Kanazawa, T., Distefano, M. & Bassham, J.A. (1983). Ammonia regulation of intermediary metabolism in photosynthesizing and respiring *Chlorella pyrenoidosa*: comparative effects of methylamine. *Plant and Cell Physiology* **24**, 979–86.

Kanazawa, T., Kirk, M.R. & Bassham, J.A. (1970). Regulatory effects of ammonia on carbon metabolism in photosynthesizing *Chlorella pyrenoidosa*. *Biochimica et Biophysica Acta* **205**, 401–8.

Kanazawa, T.K., Kanazawa, K., Kirk, M.R. & Bassham, J.A. (1972). Regulatory effects of ammonia on carbon metabolism in *Chlorella pyrenoidosa* during photosynthesis and respiration. *Biochimica Biophysica Acta* **265**, 656–69.

Kelly, G.J. & Latzko, E. (1977). Chloroplast phosphofructokinase II: partial purification, kinetic and regulatory properties. *Plant Physiology* **60**, 295–9.

Kelly, G.J., Mukherjee, U., Holtum, J.A.M. & Latzko, E. (1985). Purification, kinetic and regulatory properties of phosphofructokinases from *Chlorella pyrenoidosa*. *Plant and Cell Physiology* **26**, 301–7.

Knowles, V.L., Greyson, M.F. & Dennis, D.T. (1990). Characterization of ATP-dependent fructose 6-phosphate 1-phosphotransferase isoenzymes from leaf and endosperm tissues of *Ricinis communis*. *Plant Physiology* **92**, 155–9.

Lara, C., Romero, J.M., Coronil, R. & Guerrero, M.G. (1987). Interactions between photosynthetic nitrate assimilation and CO_2 fixation in cyanobacteria. In *Inorganic Nitrogen Metabolism*, ed. W. Ullrich,

P.J. Aparicio, P.J. Syrett & F. Castillo, pp. 45–52. New York: Springer-Verlag.

Larsson, C.M. & Larsson, M. (1987). Regulation of nitrate utilization in green algae. In *Inorganic Nitrogen Metabolism*, ed. W. Ullrich, P. J. Aparicio, P.J. Syrett & F. Castillo, pp. 203–7. New York: Springer-Verlag.

Lin, M., Turpin, D.H. & Plaxton, W. (1989a). Pyruvate kinase isozymes from the green alga, *Selenastrum minutum*. I. Purification and physical and immunological characterization. *Archives of Biochemistry and Biophysics* 269, 219–27.

Lin, M., Turpin, D.H. & Plaxton, W.C. (1989b). Pyruvate kinase isoenzymes from the green alga, *Selenastrum minutum*. II. Kinetic and regulatory properties. *Archives of Biochemistry and Biophysics* 269, 228–38.

Miyachi, S. & Miyachi, W. (1985). Ammonia induces starch degradation in *Chlorella* cells. *Plant and Cell Physiology* 26, 245–52.

Mohanty, N., Bruce, D. & Turpin, D.H. (1991). Dark ammonium assimilation reduces the plastoquinone pool of photosystems II in the green alga *Selenastrum minutum*. *Plant Physiology* 96, 513–17.

Ohmori, J. & Hattori, A. (1978). Transient change in ATP pool of *Anabaena cylindrica* associated with ammonium assimilation. *Archives of Microbiology* 117, 17–20.

Peltier, G., Ravenel, J. & Vermeglio, A. (1987). Inhibition of a respiratory activity by short saturating flashes in *Chlamydomonas*: evidence for a chlororespiration. *Biochimica et Biophysica Acta* 893, 83–90.

Peltier, G. & Thibault, P. (1985). O_2 uptake in the light in *Chlamydomonas*: evidence for persistent mitochondrial respiration. *Plant Physiology* 79, 225–30.

Quick, W.P. & Horton, P. (1986). Studies on the induction of chlorophyll fluorescence changes in the level of glycerate-3-phosphate and the patterns of fluorescence quenching. *Biochimica Biophysica Acta* 849, 1–6.

Rigano, D., Di Martino Rigano, V., Vona, V., Esposito, S. & Di Martino, C. (1990). Carbon metabolism and ammonium assimilation under light or dark conditions in N-sufficient and N-limited unicellular algae. In *Inorganic Nitrogen in Plants and Microorganisms, Uptake and Metabolism*, ed. W.R. Ullrich, C. Rigano, A. Fuggi & P.J. Aparico, pp. 131–6. New York: Springer-Verlag.

Romero, J.M., Lara, C. & Guerrero, M.G. (1985). Dependence of nitrate utilization upon active CO_2 fixation in *Anacystis nidulans*: a regulatory aspect of the interaction between photosynthetic carbon and nitrogen metabolism. *Archives of Biochemistry and Biophysics* 237, 396–401.

Schlee, J., Cho, B.H. & Komor, E. (1985). Regulation of nitrate uptake by glucose in *Chlorella*. *Plant Science* 39, 25–30.

Schreiber, U. & Bilger, W. (1985). Rapid assessment of stress effects on plant leaves by chlorophyll fluorescence measurements. *NATO Advanced Research Workshop*. Berlin: Springer-Verlag.

Schuller, K.A., Plaxton, W.C. & Turpin, D.H. (1990). Regulation of phospho*enol*pyruvate carboxylase from the green alga *Selenastrum minutum*: properties associated with replenishment of tricarboxylic acid cycle intermediates during ammonium assimilation. *Plant Physiology* **93**, 1303–11.

Schuller, K.A. & Randall, D.D. (1989). Regulation of pea mitochondrial pyruvate dehydrogenase complex: does photorespiratory ammonium influence mitochondrial carbon metabolism? *Plant Physiology* **89**, 1207–12.

Seaton, G.G.R. & Walker, D.A. (1990). Chlorophyll fluorescence as a measure of photosynthetic carbon assimilation. *Proceedings of the Royal Society of London B* **242**, 29–35.

Sivak, M.N. & Walker, D. (1985). Chlorophyll *a* fluorescence: can it shed light on fundamental questions in photosynthetic carbon dioxide fixation? *Plant, Cell and Environment* **8**, 439–48.

Smith, R.G., Vanlerberghe, G.C., Stitt, M. & Turpin, D.H. (1989). Short-term metabolite changes during transient ammonium assimilation by the N-limited green alga *Selenastrum minutum*. *Plant Physiology* **91**, 749–55.

Steup, M. (1988). Starch degradation. In *The Biochemistry of Plants: A Comprehensive Treatise*, Vol. 14, ed. P.K. Stumpf & E.E. Conn, pp. 255–96. New York: Academic Press.

Syrett, P.J. (1953). The assimilation of ammonia by nitrogen starved cells of *Chlorella vulgaris*. Part I. The correlation of assimilation with respiration. *Annals of Botany* **65**, 1–19.

Syrett, P.J. (1956*a*). The assimilation of ammonia and nitrate by nitrogen starved cells of *Chlorella vulgaris*. II. The assimilation of large quantities of nitrogen. *Physiologia Plantarum* **9**, 19–27.

Syrett, P.J. (1956*b*). The assimilation of ammonia and nitrite by nitrogen starved cells of *Chlorella vulgaris*. IV. The dark fixation of carbon dioxide. *Physiologia Plantarum* **9**, 165–71.

Syrett, P.J. (1981). Nitrogen metabolism of microalgae. *Canadian Bulletin of Fisheries and Aquatic Sciences* **210**, 182–210.

Syrett, P.J. (1988). Uptake and utilization of nitrogen compounds. In *Biochemistry of the Algae and Cyanobacteria*, ed. L.J. Rogers & J.R. Gallon, pp. 23–39. Oxford: Clarendon Press.

Thacker, A. & Syrett, P.J. (1972). The assimilation of nitrate and ammonium by *Chlamydomonas reinhardtii*. *New Phytologist* **71**, 423–33.

Turpin, D.H. (1991). Effects of inorganic N-availability on algal photosynthesis and carbon metabolism. *Journal of Phycology* **27**, 14–20.

Turpin, D.H., Botha, F.C., Smith, R.G., Feil, R., Horsey, A.K. &

Vanlerberghe, G.C. (1990). Regulation of carbon partitioning to respiration during dark ammonium assimilation by the green alga *Selenastrum minutum*. *Plant Physiology* **93**, 166–75.

Turpin, D.H. & Bruce, D. (1990). Regulation of photosynthetic light harvesting by nitrogen assimilation in the green alga *Selenastrum minutum*. *FEBS Letters* **263**, 99–103.

Turpin, D.H. & Weger, H.G. (1988). Steady-state chlorophyll *a* fluorescence transients during ammonium assimilation by the N-limited green alga *Selenastrum minutum*. *Plant Physiology* **88**, 97–101.

Ullrich, W.R., Lesch, S., Jarczyk, L., Harterich, M. & Trogisch, G.D. (1990). Transport of inorganic nitrogen compounds: physiological studies on uptake and assimilation. In *Inorganic Nitrogen in Plants and Microorganisms, Uptake and Metabolism*, ed. W.R. Ullrich, C. Rigano, A. Fuggi & P.J. Aparico, pp. 44–50. New York: Springer-Verlag.

Vanlerberghe, G.C., Feil, R. & Turpin, D.H. (1990*a*). Anaerobic metabolism in the N-limited green alga *Selenastrum minutum*. *Plant Physiology* **94**, 1116–23.

Vanlerberghe, G.C., Schuller, K.A., Smith, R.G., Feil, R., Plaxton, W.C. & Turpin, D.H. (1990*b*). Relationship between NH_4^+ assimilation rate and *in vivo* phospho*enol*pyruvate carboxylase activity. *Plant Physiology* **94**, 284–90.

Walker, D.A., Sivak, M.N., Prinsley, R.T. & Cheesbrough, J.K. (1983). Simultaneous measurement of oscillations in oxygen evolution and chlorophyll *a* fluorescence in leaf pieces. *Plant Physiology* **73**, 542–9.

Weger, H.G., Birch, D.G., Elrifi, I.R. & Turpin, D.H. (1988). Ammonium assimilation requires mitochondrial respiration in the light. A study with the green alga *Selenastrum minutum*. *Plant Physiology* **86**, 688–92.

Weger, H.G., Chadderton, A.R., Lin, M., Guy, R.D. & Turpin, D.H. (1990*a*). Cytochrome and alternative pathway respiration during transient ammonium assimilation by N-limited *Chlamydomonas reinhardtii*. *Plant Physiology* **94**, 1131–6.

Weger, H.G., Guy, R.D. & Turpin, D.H. (1990*b*). Cytochrome and alternative pathway respiration in green algae. *Plant Physiology* **93**, 356–60.

Weger, H.G. & Turpin, D.H. (1989). Mitochondrial respiration can support NO_3^- and NO_2^- reduction during photosynthesis: interactions between photosynthesis, respiration and N assimilation in the N-limited green algae *Selenastrum minutum*. *Plant Physiology* **89**, 409–15.

Weis, E. & Berry, J.A. (1987). Quantum efficiency of photosystem II in relation to energy dependent quenching of chlorophyll fluorescence. *Biochimica et Biophysica Acta* **894**, 198–208.

Wilhem, C. & Duval, J.C. (1990). Fluorescence induction kinetics as a tool to detect a chlororespiratory activity in the prasinophycean alga *Mantoniella squamata. Biochimica et Biophysica Acta* **1016**, 197–202.

Williams, W.P. & Allen, J.F. (1987). State 1/State 2 changes in higher plants and algae. *Photosynthesis Research* **13**, 19–45.

ROGER M. WALLSGROVE, ANITA C. BARON
and ALYSON K. TOBIN

Carbon and nitrogen cycling between organelles during photorespiration

Photorespiration is defined as the light-dependent release of CO_2 and uptake of O_2 in photosynthetic tissues. It is distinct from 'dark' respiration, and is a result of the intrinsic inefficiency of ribulose bisphosphate carboxylase/oxygenase (Rubisco), the enzyme responsible for the assimilation of carbon (Keys, 1986). The oxygenase reaction catalysed by Rubisco produces one molecule of 3-phosphoglycerate (PGA), and one molecule of 2-phosphoglycolate from one molecule of ribulose bisphosphate (RuBP) and oxygen, rather than the two molecules of PGA formed by the carboxylase reaction.

Carboxylase: $RuBP + CO_2 \rightarrow 2\ PGA$

Oxygenase: $RuBP + O_2 \rightarrow PGA + 2\text{-phosphoglycolate}$

Metabolism of the phosphoglycolate leads to a release of CO_2. The relative rates of carboxylase and oxygenase activity depend on the relative concentrations of oxygen and CO_2 at the active site of the enzyme. The question of CO_2 diffusion and transport into chloroplasts has recently been reviewed (Machler *et al.*, 1990). Both temperature and irradiance affect the relative rates of the two reactions *in vivo* (see for example Fuhrer & Erismann, 1984; Peterson, 1990). There are also quite distinctive differences in the $CO_2{:}O_2$ 'specificity factor' between the Rubisco enzymes from different plants (Parry *et al.*, 1989). Phosphoglycolate represents a potential 'loss' of both carbon and phosphate from the photosynthetic carbon assimilation pathway, and the photorespiratory pathway can be seen in part as a 'scavenging cycle' to minimise this loss. It has long been recognised that photorespiration represents an inherent inefficiency in photosynthesis, and considerable efforts have been made to find ways of reducing or eliminating photorespiration (see for example reviews by Keys, 1986, and Ogren, 1984).

Society for Experimental Biology Seminar Series 50: *Plant organelles*, ed. A. K. Tobin.
© Cambridge University Press 1992, pp. 79–96.

Carbon and nitrogen cycling

The currently accepted pathway for the photorespiration cycle is shown in Fig. 1. It will be seen that metabolism of phosphoglycolate proceeds via reactions in the chloroplast, cytosol, peroxisome and mitochondrion. The reaction that actually results in the release of CO_2 is the conversion of glycine to serine in the mitochondrion, catalysed by the glycine decarboxylase complex together with serine hydroxymethyltransferase:

$$2\ Gly + H_2O + NAD^+ \rightarrow Ser + CO_2 + NH_3 + NADH + H^+$$

It was not until the late 1970s that the significance of the stoichiometric release of ammonia in this reaction was appreciated, and a photorespiratory nitrogen cycle described (Keys *et al.*, 1978). Overall, 75% of the carbon in phosphoglycolate and all of the phosphate is re-

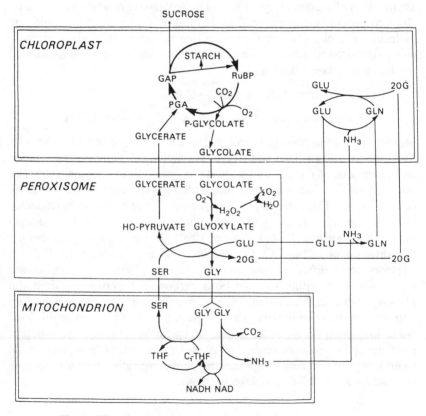

Fig. 1. The photorespiration pathway.

cycled, without allowing for any of the released CO_2 being reassimilated. The C4 plants, such as maize and sugar cane, do not exhibit photorespiration as defined above, but as will be seen below this does not mean that the photorespiratory cycle is absent from or inactive in such plants.

Glycine decarboxylation is not the only possible source of CO_2 release from phosphoglycolate metabolism. At various times other alternatives have been suggested, such as direct decarboxylation of glyoxylate by H_2O_2 (Grodzinski, 1978; Shingles *et al.*, 1984) or the reaction of H_2O_2 with either formate or hydroxypyruvate (Singh *et al.*, 1985). There is, however, now very clear genetic and biochemical evidence that the pathway as described in Fig. 1 represents the main, and possibly only, route for phosphoglycolate metabolism and photorespiration. One striking feature of all the alternative pathways suggested is that they are very much less efficient at returning carbon to the Calvin cycle, in some cases involving a 100% loss of carbon, and this in itself makes their operation *in vivo* unlikely even though the reactions can be demonstrated in isolated peroxisomes (see for example Grodzinski, 1979).

Chemical inhibition of enzymes involved in the photorespiratory cycle leads to a decrease in photosynthetic rate (Lawyer & Zelitch, 1979; Jenkins *et al.*, 1982*a*; Ikeda *et al.*, 1984) to the extent that many such compounds are effective herbicides. Interestingly, such inhibitor studies revealed a significant photorespiratory carbon and nitrogen cycle in C4 plants (Jenkins *et al.*, 1982*b*; Berger & Fock, 1983), as did a study of glycine and glycolate pools in maize leaves (Marek & Stewart, 1983). Another way of blocking the photorespiration cycle is by the selection of mutants which lack one of the enzymes. This approach was first employed by Somerville & Ogren (1979) with *Arabidopsis thaliana*. After chemical mutagenesis, M_2 lines were selected by their inability to grow in air although they were viable in high CO_2 (i.e. under conditions that suppress photorespiration). Several different classes of mutants were isolated, and similar studies with barley and pea have lead to the isolation of similar mutants in these species (see review by Blackwell *et al.*, 1988). Table 1 lists the specific enzyme lesions so far identified by such mutant screening.

The physiological and biochemical characterisation of such mutants has provided unequivocal confirmation of the photorespiratory pathway of carbon and nitrogen cycling, as shown in Fig. 1. What is immediately clear from these studies is that *any* lesion in the cycle is ultimately fatal to plants when they are grown under photorespiratory conditions. If any alternative pathways exist, they can at best be of only minor significance, and they cannot cope with the full flux of phosphoglycolate. This does not mean to say that no other biochemical processes are involved in

Table 1. *Photorespiration mutants in higher plants*

Biochemical lesion	Plant species	Reference
Rubisco protein	barley	Hall *et al.*, 1986
Rubisco activase	*Arabidopsis*	Somerville *et al.*, 1982
Phosphoglycolate phosphatase	*Arabidopsis*	Somerville & Ogren, 1979
	barley	Hall *et al.*, 1987
Catalase	barley	Kendall *et al.*, 1983
Serine: glyoxylate aminotransferase	*Arabidopsis*	Somerville & Ogren, 1980a
	barley	Murray *et al.*, 1987
	Nicotiana sylvestris	McHale *et al.*, 1988
Glycine-serine conversion	*Arabidopsis*	Somerville & Ogren, 1981a, b
	barley	Lea *et al.*, 1984, Blackwell *et al.*, 1987
Hydroxypyruvate reductase	barley	Murray *et al.*, 1989
Glutamine synthetase	barley	Wallsgrove *et al*, 1987
Fd-glutamate synthase	*Arabidopsis*	Somerville & Ogren, 1980b
	barley	Kendall *et al.*, 1986
	pea	Blackwell *et al.*, 1987
Chloroplast 2-oxoglutarate uptake	*Arabidopsis*	Somerville & Ogren, 1983
	barley	Wallsgrove *et al.*, 1986

photorespiration. Indeed, many mutants have been generated which have all the characteristics of photorespiratory mutants (i.e. air-sensitive, viable in high CO_2) but appear to have no lesions in the main pathway (Blackwell *et al.*, 1988). This suggests that there are reactions outside the main pathway that are nevertheless essential to the photorespiration cycle. There are clearly aspects of photorespiration and its interaction with other metabolic processes which remain to be elucidated.

Carbon dioxide fixation rates in most of the mutant lines are severely reduced on transfer of plants from 1% O_2 (little or no oxygenase activity) to air (Blackwell *et al.*, 1988). This is especially true of the lines deficient in phosphoglycolate phosphatase, serine:glyoxylate aminotransferase, glycine–serine conversion, and ferredoxin glutamate synthase. The implications are that blocking either carbon or nitrogen cycling rapidly depletes the chloroplast of Calvin cycle intermediates or that toxic levels of intermediates rapidly build up (though this is probably only significant in the case of phosphoglycolate). Where this rapid drop in CO_2 fixation

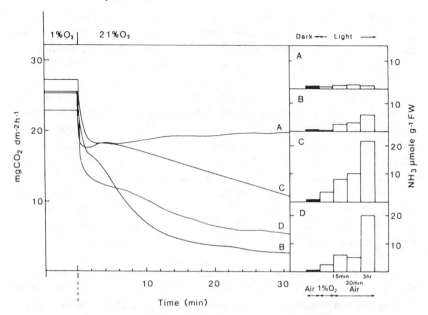

Fig. 2. CO_2 fixation and ammonia accumulation in barley leaves under photorespiratory conditions. A, wild-type (cv. Maris Mink); B, Fd-glutamate synthase deficient mutant; C, glutamine synthetase deficient mutant; D, double mutant deficient in both enzymes.

does not occur, it seems that the limitation in carbon or nitrogen flux is less severe. For example, glutamine synthetase (GS) and hydroxy-pyruvate reductase mutants maintain a significant photorespiratory flux because of the presence of unaltered isozymes of these enzymes (Walls-grove *et al.*, 1987; Murray *et al.*, 1989). The catalase mutants of barley have no short-term decline in CO_2 fixation, because there is effectively no block in carbon or nitrogen cycling. Instead, longer exposure to photorespiratory conditions leads to breakdown of cell membranes by the accumulation of H_2O_2 (Parker & Lea, 1983), after other protective mechanisms have been overloaded (Smith *et al.*, 1984).

Blocking the photorespiratory N cycle, either in the mutants (Black-well *et al.*, 1988) or by application of inhibitors (e.g. Ikeda *et al.*, 1984), can produce a very large accumulation of ammonia. This ammonia was originally thought to be the cause of the decline in photosynthesis, through 'uncoupling' of chloroplasts, but several lines of evidence suggest that this is not the case. In the mutants, there was no correlation between ammonia accumulation and inhibition of photosynthesis (Fig. 2; Black-well *et al.*, 1988), and chlorophyll fluorescence studies of these plants failed to detect the expected uncoupling of electron transport (Sivak *et al.*, 1988). In addition, feeding ammonia to leaves causes a greater accumulation of ammonia than do the inhibitors, but without any effect on photosynthetic rate (Ikeda *et al.*, 1984; Walker *et al.*, 1984). When reassimilation is blocked, ammonia appears to be sequestered in the vacuole and does not accumulate to toxic concentrations in the chloroplasts. The real effect of such blockage is in the cycling of carbon, so that there is no regeneration of the amino donors needed for conversion of glyoxylate to glycine. Feeding exogenous glutamate, glutamine or other amino acids can at least in part overcome the deficiency (Blackwell *et al.*, 1988). Studies with *Chlamydomonas* illustrate the drastic depletion of amino nitrogen, and mobilisation of reserves in protein, that occurs when photorespiratory N reassimilation is blocked in this alga (Cullimore & Sims, 1980), and a similar flow of nitrogen from protein to photorespiratory ammonia has been demonstrated in *Lemna* (Rhodes *et al.*, 1986).

Isolation of a hydroxypyruvate reductase deficient mutant in barley has clarified the position regarding isoforms of this enzyme, previously indicated by the separation of multiple forms and differential localisation. The major form of this enzyme in green leaves is the NADH-dependent isozyme, located in the peroxisomes. A minor, NADPH-specific isozyme was reported in the cytosol (Kleczkowski *et al.*, 1988). In a barley mutant devoid of the peroxisomal hydroxypyruvate reductase this latter activity is unaffected (Murray *et al.*, 1989), proving it to be a distinct isozyme. This mutant also provides evidence that NADH-glyoxylate reductase is a separate enzyme from NADH-hydroxypyruvate reductase, in that the former activity is little affected by the mutation.

Another aspect of carbon and nitrogen cycling concerns the utilisation of NADH formed during serine synthesis in the mitochondria, and the supply of NADH for hydroxypyruvate reduction in the peroxisome. A malate–oxaloacetate shuttle has been described that fulfils both these operations (Journet *et al.*, 1981). (For a more detailed discussion see Wiskich & Meidan, and Heldt & Flügge, this volume.)

Other metabolic functions for photorespiration

The effective operation of the pathway in Fig. 1 is thus essential to the metabolism of a photosynthetically active cell in air. Conversely, under conditions that minimise or abolish the oxygenase reaction and the flux into phosphoglycolate, the mutants demonstrate quite clearly that this pathway is not necessary for primary metabolism and normal growth and development. Whatever role(s) for photorespiration exist (and there have been many suggestions), it is clearly not indispensable. The most striking examples are those of mutants which lack GS2, the chloroplast GS isoenzyme (which comprises up to 90% of the total GS activity in barley leaves), and those lacking ferredoxin-glutamate synthase (>95% of total glutamate synthase). These plants grow and develop quite normally under high CO_2 (Blackwell *et al.*, 1988). There is thus sufficient cytosolic GS activity, and NADH-glutamate synthase, for primary ammonia assimilation. The two minor isoenzymes are adequate for all glutamate (and hence other amino acid) synthesis in the absence of a photorespiratory ammonia flux, as no other route of primary N assimilation has been shown to operate. It has been suggested that the photorespiration pathway provides metabolites for other pathways, particularly through the synthesis of glycine and serine. All the evidence from the mutants and with photorespiration inhibitors provides strong support for the view that no significant net flux of carbon out of the pathway is likely to occur. This does not mean that no glycine or serine are diverted from the pathway: indeed, there is experimental evidence that such a flow occurs (Madore & Grodzinski, 1984). No real quantitative estimate of such a flow exists, and the mutants clearly show that photorespiration is not absolutely required for sufficient synthesis of glycine and serine, which are also synthesised in the chloroplast from PGA (Larsson & Albertsson, 1979). There may well be significant species differences in diversion of metabolites from photorespiration. Studies with guayule (*Parthenium argentatum*) suggest that in this plant there is a major flow of carbon from photorespiratory serine into rubber (Reddy *et al.*, 1987).

What has become apparent is the interaction of the nitrogen cycle with other aspects of leaf (and plant) nitrogen metabolism, and an apparent flow of nitrogen from asparagine and alanine into the pathway (see review by Joy, 1988). The role of alanine in photorespiratory N metabolism is not clear, although it appears to be a preferred substrate for glyoxylate transamination *in vivo* (Ta & Joy, 1986). Alanine is usually assumed to be produced via transamination of pyruvate with glutamate, the major glutamate:pyruvate aminotransferase being in the peroxisome (Biekmann & Feierabend, 1982) and apparently the same enzyme as glutamate:glyoxylate aminotransferase. This makes the contribution of alanine to glyoxylate transamination somewhat puzzling. Both asparagine and alanine transfer nitrogen to glycine via serine:glyoxylate aminotransferase (SGAT), rather than the glutamate:glyoxylate aminotransferase, as clearly demonstrated by the concomitant loss of serine, asparagine and alanine transamination activity in SGAT-deficient mutants (Somerville & Ogren, 1980a; Murray *et al.*, 1987).

As drawn in Fig. 1 the photorespiratory cycle has no net demand for 2-oxoglutarate, this oxoacid being cycled from peroxisome to chloroplast. The vital role of chloroplast 2-OG uptake has been demonstrated (Somerville & Ogren, 1983; Wallsgrove et al., 1986). Any shortfall in 2-oxoglutarate was assumed to be met from the tricarboxylic acid cycle in the mitochondria. However, recent work has suggested alternative sources of 2-OG. As well as the mitochondrial NAD-dependent isocitrate dehydrogenase (IDH), NADP-dependent IDH isoenzymes have been detected in the chloroplast (Randall & Givan, 1981; Chen et al., 1989) and in the cytosol (Chen et al., 1988). Based on the relative activities of these isoenzymes, it has been suggested that the cytosolic IDH is the most likely source of the 2-OG needed for chloroplast ammonia assimilation (Chen & Gadal, 1990). The substrate for cytosolic IDH is probably provided by mobilisation of a vacuolar citrate pool (Oleski et al., 1987) which is converted to isocitrate by a cytosolic aconitase (Chen & Gadal, 1990). The possibility therefore exists for interaction of photorespiratory glutamate metabolism with other aspects of organic acid metabolism in the leaf. A further example of this is illustrated by the possible interaction between photorespiratory ammonia release and the phosphorylation state of the TCA cycle pyruvate dehydrogenase complex (PDC). PDC is regulated by reversible inactivation through phosphorylation of the enzyme by pyruvate dehydrogenase (PDH) kinase, and activation by dephosphorylation via P-PDH phosphatase (Miernyk et al., 1985). PDH kinase is stimulated by NH_4^+ and it has been suggested that the release of ammonia from glycine oxidation during photorespiration leads to inactivation of PDC in the light (Schuller & Randall, 1989). This would progressively halt the flow of carbon into the TCA cycle via PDC. The full details of this regulation remain to be determined, and as yet there is no direct evidence that this operates in vivo. (For further discussion of TCA cycle regulation, see Wiskich & Meidan, and Moore et al., this volume.)

One other possible metabolic function of the photorespiratory pathway is the transfer of reducing power and ATP from plastid to cytoplasm. As already seen, the conversion of glycine to serine involves a net production of NADH and protons, and there is direct evidence for net ATP production as a result of this reaction (Gardeström & Wigge, 1988). Under photorespiratory conditions carbon assimilation is limited, possibly reducing the capacity of chloroplasts to export triosephosphates to the cytosol for NADH and ATP synthesis; but under these same conditions it seems that glycine oxidation in the mitochondria can make up for any shortfall (see Wiskich & Meidan, Heldt & Flügge, and Moore et al., this volume).

A general role for photorespiration in the dissipation of excess light energy has frequently been suggested, although other mechanisms for this dissipation also operate. Recent studies on the effect of irradiance, using simultaneous gas exchange and fluorescence measurements, suggested that this was a real possibility at very high irradiances (Peterson, 1990), and might involve some alteration in photorespiratory metabolism. No direct biochemical evidence for such changes has yet emerged, but further work is needed to clarify this.

Control of photorespiratory metabolism

The most significant regulator of flux through the photorespiration pathway is the relative concentration of O_2 and CO_2 at the active site of Rubisco. Whilst the atmospheric concentrations of these gases do not vary significantly (fossil fuel burning and the greenhouse effect notwithstanding!), the internal concentration of CO_2 in the leaf may well vary, depending on water supply and subsequent stomatal opening or closing. Temperature also affects photorespiration, the relative solubilities of the two gases changing in such a way that oxygenase activity is favoured at higher temperatures (Brooks & Farquhar, 1985).

Other environmental effects on photorespiration have been reported, in particular the influence of nitrogen supply (quantitative and qualitative). Nitrogen starvation or limitation are reported to increase photorespiration (Hak & Natr, 1987*a*, *b*), as is the supply of nitrogen as ammonium rather than nitrate (Fuhrer & Erismann, 1984). Other studies have been unable to reproduce some of these effects (Hall *et al.*, 1984), so at the present time it is unclear to what extent such nutritional controls actually operate. No obvious biochemical explanation for any influence of nitrogen nutrition on Rubisco activity is apparent, though a possible stimulation of CO_2 release from glyoxylate by ammonia has been suggested (Fuhrer & Erismann, 1984). This suggestion is hard to reconcile with the findings of Somerville & Ogren (1981*a*, *b*), who showed that feeding ammonia to mutants blocked in glycine metabolism abolished all detectable CO_2 release.

Developmental regulation of the photorespiratory pathway has been demonstrated in a number of species. Many of the enzymes have been shown to be synthesised in coordination with the photosynthetic apparatus. These include chloroplastic Fd-glutamate synthase (Wallsgrove *et al.*, 1982) and GS2 (Tobin *et al.*, 1985), mitochondrial glycine decarboxylase (Tobin *et al.*, 1989), and the peroxisomal enzymes glycolate oxidase (Tobin *et al.*, 1988), hydroxypyruvate reductase (Bajracharya *et al.*, 1987) and glutamate:glyoxylate aminotransferase (Biekmann & Feierabend, 1982). For a more detailed discussion, see

Tobin & Rogers (this volume). In addition to this developmental regulation, there is a spatial, cell-specific control of photorespiratory enzymes. Glycine decarboxylase protein (Tobin *et al.*, 1989) and activity (Gardestrom *et al.*, 1980) are concentrated in the photosynthetic cells of C3 leaves. In C4 plants GDC activity is confined to the bundle sheath cells and is absent from the mesophyll cells which lack Rubisco (Ohnishi & Kanai 1983). In C3–C4 intermediate species, the P protein of GDC is also confined to the bundle sheath cells, although Rubisco is present in both mesophyll and bundle sheath cells (Hylton *et al.*, 1988; see Rawsthorne *et al.*, this volume).

The existence of other regulatory factors modulating enzyme activity or expression has been suggested. In pea leaves, growth in high CO_2 was reported to reduce the expression of chloroplast GS mRNA (Edwards & Coruzzi, 1989). This study did not monitor enzyme activity or protein content, however, and we have been unable to demonstrate any influence of CO_2 on the activity or content of GS in barley leaves (Fig. 3). In some experiments Northern blots indicated that there was an effect of CO_2 on GS mRNA levels, but this was not consistent or reproducible. A marked reduction in catalase activity was found (Fig. 4), as previously reported in tobacco leaves (Havir & McHale 1989), but this seemed to be a transient effect. It is important to recognise that as there are strong developmental influences on the levels of photorespiratory enzymes in leaves (as dis-

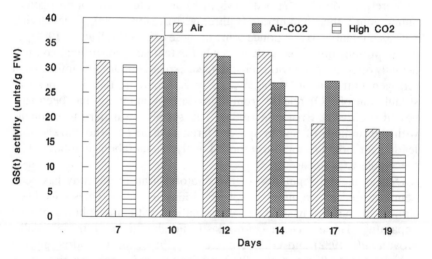

Fig. 3. Glutamine synthetase activity in first leaves of barley, grown (a) in air, (b) transferred from air to 0.7% CO_2 after 7 days, (c) continuously in 0.7% CO_2. Total GS activity (transferase assay) was measured in soluble leaf extracts.

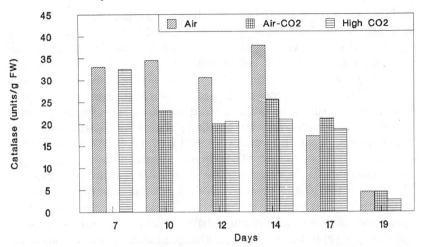

Fig. 4. Catalase activity in first leaves of barley, grown under the same conditions as in Fig. 3.

cussed above and in Tobin & Rogers, this volume), any effect of high CO_2 on the expression of these enzymes may be an indirect effect resulting from a change in plant growth, rather than a direct metabolic regulation.

Genetic analysis of the photorespiration mutants has strongly suggested that there is little or no metabolic regulation of gene expression for the enzymes of the pathway. For phosphoglycolate phosphatase, catalase, GS and glutamate synthase mutants, heterozygous plants produced by back-crosses with the parent line have only 50% of the wild-type activity of these enzymes (Blackwell *et al.*, 1988). At least in the case of the GS mutants we know that the mutations are in the structural gene for the protein (Freeman *et al.*, 1990), so it would seem that expression of the enzyme is determined only by the number of gene copies. More detailed molecular studies of gene expression, through development and following environmental changes, are required in this area. It should always be borne in mind that the most important changes are in enzyme protein content and enzyme activity: merely following changes in mRNA levels may not accurately reflect the *in vivo* capacity or capability of the metabolic pathway.

Conclusions

The photorespiration cycle represents a major metabolic pathway in photosynthetic tissues. The flux of both carbon and nitrogen through this cycle is considerable, yet the sole clearly defined role for the pathway is to scavenge carbon that would be potentially lost to the Calvin cycle as a result of Rubisco oxygenase activity. Operation of the cycle involves many organelles within the cell, and the enzymes of the pathway form a major constituent of these organelles in photosynthetic tissues. Although we have a reasonable understanding of the biochemistry of photorespiration, there are still some unanswered questions and problems. To what extent do photorespiratory intermediates contribute to other metabolic pathways, and what other pathways feed into the cycle? Are there any environmental or metabolic influences or controls on the operation of the photorespiration cycle, other than the relative concentrations of CO_2 and O_2 at the active site of Rubisco? Although some membrane transport systems important in the cycle have been identified, a clear view of the transport/diffusion of all the intermediates, across the many membrane barriers, has yet to be established. What are the metabolic lesions in the many conditional-lethal, CO_2-sensitive mutants selected that do *not* have defects in any known enzyme of the cycle? More work in this area is undoubtedly needed, and there are complexities yet to be uncovered.

There were originally optimistic ideas that some simple way of blocking photorespiration might be found, leading to significant increases in agricultural productivity. Alas, these hopes have been dashed. The photorespiration pathway is certainly a useful target for chemical inhibitors, but only as potent herbicides. To reduce photorespiration, we must either increase the CO_2 concentration in the atmosphere (though the other consequences of such action are likely to be unwelcome), or change the active site of Rubisco to reduce the oxygenase reaction. Research in several laboratories is investigating the possibility of manipulating Rubisco genes to this end (see McFadden & Small, 1988; Kettleborough et al., 1991). The significant improvements in the yield of some glasshouse crops grown under enhanced CO_2 concentrations provide a pointer to what might be possible if photorespiratory flux could be reduced.

Acknowledgements

We thank the Agricultural and Food Research Council for financial support.

References

Bajracharya, D., Bergfeld, R., Hatzfeld, W.-D., Klein, S. & Schopper, P. (1987). Regulatory involvement of plastids in the development of peroxisomal enzymes in the cotyledons of mustard seedlings. *Journal of Plant Physiology* **126**, 421–436.

Berger, M.G. & Fock, H.P. (1983). [15]N and inhibitor studies on the photorespiratory nitrogen cycle in maize leaves. *Photosynthesis Research* **4**, 3–7.

Biekmann, S. & Feierabend, J. (1982). Subcellular distribution, multiple forms and development of glutamate-pyruvate (glyoxylate) aminotransferase in plant tissues. *Biochimica et Biophysica Acta* **721**, 268–79.

Blackwell, R.D., Murray, A.J.S. & Lea, P.J. (1987). The isolation and characterisation of photorespiratory mutants of barley and pea. In *Progress in Photosynthesis Research*, Vol. 3, ed. J. Biggins, pp. 625–8. Dordrecht: Martinus Nijhoff.

Blackwell, R.D., Murray, A.J.S., Lea, P.J., Kendall, A.C., Hall, N.P., Turner, J.C. & Wallsgrove, R.M. (1988). The value of mutants unable to carry out photorespiration. *Photosynthesis Research* **16**, 155–76.

Brooks, A. & Farquhar, G.D. (1985). Effect of temperature on the CO_2/O_2 specificity of ribulose-1,5-bisphosphate carboxylase/oxygenase and the rate of respiration in the light. *Planta* **165**, 397–406.

Chen, R., Bismuth, E., Champigny, M.L. & Gadal, P. (1989). Chromatographic and immunological evidence that chloroplastic and cytosolic pea (*Pisum sativum* L.) NADP isocitrate dehydrogenases are distinct isoenzymes. *Planta* **178**, 157–63.

Chen, R. & Gadal, P. (1990). Do the mitochondria provide the 2-oxoglutarate needed for glutamate synthesis in higher plant chloroplasts? *Plant Physiology and Biochemistry* **28**, 141–5.

Chen, R., Le Marechal, P., Vidal, J., Jacquot, J.P. & Gadal, P. (1988). Purification and comparative properties of the cytosolic isocitrate dehydrogenases (NADP) from pea (*Pisum sativum*) roots and green leaves. *European Journal of Biochemistry* **175**, 565–72.

Cullimore, J.V. & Sims, A.P. (1980). An association between photorespiration and protein catabolism: studies with *Chlamydomonas*. *Planta* **150**, 392–6.

Edwards, J.W. & Coruzzi, G.M. (1989). Photorespiration and light act in concert to regulate the expression of the nuclear gene for chloroplast glutamine synthetase. *The Plant Cell* **1**, 241–8.

Freeman, J., Marquez, A.J., Wallsgrove, R.M., Saarelainen, R. & Forde, B.G. (1990). Molecular analysis of barley mutants deficient in chloroplast glutamine synthetase. *Plant Molecular Biology* **14**, 297–311.

Fuhrer, J. & Erismann, K.H. (1984). Steady-state carbon flow in photosynthesis and photorespiration in *Lemna minor* L.: the effect of temperature and ammonium nitrogen. *Photosynthetica* **18**, 74–83.

Gardeström, P., Bergmann, A. & Ericson, I. (1980). Oxidation of glycine via the respiratory chain in mitochondria prepared from different parts of spinach. *Plant Physiology* **65**, 389–91.

Gardeström, P. & Wigge, B. (1988). Influence of photorespiration on ATP/ADP ratios in the chloroplasts, mitochondria and cytosol, studied by rapid fractionation of barley protoplasts. *Plant Physiology* **88**, 69–76.

Grodzinski, B. (1978). Glyoxylate decarboxylation during photorespiration. *Planta* **144**, 31–7.

Grodzinski, B. (1979). A study of formate production and oxidation in leaf peroxisomes during photorespiration. *Plant Physiology* **63**, 289–93.

Hak, R. & Natr, L. (1987a). Effect of nitrogen starvation and recovery on gas exchange characteristics of young barley leaves. *Photosynthetica* **21**, 9–14.

Hak, R. & Natr, L. (1987b). Effect of nitrogen starvation and recovery on carbon fluxes in photosynthetic carbon reduction and oxidation cycles in young barley leaves. *Photosynthetica* **21**, 15–22.

Hall, N.P., Franklin, J., Keys, A.J., Reggiani, R. & Lea, P.J. (1984). An investigation into the interaction between nitrogen nutrition, photosynthesis and photorespiration. *Photosynthesis Research* **5**, 361–9.

Hall, N.P., Kendall, A.C., Lea, P.J., Turner, J.C. & Wallsgrove, R.M. (1987). Characteristics of a photorespiratory mutant of barley deficient in phosphoglycolate phosphatase. *Photosynthesis Research* **11**, 89–96.

Hall, N.P., Kendall, A.C., Turner, J.C., Wallsgrove, R.M. & Keys, A.J. (1986). A barley mutant deficient in RuBP carboxylase. *Plant Physiology* **80**, 281S.

Havir, E.A. & McHale, N.A. (1989). Regulation of catalase activity in leaves of *Nicotiana sylvestris* by high CO_2. *Plant Physiology* **89**, 952–7.

Hylton, C.M., Rawsthorne, S., Smith, A.M., Jones, D.A. & Woolhouse, H.W. (1988). Glycine decarboxylase is confined to the bundle sheath cells of leaves of C3–C4 intermediate species. *Planta* **175**, 452–9.

Ikeda, M., Ogren, W.L. & Hageman, R.H. (1984). Effect of methionine sulphoximine on photosynthetic carbon metabolism in wheat leaves. *Plant and Cell Physiology* **25**, 447–52.

Jenkins, C.L.D., Rogers, L.J. & Kerr, M.W. (1982a). Glycolate oxidase inhibition and its effect on photosynthesis and pigment formation in *Hordeum vulgare*. *Phytochemistry* **21**, 1849–58.

Jenkins, C.L.D., Rogers, L.J. & Kerr, M.W. (1982b). Glycolate oxidase inhibition and its effect on photosynthesis and pigment formation in *Zea mays*. *Phytochemistry* **21**, 1859–63.

Journet, E.-P., Neuberger, M. & Douce, R. (1981). Role of glutamate-oxaloacetate transaminase and malate dehydrogenase in the regeneration of NAD for glycine oxidation by spinach leaf mitochondria. *Plant Physiology* **67**, 467–9.

Joy, K.W. (1988). Ammonia, glutamine, and asparagine: a carbon–nitrogen interface. *Canadian Journal of Botany* **66**, 2103–9.

Kendall, A.C., Keys, A.J., Turner, J.C., Lea, P.J. & Miflin, B.J. (1983). The isolation and characterisation of a catalase-deficient mutant of barley. *Planta* **159**, 505–11.

Kendall, A.C., Wallsgrove, R.M., Hall, N.P., Turner, J.C. & Lea, P.J. (1986). Carbon and nitrogen metabolism in barley mutants lacking ferredoxin-dependent glutamate synthase. *Planta* **168**, 316–23.

Kettleborough, C.A., Phillips, A.L., Keys, A.J. & Parry, M.A. (1991). A point mutation in the N-terminus of ribulose-1,5-bisphosphate carboxylase affects ribulose-1,5-bisphosphate binding. *Planta* **184**, 35–9.

Keys, A.J. (1986). Rubisco, its role in photorespiration. *Philosophical Transactions of the Royal Society of London, B* **313**, 325–36.

Keys, A.J., Bird, I.F., Cornelius, M.J., Lea, P.J., Wallsgrove, R.M. & Miflin, B.J. (1978). Photorespiratory nitrogen cycle. *Nature* **275**, 741–3.

Kleczkowski, L.A., Givan, C.V., Hodgson, J.M. & Randall, D.D. (1988). Subcellular location of NADPH-dependent hydroxypyruvate reductase activity in leaf protoplasts of *Pisum sativum* and its role in photorespiratory metabolism. *Plant Physiology* **88**, 1182–5.

Larsson, C. & Albertsson, E. (1979). Enzymes related to serine synthesis in spinach chloroplasts. *Physiologia Plantarum* **45**, 7–10.

Lawyer, A.L. & Zelitch, I. (1979). Inhibition of glycine decarboxylation and serine formation in tobacco by glycine hydroxamate and its effect on photorespiratory carbon flow. *Plant Physiology* **64**, 706–11.

Lea, P.J., Kendall, A.C., Keys, A.J. & Turner, J.C. (1984). The isolation of a photorespiratory mutant of barley unable to convert glycine to serine. *Plant Physiology* **75**, 881S.

McFadden, B.A. & Small, C.L. (1988). Cloning, expression and directed mutagenesis of the genes for ribulose bisphosphate carboxylase/oxygenase. *Photosynthesis Research* **18**, 245–60.

McHale, N.A., Havir, E.A. & Zelitch, I. (1988). A mutant of *Nicotiana sylvestris* deficient in serine glyoxylate aminotransferase activity. *Theoretical and Applied Genetics* **76**, 71–5.

Machler, F., Muller, J. & Dubach, M. (1990). RuBPCO kinetics and the mechanism of CO_2 entry in C_3 plants. *Plant, Cell & Environment* **13**, 881–99.

Madore, M. & Grodzinski, B. (1984). Effect of oxygen concentration on the ^{14}C-photoassimilate transport from leaves of *Salvia splendens*. *Plant Physiology* **76**, 782–6.

Marek, L.F. & Stewart, C.R. (1983). Photorespiratory glycine metabolism in corn leaf discs. *Plant Physiology* **73**, 118–20.

Miernyk, J.A., Camp, P.J. & Randall, D.D. (1985). Regulation of plant pyruvate dehydrogenase complexes. *Current Topics in Plant Biochemistry and Physiology* **4**, 175–90.

Murray, A.J.S., Blackwell, R.D., Joy, K.W. & Lea, P.J. (1987). Photorespiratory N donors, aminotransferase specificity and photosynthesis in a mutant of barley deficient in serine:glyoxylate aminotransferase activity. *Planta* **172**, 106–13.

Murray, A.J.S., Blackwell, R.D. & Lea, P.J. (1989). Metabolism of hydroxypyruvate in a mutant of barley lacking NADH-dependent hydroxypyruvate reductase, an important photorespiratory enzyme activity. *Plant Physiology* **91**, 395–400.

Ogren, W.L. (1984). Photorespiration: pathways, regulation and modification. *Annual Review of Plant Physiology* **35**, 415–42.

Ohnishi, J. & Kanai, R. (1983). Differentiation of photorespiratory activity between mesophyll and bundle sheath cells of C4 plants. I. Glycine oxidation by mitochondria. *Plant and Cell Physiology* **24**, 1411–20.

Oleski, N., Mahdavi, P. & Bennett, A.B. (1987). Transport properties of the tomato fruit tonoplast. *Plant Physiology* **84**, 997–1000.

Parker, M.L. & Lea, P.J. (1983). Ultrastructure of the mesophyll cells of leaves of a catalase-deficient mutant of barley. *Planta* **159**, 512–17.

Parry, M.A.J., Keys, A.J. & Gutteridge, S. (1989). Variation in the specificity factor of C_3 higher plant Rubiscos determined by the total consumption of ribulose-P_2. *Journal of Experimental Botany* **40**, 317–20.

Peterson, R.B. (1990). Effects of irradiance on the *in vivo* $CO_2:O_2$ specificity factor in tobacco using simultaneous gas exchange and fluorescence techniques. *Plant Physiology* **94**, 892–8.

Randall, D.D. & Givan, C.V. (1981). Subcellular location of NADP-isocitrate dehydrogenase in *Pisum sativum* leaves. *Plant Physiology* **68**, 70–3.

Reddy, A.R., Suhasini, M. & Rama Das, V.S. (1987). Impairment of photorespiratory carbon flow into rubber by the inhibition of the glycolate pathway in guayule. *Plant Physiology* **84**, 1447–50.

Rhodes, D., Deal, L., Haworth, P., Jamieson, J.C., Reuter, C.C. & Ericson, M.C. (1986). Amino acid metabolism of *Lemna minor* L. *Plant Physiology* **98**, 1057–62.

Rogers, W.J., Jordan, B.R., Rawsthorne, S. & Tobin, A.K. (1991). Changes to the stoichiometry of glycine decarboxylase subunits during wheat (*Triticum aestivum* L.) and pea (*Pisum sativum* L.) leaf development. *Plant Physiology* **96**, 952–6.

Schuller, K.A. & Randall, D.D. (1989). Regulation of pea mitochondrial pyruvate dehydrogenase complex. *Plant Physiology* **89**, 1207–12.

Shingles, R., Woodrow, L. & Grodzinski, B. (1984). Effects of glycolate pathway intermediates on glycine decarboxylation and serine synthesis in pea. *Plant Physiology* **74**, 705–10.

Singh, P., Kumar, P.A., Abrol, Y.P. & Naik, M.S. (1985).

Photorespiratory nitrogen cycle – a critical evaluation. *Physiologia Plantarum* **66**, 169–76.

Sivak, M.N., Lea, P.J., Blackwell, R.D., Murray, A.J.S., Hall, N.P., Kendall, A.C., Turner, J.C. & Wallsgrove, R.M. (1989). Some effects of oxygen on photosynthesis by photorespiratory mutants of barley. I. Response to changes in oxygen concentration. *Journal of Experimental Botany* **39**, 655–66.

Smith, I.K., Kendall, A.C., Keys, A.J., Turner, J.C. & Lea, P.J. (1984). Increased levels of glutathione in a catalase deficient mutant of barley. *Plant Science Letters* **37**, 29–33.

Somerville, C.R. & Ogren, W.L. (1979). A phosphoglycolate phosphatase deficient mutant of *Arabidopsis*. *Nature* **280**, 833–6.

Somerville, C.R. & Ogren, W.L. (1980*a*). Photorespiration mutants of *Arabidopsis thaliana* deficient in serine–glyoxylate aminotransferase activity. *Proceedings of the National Academy of Sciences, USA* **77**, 2684–7.

Somerville, C.R. & Ogren, W.L. (1980*b*). Inhibition of photosynthesis in *Arabidopsis* mutants lacking leaf glutamate synthase activity. *Nature* **286**, 257–9.

Somerville, C.R. & Ogren, W.L. (1981*a*). Photorespiration-deficient mutants of *Arabidopsis thaliana* lacking mitochondrial serine transhydroxymethylase activity. *Plant Physiology* **67**, 666–71.

Somerville, C.R. & Ogren, W.L. (1981*b*). Mutants of the cruciferous plant *Arabidopsis thaliana* lacking glycine decarboxylase activity. *Biochemical Journal* **202**, 373–80.

Somerville, C.R. & Ogren, W.L. (1983). An *Arabidopsis thaliana* mutant defective in chloroplast dicarboxylate transport. *Proceedings of the National Academy of Sciences, USA* **80**, 1290–4.

Somerville, C.R., Portis, A.R. & Ogren, W.L. (1982). A mutant of *Arabidopsis thaliana* which lacks activation of RuBP carboxylase *in vivo*. *Plant Physiology* **70**, 381–7.

Ta, T.C. & Joy, K.W. (1986). Metabolism of some amino acids in relation to the photorespiratory nitrogen cycle of pea leaves. *Planta* **169**, 117–22.

Tobin, A.K., Ridley, S.M. & Stewart, G.R. (1985). Changes in the activities of chloroplast and cytosolic isoenzymes of glutamine synthetase during normal leaf growth and plastid development in wheat. *Planta* **163**, 544–8.

Tobin, A.K., Sumar, N., Patel, M., Moore, A.L. & Stewart, G.R. (1988). Development of photorespiration during chloroplast biogenesis in wheat leaves. *Journal of Experimental Botany* **39**, 833–43.

Tobin, A.K., Thorpe, J.R., Hylton, C.M. & Rawsthorne, S. (1989). Spatial and temporal influences on the cell-specific distribution of glycine decarboxylase in leaves of wheat (*Triticum aestivum* L.) and pea (*Pisum sativum* L.). *Plant Physiology* **91**, 1219–25.

Walker, K.A., Givan, C.V. & Keys, A.J. (1984). Glutamic acid metabolism and the photorespiratory nitrogen cycle in wheat leaves. Metabolic consequences of elevated ammonia concentrations and of blocking ammonia assimilation. *Plant Physiology* **75**, 60–6.

Wallsgrove, R.M., Lea, P.J. & Miflin, B.J. (1982). The development of NAD(P)H-dependent and ferredoxin dependent glutamate synthase in greening pea and barley leaves. *Planta* **154**, 473–6.

Wallsgrove, R.M., Kendall, A.C., Hall, N.P., Turner, J.C. & Lea, P.J. (1986). Carbon and nitrogen metabolism in a barley mutant with impaired chloroplast dicarboxylate transport. *Planta* **168**, 324–9.

Wallsgrove, R.M., Turner, J.C., Hall, N.P., Kendall, A.C. & Bright, S.W.J. (1987). Barley mutants lacking chloroplast glutamine synthetase – biochemical and genetic analysis. *Plant Physiology* **83**, 155–8.

G. E. EDWARDS and J. P. KRALL

Metabolic interactions between organelles in C4 plants

In a simplistic, textbook interpretation of plant cell biology, the organelles of cells and different tissues have specific functions, and house metabolic machinery to catalyse independent metabolic pathways. Thus, chloroplasts are responsible for photosynthesis, mitochondria for respiration, etc. With respect to photosynthesis, there is a combination of biochemical and environmental factors that belie this simple theory and that have affected the course of evolution. The most apparent is that Rubisco (ribulose 1,5-bisphosphate carboxylase/oxygenase), a key enzyme in the pathway for photosynthesis, functions both in carbon assimilation and in a photorespiratory process, with the magnitude of each being dependent upon the relative concentrations of carbon dioxide and oxygen available to the chloroplast. CO_2 and O_2 each serve as substrates and react with RuBP (ribulose 1,5-bisphosphate), and each is a competitive inhibitor with respect to the other. In C3 plants, where the supply of CO_2 from the atmosphere to the chloroplast is dependent on simple diffusion, the ratio of carboxylase to oxygenase activity is about 2.5:1 under current atmospheric conditions (Sharkey, 1988). Whether or not photorespiration has ever been of benefit to plants, it is known that photosynthesis in C3 plants can be increased by increasing CO_2 in the atmosphere, and that some plants, called C4 plants, have evolved a mechanism to concentrate CO_2 around Rubisco and minimise photorespiration. This evolutionary development in C4 plants resulted in an obvious benefit in carbon assimilation under environmental conditions where CO_2 is most limiting. However, this benefit comes with requirements, including (i) a cost in photochemically derived energy to drive the CO_2 concentrating mechanism, (ii) an investment in proteins associated with the CO_2 concentrating mechanism (catalysts and translocators), and (iii) cooperation between the two photosynthetic cell types in the leaf, mesophyll and bundle sheath, and between organelles in these cell types with respect to the C4 cycle, the C3 pathway, and photorespiration (to the limited extent that this latter process occurs in C4 plants).

Society for Experimental Biology Seminar Series 50: *Plant organelles*, ed. A. K. Tobin.
© Cambridge University Press 1992, pp. 97–112.

Intercellular transport in C4 plants during photosynthesis

Figure 1 illustrates the cooperation between mesophyll and bundle sheath cells in C4 plants, whereby solar energy is used to drive the C4 cycle, the CO_2 concentration is elevated in the bundle sheath cells, and then is assimilated via the C3 pathway. In the C4 cycle, the CO_2 is fixed in the mesophyll cells via phosphoenolpyruvate (PEP) carboxylase and then released in the bundle sheath cells via a C4 acid decarboxylase. The amount of energy required per CO_2 fixed is the sum of that needed by the C4 cycle and the C3 pathway, plus any losses which may occur through the low levels of photorespiration.

There are three subgroups of C4 plants, based on differences in their C4 cycle and in the associated C4 acid decarboxylases (Gutierrez *et al.*, 1974; Hatch *et al.*, 1975). Figure 2 illustrates the operation of the C4 cycles in these photosynthetic subgroups and the energy requirements per CO_2 moved through the cycle to the bundle sheath compartment. In each subgroup the generation of PEP, the substrate for CO_2 fixation via PEP carboxylase, from pyruvate in the mesophyll cell, requires 2 ATP

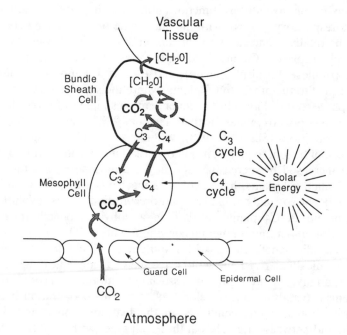

Fig. 1. Simplified scheme illustrating the C4 cycle in C4 plants. [CH₂O] is an abbreviation for carbohydrates.

NADP-ME TYPE

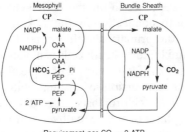

Requirement per CO_2 = 2 ATP

NAD-ME TYPE

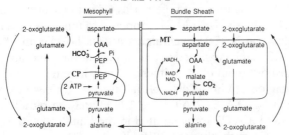

Requirement per CO_2 = 2 ATP

PEP-CK TYPE

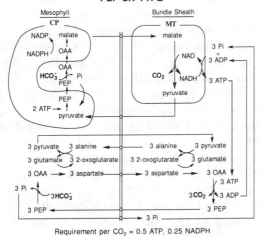

Requirement per CO_2 = 0.5 ATP, 0.25 NADPH

Fig. 2. Proposed schemes of intercellular transport associated with the C4 cycles in the three C4 subgroups of plants. CP, chloroplast; MT, mitochondrion; OAA, oxaloacetate; PEP, phosphoenolpyruvate.

(through the combined action of pyruvate, P_i dikinase and adenylate kinase). This conversion takes place in the mesophyll chloroplast where the ATP is generated photochemically. In the NADP-malic enzyme (NADP-ME) type species, the primary shuttle of metabolites in the C4 cycle between mesophyll and bundle sheath cells is malate/pyruvate, and the net energy requirement is 2 ATP per CO_2 fixed. The NADPH utilised in the mesophyll cells for conversion of oxaloacetate to malate is regenerated in the bundle sheath chloroplasts via NADP-ME, where it is then used in the reductive phase of the C3 pathway. In the NAD-malic enzyme (NAD-ME) species the primary shuttle between mesophyll and bundle sheath cells is aspartate/alanine and the net requirement per turn of the cycle is 2 ATP. In the PEP carboxykinase (PEP-CK) type species, PEP-CK and NAD-ME function together as C4 acid decarboxylases. In this case, the C4 cycle for NAD-ME is proposed to function via a malate/ pyruvate shuttle. The NADH generated via decarboxylation of malate through NAD-ME is then oxidised in the mitochondria, generating ATP. The ATP is transported out of the mitochondria, where it is utilised in the PEP-CK reaction located in the cytosol. In this case, for each turn of the C4 cycle through NAD-ME, 2 ATP and 1 NADPH are required in the mesophyll cell, and 3 ATP are generated in the bundle sheath cell. For each turn of the C4 cycle through PEP-CK, 1 ATP is required in the bundle sheath cell. Although the exact stoichiometry between these two cycles in different PEP-CK species is not known, the maximum *in vitro* activity of PEP-CK is generally higher than that of NAD-ME. Also, the primary initial product of $^{14}CO_2$ fixation by leaves of these species is aspartate, suggesting a major shuttle of aspartate/alanine, which may be associated with a higher flux through the PEP-CK cycle than through the NAD-ME cycle. An efficient operation as far as energy output per CO_2 fixed would be 1 turn of the NAD-ME cycle per 3 turns of the PEP-CK cycle, provided 3 ATP are generated per NADH oxidised by the bundle sheath mitochondria. With this stoichiometry, the average energy requirement per turn of the C4 cycle in these species would be 0.5 ATP, and 0.25 NADPH.

The energy requirements in the C3 pathway per CO_2 fixed to the level of triose-P are 3 ATP and 2 NADPH in the absence of photorespiration. Therefore, the total energy requirements per CO_2 fixed in C4 plants can be taken as the sum of that required in the C4 cycle plus that required in the C3 pathway, assuming (for the moment) that there are no losses attributable to photorespiration (Table 1). This results in 5 ATP + 2 NADPH per CO_2 fixed in NADP-ME type species and NAD-ME type species, and 3.5 ATP + 2.25 NADPH per CO_2 fixed in PEP-CK type species. Despite the differences in the C4 cycles, the calculated energy

Table 1. *Minimum energy requirements in C4 plants per CO$_2$ fixed to the level of triose-P*

Type	Energy	Requirements in C4 cycle	Requirements in C3 pathway	Total per CO$_2$
NADP–ME	ATP	2	3	5
	NADPH	0	2	2
NAD–ME	ATP	2	3	5
	NADPH	0	2	2
PEP–CK	ATP	0.5	3	3.5
	NADPH	0.25	2	2.25

For discussions of energy requirements see also Edwards & Walker 1983; Hatch 1987; Furbank *et al.* 1990*b*.

requirements per CO$_2$ fixed are very similar, with the PEP-CK types having a lower requirement for ATP and a slightly higher requirement for NADPH (which may account for this subgroup having a lower chlorophyll *a/b* ratio). There is a considerable amount of transport of metabolites between mesophyll and bundle sheath cells, and by organelles in these cells during the operation of C4 photosynthesis. However, there is no known requirement for ATP or NADPH for these transport processes (although light enhances uptake of pyruvate by mesophyll chloroplasts; for carrier-mediated transport see the following section). The transport of metabolites between cells occurs along diffusion gradients, with a size exclusion limit of substances of molecular mass greater than approximately 800–900 Da (Hatch 1987).

The level of photorespiration in C4 plants depends on the CO$_2$/O$_2$ ratio in bundle sheath cells under a given environmental condition. Factors which will influence this ratio include the degree of overcycling of the C4 pathway per CO$_2$ fixed in the C3 pathway, the conductance of the bundle sheath cell to CO$_2$ and bicarbonate, and the level of carbonic anhydrase in the bundle sheath compartment. Overcycling is necessary to build up the CO$_2$ pool in the bundle sheath compartment and to compensate for any leakage of CO$_2$ out of the bundle sheath. Considering these factors and through a model, Jenkins *et al.* (1989) have estimated overcycling of approximately 15%, which would add to the energy requirements for the C4 cycle in Table 1. From the model, the ratio of [CO$_2$]/[O$_2$] was calculated in the bundle sheath cells and the ratio of carboxylase/oxygenase determined from the equation $v_c/v_o = S_{rel}$ [CO$_2$]/[O$_2$], where v_c is velocity of carboxylase, v_o is velocity of oxygenase, and S_{rel} is the relative specificity factor for Rubisco to perform as a carboxylase vs as an oxygenase (taken as 100). Considering that the rate of release of CO_2 in photorespiration is $\frac{1}{2} v_o$, they estimated that the rate of photorespiration is only 2–3% of the rate of photosynthesis in C4 plants under atmospheric levels of CO_2. By comparison, the rate of photorespiration in C3 plants is about 25% of the net rate of photosynthesis under atmospheric levels of CO_2 (Sharkey, 1988).

Besides the intercellular transport between mesophyll and bundle sheath cells associated with the C4 cycle, there are two other cases of intercellular transport linked to C4 photosynthesis, one associated with the C3 pathway and the other with photorespiration (Edwards & Walker, 1983; Edwards, 1986; Hatch, 1987; Edwards & Ku, 1990; Leegood & Osmond, 1990). With respect to the C3 pathway, the reductive phase (enzymes for conversion of PGA [3-phosphoglycerate] to triose-P [glyceraldehyde 3-P and dihydroxyacetone-P]) is located in both mesophyll and bundle sheath cells of C4 plants representing all three subgroups. Thus, the mesophyll chloroplasts are considered to share with the bundle sheath chloroplasts in the reduction of PGA, the product of RuBP carboxylase. Since PGA is formed in the bundle sheath chloroplast, part of it is shuttled to the mesophyll chloroplasts for reduction. The triose-P formed may be metabolised to sucrose in the mesophyll cells, or shuttled back to the bundle sheath chloroplasts.

Photorespiratory metabolism in C4 plants is largely restricted to the bundle sheath cells, where Rubisco and glycine decarboxylase of the glycolate pathway are located. However, glycerate kinase is exclusively located in the mesophyll chloroplasts of C4 plants. This enzyme is the link for returning glycerate, the product of the glycolate pathway, to the triose-P pool associated with the C3 pathway. Thus, to the extent that glycerate is formed via photorespiration, it is transported from bundle sheath to mesophyll cells.

Intracellular transport in C4 plants during photosynthesis

In leaves, the rate of photosynthesis is about 10-fold higher than the rate of dark respiration, which illustrates the potential for high rates of metabolite flux between cellular compartments during photosynthesis. For example, in C3 plants there is a high rate of movement of CO_2 into the chloroplast, and of O_2 out of the chloroplast by free diffusion, equivalent to the rate of photosynthesis. The next most rapid exchange across the C3 chloroplast envelope is P_i/triose-P on the phosphate translocator, which needs to function at rates near one-third of the rate of photosynthesis when sucrose is the major photosynthetic product. In C4 photosynthesis, movement of metabolites across organelles occurs in association with the C4 cycle (illustrated in Fig. 2), the reductive phase of the C3 pathway and photorespiration. As much as half of the PGA, the product of RuBP carboxylase, may be shuttled from bundle sheath to mesophyll chloroplasts for reduction, in which case this shuttle would be operating at the same rate as CO_2 fixation. In the C4 cycle the following metabolites, in addition to O_2, must be transported across membranes at rates similar to the rates of photosynthesis (in some cases actually exceeding rates of CO_2 assimilation, since the CO_2 cycle is proposed to turn at rates about 15% above rates of net carbon fixation).

NADP-ME species
Mesophyll chloroplasts: pyruvate, PEP, P_i, oxaloacetate and malate.
Bundle sheath chloroplasts: malate and pyruvate.

NAD-ME species
Mesophyll chloroplasts: pyruvate, PEP and P_i.
Bundle sheath mitochondria: aspartate, 2-oxoglutarate, pyruvate, glutamate, malate and P_i (see Furbank *et al.*, 1990*a*).

PEP-CK species
In this case, metabolite transport by organelles is associated with the NAD-ME component, whose cycle is illustrated in Fig. 2, and is proposed to turn at about one-quarter the rate of photosynthesis.
Mesophyll chloroplasts: pyruvate, PEP, P_i, oxaloacetate and malate.
Bundle sheath mitochondria: malate, pyruvate, ATP, ADP and P_i.

Evidence that this transport must be occurring is based on studies of enzyme compartmentation and metabolism by isolated organelles. However, direct studies on metabolite transport, in order to identify carrier-mediated mechanisms and their kinetic properties, has been limited mainly to the mesophyll chloroplasts. The phosphate translocator is the only carrier protein identified in C4 photosynthesis. A summary of the main translocators which have been found in the mesophyll chloroplast is presented in Fig. 3 and discussed below. For additional information on metabolite transport in C4 plants see Hatch (1987), Heldt *et al.* (1990, 1991) and Heldt & Flügge (this volume).

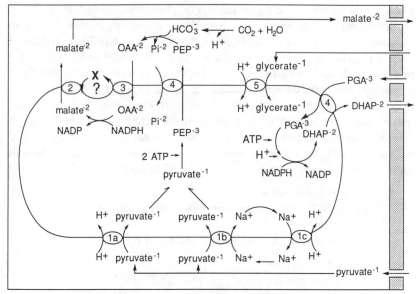

Translocators in C_4 Mesophyll Chloroplasts

Fig. 3. Scheme illustrating the transporters of the C4 mesophyll chloroplast during C4 photosynthesis. 1a, pyruvate translocator as proposed for NADP-ME type species; 1b, pyruvate translocator linked to sodium transport (1c) as proposed for NAD-ME and PEP-CK type species (see text); 2, dicarboxylate translocator; 3, oxaloacetate translocator; 4, phosphate translocator; 5, glycerate translocator. Carriers 2 and 3 may function together in uptake of oxaloacetate and export of malate, but the mechanism is uncertain. The dicarboxylate carrier functions as an exchange, so transport of a dicarboxylic acid (X) as illustrated is one possibility. DHAP, dihydroxyacetone phosphate, PGA, 3-phosphoglycerate, PEP, phosphoenolpyruvate.

Phosphate translocator

A single phosphate translocator in the mesophyll chloroplast is considered to mediate PEP/P_i exchange associated with the C4 cycle and PGA/triose-P exchange associated with the C3 pathway. Thus, the specificity of this translocator is different from that in C3 chloroplasts such as those of spinach, which can effectively exchange P_i/triose-P but not PEP/P_i (Huber & Edwards, 1977a; Hatch, 1987; Rumpho et al., 1987; Heldt et al., 1990, 1991; Ohnishi et al., 1990a). DIDS (4,4'-diisothiocyanatostilbene-2,2'-disulphonic acid), a disulphonic acid stilbene derivative, is a potent, irreversible inhibitor of the phosphate translocator. Incubation of chloroplasts with tritiated DIDS results in specific labelling of a polypeptide of *c.* 29 kDa in the envelopes of C3 and C4 mesophyll chloroplasts, which is considered to be the phosphate translocator protein (Rumpho *et al.*, 1988).

Pyruvate translocator

This translocator facilitates the uptake of pyruvate into the mesophyll chloroplast. It shows Michaelis–Menten type kinetics and is inhibited by analogues of pyruvate and by low temperature (Huber & Edwards, 1977*b*). There is light-induced uptake and accumulation of pyruvate by C4 mesophyll chloroplasts, which is driven by a proton or sodium gradient, depending on the species (Ohnishi & Kanai, 1990; Ohnishi *et al.*, 1990*b*). In mesophyll chloroplasts from monocots of NAD-ME and PEP-CK species there is pyruvate uptake in the light; addition of Na^+ can substitute for the light requirement. This light-dependent accumulation of pyruvate may be associated with an active Na^+/H^+ exchange, or active Na^+ efflux from the chloroplast which drives co-transport of Na^+ and pyruvate into the chloroplast. In NADP-ME type monocots there is enhanced uptake of pyruvate by mesophyll chloroplasts following a pH drop, suggesting a H^+/pyruvate import. Sodium does not stimulate uptake of pyruvate by chloroplasts of these species. In this case, the light-enhanced pyruvate uptake may be linked to light-dependent alkalisation of the stromal compartment (Ohnishi & Kanai, 1990).

Oxaloacetate translocator

There is a specific carrier in C4 mesophyll chloroplasts for uptake of oxaloacetate. It has a high affinity for oxaloacetate ($K_m = 45$ μM for maize chloroplasts), which is considered important since oxaloacetate is present at low levels in the cell (Hatch *et al.*, 1984).

Dicarboxylate translocator

This translocator catalyses the exchange of dicarboxylic acids by the C4 mesophyll chloroplast, as seen with *Digitaria sanguinalis* and maize (Huber, 1977; Day & Hatch, 1981). *In vivo*, its function may be linked in some way to the oxaloacetate translocator and involves the uptake of oxaloacetate and export of malate by the C4 mesophyll chloroplast (Fig. 3). The dicarboxylate translocator(s) may also have other functions in association with movement of other dicarboxylic acids across the chloroplast envelope (e.g. aspartate, 2-oxoglutarate and glutamate).

Glycerate translocator

There is evidence for a glycerate/glycolate translocator in C4 mesophyll and bundle sheath chloroplasts which may function in the export of glycolate from the bundle sheath chloroplasts and import of glycerate into the mesophyll chloroplasts (Ohnishi & Kanai, 1988). Some stimulation of uptake of glycerate by light may be attributed to an alkalisation of the stroma and a H^+/glycerate$^-$ import.

The function of these translocators in the C4 mesophyll chloroplast is shown in Fig. 3. Interestingly, if half of the PGA formed via RuBP carboxylase in bundle sheath cells were translocated to mesophyll cells for reduction, then there would be a similar rate of transfer of PGA/ triose-P and malate/pyruvate between mesophyll and bundle sheath cells. In that case, charge balance could be maintained as indicated. For simplicity, the exchange of aspartate/alanine between mesophyll/bundle sheath cells is not shown.

Energetics of C4 photosynthesis: linkage of noncyclic electron flow to CO_2 fixation

It is apparent that the different C4 subgroups have considerable differences in their C4 cycles and in metabolite transport. The question of whether these differences are simple alternatives to C4 photosynthesis or whether they confer some photosynthetic advantage under certain environmental conditions is unresolved. Recently, we made simultaneous measurements on photochemistry and carbon fixation in leaves of C4 plants representing different subgroups. Specifically, the relationship between photosystem II (PS II) activity and CO_2 fixation was examined. Do differences exist in PS II activity per CO_2 fixed because of differences in requirement for reductant (NADPH) in the different subgroups? Is there an increase in PS II activity per CO_2 fixed owing to increased photorespiration in C4 plants under limiting CO_2 (e.g. under stress with decreased stomatal conductance and decreased intercellular levels of CO_2)? Another factor to consider is that PS II activity in C4 plants might be linked to pseudocyclic or cyclic electron flow around PS II to provide additional ATP (there is currently speculation about cyclic electron flow around PS II, but there is no evidence for this in C4 plants; see Krall & Edwards, 1991).

We have measured the quantum yield of PS II (Φ_e) in leaves of C4 plants (electrons transported via PS II per photon absorbed by PS II) using a modulated fluorescence system based on the method developed by Genty et al. (1989). Simultaneously, rates of CO_2 fixation have been measured and the incident quantum yield for CO_2 fixation (Φ_{CO_2}) has been calculated under any given condition by dividing the rate of photosynthesis by the light intensity (A/I). Figure 4 shows the results with C4 monocots representing the three C4 subgroups where different

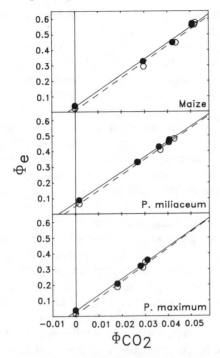

Fig. 4. The relationship between the quantum yield of photosystem II activity (Φ_e) and the incident quantum yield of CO_2 fixation (Φ_{CO_2}) under different levels of CO_2 with C4 species representing the three photosynthetic groups. Maize, NADP-ME type; *Panicum miliaceum*, NAD-ME type; *P. maximum*, PEP-CK type (adapted from Krall & Edwards, 1990). Closed circles, 21% O_2; open circles, 2% O_2.

quantum yields were obtained under varying concentrations of CO_2. As the atmospheric level of CO_2 is decreased, the quantum yields of PS II and of CO_2 fixation decrease. There is a linear relationship between Φ_e and Φ_{CO_2} which is similar under 2 vs 21% O_2. The slopes of the Φ_e vs Φ_{CO_2} plots are essentially the same for species representing all three subgroups (see also Krall & Edwards, 1990). One major conclusion from these experiments is that the efficiency in utilising PS II for CO_2 fixation in C4 plants is very similar in C4 monocots representing all three subgroups. An equation which best fits this relationship for the three species under 21% O_2, from the data with varying CO_2, is

$$A = \frac{(\Phi_e - 0.046)I}{10.12}$$

where A is the rate of photosynthesis, 0.046 is the intercept, I is the light intensity and 10.12 is the slope. This suggests Φ_e may be used as a gauge of photosynthetic rates in C4 plants. Despite marked differences in the C4 cycles among subgroups, uncertainties about bundle sheath conductance (leakiness of bundle sheath), differences in the degree of overcycling of the C4 pathway, and the level of photorespiration under limiting CO_2 (see below), representatives of these subgroups show a remarkable similarity in efficiency of utilisation of PS II in carbon fixation. The similarity in the relationship of Φ_e to Φ_{CO_2} in the three subgroups also supports calculations from the metabolic pathways of C4 photosynthesis which show a similar NADPH requirement per CO_2 fixed (Table 1).

Another major conclusion which could be drawn from the results shown in Fig. 4 is that photorespiration is relatively low in C4 plants, even when CO_2 is limiting. Although C4 plants have long been known to have low apparent levels of photorespiration (e.g. reflected in CO_2 compensation points being near zero) there has been uncertainty about the rate of true photorespiration and refixation of photorespired CO_2 in the bundle sheath compartment. As shown in Fig. 4, under 2% O_2 the intercept on the y-axis is very low (particularly in maize and *Panicum maximum*). The slightly positive intercept on the y-axis when Φ_{CO_2} is zero may occur because of other electron sinks (e.g. nitrogen assimilation). Under 21% O_2 there is a slight increase in Φ_e, indicating a minimal effect of O_2 on PS II activity. This suggests there is limited photorespiration and refixation of photorespired CO_2 in C4 plants, even under low CO_2. In contrast, in C3 plants under low CO_2 and 21% O_2, Φ_e remains high as a result of photorespiration becoming a major sink for electrons (Krall & Edwards, 1990). Also, with C3–C4 intermediate species, a progressive development of C4 characteristics results in an increased efficiency in utilisation of PS II derived energy for CO_2 assimilation, owing to reduced photorespiratory losses under low CO_2 levels (Krall *et al.*, 1991).

As noted earlier, Jenkins *et al.* (1989) recently developed a model for C4 photosynthesis which predicts the CO_2 and O_2 concentrations in bundle sheath cells at a given rate of photosynthesis and a given pool of inorganic carbon. Considerations in developing the model were the conductance across bundle sheath cells, the carbonic anhydrase content of bundle sheath cells, and the volume of bundle sheath cells and intracellular compartments. They also used the model to predict how the $[CO_2]/[O_2]$ ratio decreases in bundle sheath cells with decreasing CO_2 in the atmosphere, based on a study with the PEP-CK species *Urochloa panicoides* (Fig. 7 in Jenkins *et al.*, 1989). As noted above, the ratio of carboxylase/oxygenase can be calculated based on $v_c/v_o = S_{rel}[CO_2]/[O_2]$. For each CO_2 and for each O_2 reacting with RuBP a similar amount of

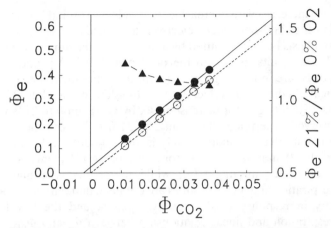

Fig. 5. Predicted relationship between Φ_e vs Φ_{CO_2} for a C4 species with varying CO_2. Rates of photosynthesis (A) and the modelled $[CO_2]/[O_2]$ ratio were obtained with the species *Urochloa panicoides* under varying levels of CO_2 (Furbank & Hatch, 1987; Jenkins *et al.*, 1989), and from this, v_c/v_o can be calculated. To calculate Φ_{CO_2} (A/I), rates of photosynthesis (rates were converted to an area basis, assuming 300 mg Chl m^{-2}) were divided by light intensity (900 µmol m^{-2} s^{-1}) at arbitrary points over the CO_2 response curve. Φ_e values under 0% O_2 were estimated using a slope of 10 (based on experimental values of Krall & Edwards, 1990) and an intercept of zero. The ratio of Φ_e 21% O_2/Φ_e 0% O_2 was then determined based on the calculated v_c/v_o ratios. For each CO_2 reacting with RuBP, 2 NADPH are required. Also, *c.* 2 NADPH equivalents are required per O_2 reacting with RuBP (1 NADPH to reduce the PGA formed as a product of oxygenase, 0.5 NADPH to reduce the PGA formed via the glycolate pathway, and equivalent of 0.5 NADPH to reduce the ammonia formed via glycine decarboxylation). Thus, for photosynthesis in the absence of O_2, $A = v_c$, and in the presence of O_2, $A = v_c - 0.5\ v_o$ (ignoring dark respiration). The degree of increase in requirement of reductant in the presence of O_2 is then equal to $(v_c + v_o)/A$. Closed circles, 21% O_2; open circles, 0% O_2; closed triangles, ratio of Φ_e 21% O_2/Φ_e 0% O_2.

reductant is required (see legend to Fig. 5). Therefore, at a given CO_2 concentration in the atmosphere, and a given $[CO_2]/[O_2]$ ratio in the bundle sheath cells as predicted from their model, the additional electron flow resulting from O_2 reacting with RuBP can be predicted (Fig. 5). This analysis indicates that the Φ_e vs Φ_{CO_2} plots, generated from varying CO_2 levels, have a very similar slope in the presence or absence of O_2. According-ing to our analysis, O_2 would cause a relatively small increase in Φ_e, similar in magnitude over the range of Φ_e values. The ratio of Φ_e 21% $O_2/$

Φ_e 0% O_2 increases with decreasing Φ_{CO_2} values. This indicates the extent to which the PS II-dependent electron flow increases per CO_2 fixed, as photosynthesis becomes limited by a decrease in the atmospheric level of CO_2. These results, based on the predicted $[CO_2]/[O_2]$ ratio in bundle sheath cells, are in good agreement with the experimental results of Fig. 5, particularly the lack of effect of O_2 on the slope of the plots. The model predicts a slightly greater increase in Φ_e by O_2 (from 0 to 21% O_2) than was observed experimentally (from 2 to 21% O_2). Both the model and experimental data (including analysis of $^{18}O_2$ uptake by Furbank & Badger, 1982) indicate that photorespiration in C4 plants is relatively low, even under limiting CO_2, while in C3 plants a high rate of photorespiration and high Φ_e values are obtained under limiting CO_2. The similarity in response of different C4 plants, and the low levels of photorespiration and linear relationship between Φ_e and Φ_{CO_2}, suggest that measurements of Φ_e may be a very useful indicator of photosynthetic activity of C4 species. More study is needed with various species under different conditions. However, considering the complexity and diversity of the C4 cycle, the similarity of the relationship between Φ_e and Φ_{CO_2} which has been documented so far may be fortunate for plant physiologists who wish to measure photochemistry and carbon fixation of C4 species *in vivo*.

Acknowledgements

G.E. acknowledges the support of the National Science Foundation (DCB-8816322) and the USDA Competitive Grants Program (90-37280–5706) for recent support for this work, and thanks Mr Dai for assistance on the figures. Thanks also to Dr A.K. Tobin and the Society for Experimental Biology, for the opportunity to present this paper.

References

Day, D. & Hatch, M.D. (1981). Dicarboxylate transport in maize mesophyll chloroplasts. *Archives of Biochemistry and Biophysics* **211**, 738–42.

Edwards, G.E. (1986). Carbon fixation and partitioning in the leaf. In *Regulation of Carbon and Nitrogen Reduction and Utilization in Maize*, ed. J.C. Shannon, D.P. Knievel & C.D. Boyer, pp. 51–65. Beltsville, MD: American Society of Plant Physiologists.

Edwards, G.E. & Ku, M.S.B. (1990). Regulation of the C_4 pathway of photosynthesis. In *Perspectives in Biochemical and Genetic Regulation of Photosynthesis*, ed. I. Zelitch, pp. 175–90. New York: Alan R. Liss.

Edwards, G.E. & Walker, D.A. (1983). C_3, C_4. *Mechanisms, and Cellu-*

lar and Environmental Regulation, of Photosynthesis. Oxford: Blackwell Scientific Publications.

Furbank, R.T., Agostino, A. & Hatch, M.D. (1990*a*). C_4 acid decarboxylation and photosynthesis in bundle sheath cells of NAD-malic enzyme-type C_4 plants: Mechanism and the role of malate and orthophosphate. *Archives of Biochemistry and Biophysics* **276**, 347–81.

Furbank, R.T. & Badger, M.R. (1982). Photosynthetic oxygen exchange in attached leaves of C_4 monocotyledons. *Australian Journal of Plant Physiology* **9**, 553–8.

Furbank, R.T. & Hatch, M.D. (1987). Mechanism of C_4 photosynthesis. The size and composition of the inorganic carbon pool in bundle sheath cells. *Plant Physiology* **85**, 958–64.

Furbank, R.T., Jenkins, C.L.D. & Hatch, M.D. (1990*b*). C_4 photosynthesis: Quantum requirement, C_4 acid overcycling and Q-cycle involvement. *Australian Journal of Plant Physiology* **17**, 1–7.

Genty, B., Briantais, J.-M. & Baker, N.R. (1989). The relationship between the quantum yield of photosynthetic electron transport and quenching of chlorophyll fluorescence. *Biochimica et Biophysica Acta* **990**, 87–92.

Gutierrez, M., Gracen, V.E. & Edwards, G.E. (1974). Biochemical and cytological relationships in C_4 plants. *Planta* **119**, 279–300.

Hatch, M.D. (1987). C_4 photosynthesis: a unique blend of modified biochemistry, anatomy and ultrastructure. *Biochimica et Biophysica Acta* **895**, 81–106.

Hatch, M.D., Droscher, L., Flugge, U.I. & Heldt, H.W. (1984). A specific translocator for oxaloacetate transport in chloroplasts. *FEBS Letters* **178**, 15–19.

Hatch, M.D., Kagawa, T. & Craig, S. (1975). Subdivision of C_4-pathway species based on differing C_4 acid decarboxylating systems and ultrastructural features. *Australian Journal of Plant Physiology* **2**, 111–28.

Heldt, H.W., Flugge, U.-I. & Borchert, S. (1991). Diversity of specificity and function of phosphate translocators in various plastids. *Plant Physiology* **95**, 341–3.

Heldt, H.W., Flugge, U.-I., Borchert, S., Bruckner, G. & Ohnishi, J. (1990). Phosphate translocators in plastids. In *Perspectives in Biochemical and Genetic Regulation of Photosynthesis*, ed. I. Zelitch, pp. 39–54. New York: Alan R. Liss.

Huber, S.C. (1977). The photosynthetic function of the C_4 mesophyll cell: Integration of carbon metabolism, energy provision and metabolite transport. PhD thesis, University of Wisconsin, Madison, USA.

Huber, S.C. & Edwards, G.E. (1977*a*). Transport in C_4 mesophyll chloroplasts. Evidence for an exchange of inorganic phosphate and phosphoenolpyruvate. *Biochimica et Biophysica Acta* **462**, 603–12.

Huber, S.C. & Edwards, G.E. (1977b). Transport in C_4 mesophyll chloroplasts. Characterization of the pyruvate carrier. *Biochimica et Biophysica Acta* **462**, 583–602.

Jenkins, C.L.D., Furbank, R.T. & Hatch, M.D. (1989). Mechanism of C_4 photosynthesis. A model describing the inorganic carbon pool in bundle sheath cells. *Plant Physiology* **91**, 1372–81.

Krall, J.P. & Edwards, G.E. (1990). Quantum yields of photosystem II electron transport and carbon dioxide fixation in C_4 plants. *Australian Journal of Plant Physiology* **17**, 579–88.

Krall, J. & Edwards, G.E. (1991). Environmental effects on the relationship between the quantum yields of carbon assimilation and *in vivo* PSII electron transport in maize. *Australian Journal of Plant Physiology* **8**, 267–78.

Krall, J., Edwards, G.E. & Ku, M.S.B. (1991). Quantum yield of photosystem II and efficiency of CO_2 fixation in *Flaveria* species under varying light and CO_2. *Australian Journal of Plant Physiology* **8**, 369–83.

Leegood, R. & Osmond, C.B. (1990). The flux of metabolites in C_4 and CAM plants. In *Plant Physiology, Biochemistry and Molecular Biology*, ed. D.T. Dennis & D.H. Turpin, pp. 274–98. New York: Longman Sci & Tech. John Wiley.

Ohnishi, J., Flugge, U.I. & Heldt, H.W. (1990a). Phosphate translocator of mesophyll and bundle sheath chloroplasts of a C_4 plant, *Panicum miliaceum* L. Identification and kinetic characterization. *Plant Physiology* **91**, 1507–11.

Ohnishi, J., Flugge, U.I., Heldt, H.W. & Kanai, R. (1990b). Involvement of Na^+ in active uptake of pyruvate in mesophyll chloroplasts of some C_4 plants. *Plant Physiology* **94**, 950–9.

Ohnishi, J. & Kanai, R. (1988). Glycerate uptake into mesophyll and bundle sheath chloroplasts of a C_4 plant, *Panicum miliaceum*. *Journal of Plant Physiology* **133**, 119–21.

Ohnishi, J. & Kanai, R. (1990). Pyruvate uptake induced by a pH jump in mesophyll chloroplasts of maize and sorghum, NADP-malic enzyme type C_4 species. *FEBS Letters* **69**, 122–4.

Rumpho, M., Edwards, G.E., Yousif, A.E. & Keegstra, K. (1988). Specific labelling of the phosphate translocator in C_3 and C_4 mesophyll chloroplasts by triated dihydro-DIDS (1,2-ditritio-1,2-[2,2′-disulfo-4,4′-diisothiocyano] diphenylethane). *Plant Physiology* **86**, 1193–8.

Rumpho, M., Wessinger, M.E. & Edwards, G.E. (1987). Influence of organic-phosphates on 3-phosphoglycerate dependent O_2 evolution in C_3 and C_4 mesophyll chloroplasts. *Plant and Cell Physiology* **28**, 805–13.

Sharkey, T.D. (1988). Estimating the rate of photorespiration in leaves. *Physiologia Plantarum* **73**, 147–52.

STEPHEN RAWSTHORNE, SUSANNE von
CAEMMERER, ANDREW BROOKS and
RICHARD C. LEEGOOD

Metabolic interactions in leaves of C3–C4 intermediate plants

The phenomenon of C3–C4 intermediate photosynthesis has now been recognised for about 25 years and has received a considerable amount of interest in that time, much of it aimed at the link that these plants might represent in the evolution of the C4 photosynthetic system from the older C3 one. A number of recent reviews have addressed evolutionary and adaptive aspects of C3–C4 photosynthesis together with the biochemistry (Peisker, 1986; Edwards & Ku, 1987; Monson, 1989; Monson & Moore, 1989; Araus et al., 1991). We intend, therefore, to confine our discussion in this article to those aspects of the C3–C4 character that have a major influence on the metabolic interactions within and between photosynthetic cells in the leaves of these plants and to compare and contrast this with what is known about photosynthetic metabolism in C3 and C4 plants.

Plants which have C3–C4 intermediate photosynthesis have been identified in seven genera across five families, including representatives from the Monocotyledoneae and the Dicotyledoneae (Edwards & Ku, 1987). Despite their wide distribution in the higher plant kingdom there are a number of well-conserved features across this group. We will describe these and then discuss their implications for inter- and intracellular metabolite transport and the metabolic regulation of photosynthesis and photorespiration.

Characteristics of C3–C4 photosynthesis

Leaf anatomy

A visually observable characteristic of plants which have C3–C4 photosynthesis is the increase in the number of organelles within the bundle sheath cells relative to that in both adjacent mesophyll cells and in bundle sheath cells of related C3 species. This anatomy has been described as C4- or Kranz-like. Numerous mitochondria, the peroxisomes,

Society for Experimental Biology Seminar Series 50: *Plant organelles*, ed. A. K. Tobin.
© Cambridge University Press 1992, pp. 113–39.

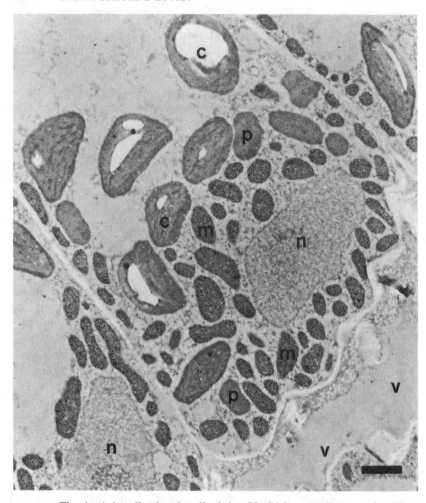

Fig. 1. A bundle sheath cell of the C3–C4 intermediate species *Moricandia spinosa* showing the arrangement of large numbers of mitochondria (m) at the centripetal end of the cell adjacent to the vascular tissue (v). The mitochondria have been immunogold labelled (visible as small electron-dense particles) with an antibody raised against the P subunit of glycine decarboxylase. Note that the peroxisomes (p) and the nucleus (n) are also in the same region of the cell and that all of these organelles are overlain by chloroplasts (c). Scale bar = 0.5 μm.

many of the chloroplasts (Holaday *et al.*, 1981; Brown *et al.*, 1983; Hylton *et al.*, 1988) and even the nuclei (C. M. Hylton and S. Rawsthorne, unpublished data from a range of C3–C4 species) are located centripetally in the bundle sheath cells (Fig. 1). The mitochondria are found along the cell wall immediately adjacent to the vascular cells and they are overlain by the chloroplasts. This anatomical development of bundle sheath cells has been reported for every C3–C4 species identified to date, except in *Neurachne minor* where the bundle sheath cells do not contain a major vacuole and the organelles are distributed throughout the cytoplasm (Hattersley *et al.*, 1986). Brown & Hattersley (1989) have quantified the partitioning of organelles (chloroplasts, mitochondria and peroxisomes) between the bundle sheath and mesophyll cells and the numbers of organelles in cell profiles in the leaves of a range of C3, C3–C4, and C4 species from the genera *Panicum, Neurachne, Flaveria* and *Moricandia*. In the C3–C4 species the combined number of mitochondria and peroxisomes per unit cell area in bundle sheath cells was on average four times that in mesophyll cells whereas the number of chloroplasts per unit cell area was only slightly increased (by 1.4-fold on average) in the bundle sheath cells (Brown & Hattersley, 1989). Other measurements have shown that the profile area of individual mitochondria in bundle sheath cells is consistently about two-fold larger than that of mitochondria in mesophyll cells in leaves of twelve C3–C4 and C4 species (Hylton *et al.*, 1988). Profile areas of mitochondria in both cell types were similar for the two C3 *Moricandia* species included in this study (Hylton *et al.*, 1988). It is interesting to note that a larger profile area, and by implication a larger mitochondrial size, is positively correlated with the presence of glycine decarboxylase in the mitochondria (see below; Hylton *et al.*, 1988). The only intermediate to show further C4-like development of the bundle sheath is *Neurachne minor* which has a suberised lamella surrounding the inner, organelle-containing, bundle sheath cells (Hattersley *et al.*, 1986).

These anatomical arrangements in the leaves of C3–C4 species are unaffected by variations in the gaseous composition of the aerial environment which alter the rate of photorespiration (Byrd & Brown, 1989), nor are they changed during leaf development (Fladung & Hesselbach, 1987). That the anatomy is determined genetically is clearly shown by segregation of the C3–C4 anatomy away from a C3–C4-like CO_2 compensation point in an F_2 population derived from a C3 $\times$ C3–C4 cross in the genus *Panicum* (Bouton *et al.*, 1986). Thus, the anatomy can be inherited independently of other C3–C4 characters which influence the leaf biochemistry, and hence the compensation point, and it cannot therefore be a consequence of the expression of those characters in a C3–C4 intermediate species. It now seems probable that the *Panicum* F_2 hybrid plant which had an intermediate anatomy but a C3-like CO_2 compensation point (Bouton *et al.*, 1986) did not have the differential distribution of glycine decarboxylase which leads to the improvement in recapture of photorespired CO_2, and hence to a lower compensation point, as discussed below.

CO_2 compensation point

The CO_2 compensation point (Γ) is a measure of the internal balance point in a leaf between photosynthetic, photorespiratory and other respiratory processes. In C3–C4 intermediate species the CO_2 compensation points are between those of C3 and C4 species, which are typically 50 and 0–5 μbar CO_2, respectively (Edwards & Ku, 1987). This intermediacy of Γ was the means by which this group of plants was first identified and named. However, there is a considerable range of values reported for all C3–C4 species and even variation within a single species. This variation is likely to reflect differences in measurement techniques, leaf age (Apel *et al.*, 1978; Peisker *et al.*, 1988) and growing conditions (Edwards & Ku, 1987; Apel & Peisker, 1988) between different laboratories. Notwithstanding such differences, the values reported at high light intensity and at atmospheric oxygen concentrations are generally between 30 and 8 μbar CO_2. This range of CO_2 compensation points between different species and different genera may reflect some modulation by leaf anatomy of the interaction between the biochemistry involved in the release and recapture of photorespired CO_2. For example, Brown and co-workers have shown good negative correlations between the percentage of the total leaf organelles in the bundle sheath cells and Γ in a range of C3 and intermediate *Panicum* species and their F_1 hybrids (Brown *et al.*, 1984) and in a range of C3, C3–C4 and C4 species from different genera (Brown & Hattersley, 1989).

A characteristic of Γ in many C3–C4 species is its strong light dependence (e.g. Brown & Morgan, 1980; Holaday *et al.*, 1982). At photon flux densities which approach the light compensation point for photosynthesis (80–150 μmol quanta m^{-2} s^{-1}) Γ can be almost as high as that of a C3 species but there is a steep decline as the light intensity increases. This also occurs in intermediate species which are reported to have elements of C4-like biochemistry (*Flaveria floridana*: Cheng *et al.*, 1989; and see below). In addition, Γ shows a biphasic response to the O_2 concentration imposed during the measurement with only limited increases occurring as the O_2 concentration is raised to 10–20% and then a C3-like response beyond this range (Edwards & Ku, 1987). In C3 species Γ increases as a linear function of the concentration of O_2.

Photorespiratory metabolism

The maximum catalytic activities of glycolate oxidase, hydroxypyruvate dehydrogenase, and the glyoxylate aminotransferases in leaves of C3 and C3–C4 *Moricandia* species are similar (Rawsthorne *et al.*, 1988*a*). This indicates that the capacities of the leaves of both photosynthetic types for photorespiratory metabolism are also likely to be similar. The maximum catalytic activities of the same enzymes are also similar in protoplast fractions enriched in either mesophyll or bundle sheath cells of the C3–C4 species *M. arvensis* (Rawsthorne *et al.*, 1988*b*). The only major difference in the photorespiratory enzymes between the C3 and all C3–C4 species reported to date is that glycine decarboxylase is confined to the mitochondria of the bundle sheath cells of C3–C4 intermediate species while it is present in the mitochondria of all the photosynthetic cells in C3 species (Rawsthorne *et al.*, 1988*a*; Hylton *et al.*, 1988). Immunogold labelling experiments have shown that the P subunit of glycine decarboxylase is not detectable in the mitochondria of mesophyll cells but is present in large amounts in bundle sheath cells (Rawsthorne *et al.*, 1988*a*; Hylton *et al.*, 1988). The absence of this subunit from the glycine decarboxylase complex prevents completely the decarboxylation of glycine (Walker & Oliver, 1986). This differential distribution of the P subunit has been confirmed by the finding that glycine decarboxylase activity is enriched in protoplast fractions derived from bundle sheath cells relative to that in protoplasts from mesophyll cells of *M. arvensis* (Rawsthorne *et al.*, 1988*b*). Some enrichment of glycine decarboxylase activity in the bundle sheath cells was also reported for another intermediate species, *Flaveria ramosissima* (Moore *et al.*, 1988). The activity of serine hydroxymethyltransferase, which is associated with glycine decarboxylase in a large multi-enzyme complex (Bourguignon *et al.*, 1988), is also enriched in bundle sheath cells of *M. arvensis* but to a lesser extent than glycine decarboxylase (Rawsthorne *et al.*, 1988*b*). The glycine decarboxylase complex has been shown to comprise about 50% of the matrix protein in mitochondria from leaves of *Pisum sativum*, a C3 species (Bourguignon *et al.*, 1988) and the larger size of the mitochondria in bundle sheath cells of C3–C4 intermediates (as described above) may be required to accommodate all of the glycine decarboxylase needed for the metabolism of photorespiratory glycine which is produced throughout the leaf.

The fact that glycine decarboxylation is confined to the bundle sheath cells but the mesophyll cells have the ability to form glycine during photorespiratory metabolism implies that the glycine must diffuse from the mesophyll to the bundle sheath cells to be decarboxylated (Rawsthorne *et al.*, 1988*a*). Carbon must then return from the bundle sheath cells to the Calvin cycle in the mesophyll cells or carbon fixation in the mesophyll could not continue. A proposed scheme for photorespiratory metabolism in the leaf of *M. arvensis* is shown in Fig. 2. The need for return of carbon committed to the photorespiratory pathway back to the Calvin cycle in the chloroplast is clearly demonstrated by the marked

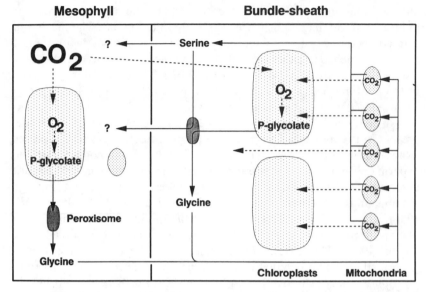

Fig. 2. A proposed pathway for photorespiration in the leaves of C3–C4 intermediate species (after Rawsthorne *et al.*, 1988*a*). Movement of metabolites is indicated by solid arrows while that of CO_2 and O_2 is shown by the dashed arrows.

inhibition of photosynthesis in photorespiratory mutants of the C3 species, *Arabidopsis* or barley, when they are grown under conditions which allow photorespiration to occur (Artus *et al.*, 1986). Such mutations can be overcome and wild-type rates of photosynthesis restored by feeding metabolites which are distal to the lesion in the photorespiratory pathway to whole leaves (Somerville & Somerville, 1983).

Rawsthorne *et al.* (1988*a*) have suggested that the differential distribution of glycine decarboxylase is critical to the C3–C4 character and could account for the low Γ of many if not *all* C3–C4 intermediate species for the following reasons. First, *all* of the release of photorespiratory CO_2 in the leaf will occur in the bundle sheath cells, and secondly, the mitochondria in these cells are in close association with, and overlain by, the chloroplasts. The potential for recapture of photorespired CO_2 before it leaves the leaf must therefore be much greater in the C3–C4 species than in C3 species. In accordance with this proposal, Hunt *et al.* (1987) had previously shown that 75% of the photorespired CO_2 is recaptured in the leaves of *M. arvensis* (C3–C4) whereas only 50% is recaptured in *M. moricandioides* (C3). Thirdly, an F_2 interspecific *Panicum* hybrid (C3 × C3–C4) has the anatomy of its intermediate parent but a C3-like Γ

(Bouton *et al.*, 1986) which shows that the anatomy alone cannot account for the low Γ of the intermediate species.

C4-like biochemistry

Despite earlier studies which suggested that there may be some C4-like biochemistry in the intermediate species in the genera *Panicum* and *Moricandia*, more recent enzyme activity and $^{14}CO_2$ labelling work has clearly shown that there is no evidence for this (Edwards *et al.*, 1982; Hunt *et al.*, 1987). There is also no evidence for C4-like metabolism in C3–C4 intermediate species of the genera *Parthenium* and *Alternanthera* (Rajendrudu *et al.*, 1986; Moore *et al.*, 1987). It had also been proposed that the low Γ of the C3–C4 intermediate species *Mollugo verticillata* was attributable to C4-like biochemistry on the basis of $^{14}CO_2$ pulse-chase labelling experiments and enzymological studies (Sayre & Kennedy, 1977; Sayre *et al.*, 1979). A more recent and critical assessment of the data and the evidence of a differential distribution of glycine decarboxylase in *M. verticillata* now suggest that this is unlikely (Edwards & Ku, 1987; Hylton *et al.*, 1988). There is also no differential distribution of ribulose 1,5-bisphosphate (RuBP) carboxylase/oxygenase (Rubisco) between bundle sheath and mesophyll cells in the leaves of *Panicum milioides*, *Moricandia arvensis* or *Mollugo verticillata* (Sayre *et al.*, 1979; Perrot-Rechenmann *et al.*, 1984; Rawsthorne *et al.*, 1988a). Phosphoenolpyruvate (PEP) carboxylase has been reported to be localised to the mesophyll cells of *P. milioides* but present in both bundle sheath and mesophyll cells of *P. decipiens* (Perrot-Rechenmann *et al.*, 1984). Whilst these are both intermediate species the minor increases reported for activities of PEP carboxylase relative to C3 species and the lack of C4-like $^{14}CO_2$ labelling patterns (Edwards *et al.*, 1982) precludes the involvement of this apparent cell-specific localisation of the enzyme in C4 metabolism in *P. milioides*.

In the genus *Flaveria* a considerable amount of data from $^{14}CO_2$ labelling experiments and measurements of enzyme activity and $\delta^{13}C$ has been used to suggest that some C4-like biochemistry may be occurring in the C3–C4 intermediate species. Significant incorporation of $^{14}CO_2$ into C4 acids occurs in a range of intermediate *Flaveria* species (Monson *et al.*, 1986; Cheng *et al.*, 1988). Within this range there is a progressive increase through the species in the degree of incorporation of $^{14}CO_2$ into C4 acids and a concomitant decrease in the proportion of CO_2 which is assimilated directly by Rubisco (Monson *et al.*, 1986; Cheng *et al.*, 1988; Moore *et al.*, 1989). This increase suggests a continuous trend in the degree of C4-ness in the Flaverias from the lowest C3–C4 type, *F. linearis*, through species like *F. floridana* and *F. ramosissima*, to *F. brownii*, a species described as 'advanced C3–C4' or C4-like, to others such as *F. palmeri* also described as C4-like, and finally to *F. trinervia* which is a true C4 species. However, in all the C3–C4 species studied, transfer of ^{14}C label from C4 acids to the Calvin cycle during the $^{12}CO_2$ chase did not occur at a rate and with a pattern similar to that in a true C4 *Flaveria* species (Monson *et al.*, 1986). Furthermore, the ^{14}C-labelled metabolites generated during the chase included significant amounts of glycine and serine (25–40% of the total label recovered), and fumarate (10–30% of the total label recovered), none of which were substantially labelled in equivalent experiments on C4 or C4 *Flaveria* species (Monson *et al.*, 1986; Moore *et al.*, 1986). Glycine and serine also become major ^{14}C-labelled products during pulse-chase measurements on leaves of *F. brownii* ('advanced C3–C4' or C4-like: Cheng *et al.*, 1988). This species has been reported to fix approximately 20% of its CO_2 directly through Rubisco (Cheng *et al.*, 1988).

The maximum catalytic activities of enzymes involved in C4 photosynthesis (PEP carboxylase, NADP-malate dehydrogenase, NADP-malic enzyme [NADP-ME] and pyruvate, P_i dikinase [PPDK]) are reported to be greater in a number of C3–C4 *Flaveria* species than in C3 species or intermediates from other genera which do not have any C4 fixation (Ku *et al.*, 1983; Moore *et al.*, 1988). However, these activities are only 10–15% of those of *Flaveria* species which have been classified as C4 or C4-like (Ku *et al.*, 1983; Cheng *et al.*, 1988; Moore *et al.*, 1989). In addition, compartmentation between bundle sheath and mesophyll cells of the enzymes involved in the carboxylation and decarboxylation stages of the C4 cycle is not complete or does not occur in leaves of *F. ramosissima* (C3–C4) or *F. brownii* (C4-like) as judged from protoplast fractionation experiments (Moore *et al.*, 1988, 1989; Cheng *et al.*, 1988, 1989). The latter species has maximum catalytic activities of PEP carboxylase, PPDK and NADP-ME that approach those in true C4 or the other C4-like *Flaveria* species which have more limited (5–10%) direct fixation of CO_2 by Rubisco (Moore *et al.*, 1989). There is no marked differential distribution of Rubisco or PEP carboxylase between mesophyll and bundle sheath in any C3–C4 *Flaveria* species examined to date (Bauwe, 1984; Reed & Chollet, 1985; Moore *et al.*, 1988). Enrichment of Rubisco in the bundle sheath relative to PEP carboxylase and vice versa for the mesophyll has been reported for *F. brownii* (Cheng *et al.*, 1988, 1989). However, without complete separation of the enzymes involved in the carboxylation and decarboxylation phases of the C4 pathway, futile cycling of CO_2 through C4 acids and hence extra energy consumption (see below) may well occur (Cheng *et al.*, 1988). All data on relative enrichments of enzyme activity between bundle sheath and mesophyll cells of the C3–C4 and C4-like *Flaveria* species should be treated with caution. This is because the protoplast separation experiments are based on (i) fractions which are reported to be enriched in bundle sheath cells and various subclasses of mesophyll cells, based upon comparisons between protoplast size and the size of cells *in vivo*, and/or (ii) assumptions as to the completeness of separation of marker enzymes between the cell types. Without distinct marker enzymes for cell subclasses, as occurs in true C4 species, the degree of cross-contamination between cell types in the protoplast fractions of leaves of C3–C4 and C4-like species is therefore impossible to determine and enrichment values for enzymes cannot be accurately measured. The use of immunogold localisation, which can be carried out on intact tissues and which provides a quantifiable labelling density, would be of considerable benefit to the understanding of C4 enzyme compartmentation in these *Flaveria* species.

Organic acids in C3–C4 intermediates

As discussed above, the C4-like biochemistry in the C3–C4 intermediate species of *Flaveria* involves the labelling by ^{14}C of organic acids such as fumarate, succinate and citrate and accumulation of label in fumarate (Monson *et al.*, 1986). Fumarate thus appears to be an end product of photosynthesis, in so far as label accumulates in it during a chase, in much the same way as it accumulates in sucrose. In C3 and C4 Flaverias, labelling of fumarate is negligible (Monson *et al.*, 1986), as it is in other studies of organic acid metabolism in plant tissues (MacLennan *et al.*, 1963) and in CAM plants. In *Opuntia*, for example, dark fixation of $^{14}CO_2$ results in the appearance of about 0.5% of the total label in fumarate (Ting & Dugger, 1968).

Measurements of organic acids in the leaves of a wide range of plants (Table 1) reveal a striking difference in the content of organic acids between certain of the C3–C4 intermediate Flaverias and other C3–C4 intermediates and C3, C4 and CAM plants. The C3–C4 intermediate *F. floridana* showed a high content of fumarate and an elevated amount of succinate, the latter comparable to amounts found in *Kalanchoe daigremontiana*. In other respects, its acid contents were comparable to those of other plants. The leaves of intermediate and C3 species of *Moricandia* and of other C3, C4 and CAM plants did not possess appreciable amounts of fumarate. In addition to the plants listed in Table 1, fumarate was low or undetectable in the C4 plants *F. brownii*, *F. palmeri* and *Sorghum bicolor*, and in the C3 plants rice, cassava and lucerne. The amounts of organic acids, and of fumarate in particular, were not much affected by illumination, as the contents were very similar in darkened

Table 1. Contents of organic acids in leaves illuminated in air of a range of species representing various photosynthetic types

Plant	Metabolite (μmol mg^{-1} chlorophyll)						
	fumarate	succinate	malate	2-oxoglutarate	glutamate	citrate	iso-citrate
C3–C4							
Flaveria floridana	10.56±1.59	2.63±0.44	15.41±3.02	0.39±0.09	2.76±0.05	0.87±0.19	0.18±0.01
Moricandia arvensis	n.d.	0.58±0.09	11.12±1.09	0.13±0.03	4.83±0.28	3.05±0.61	1.10±0.07
M. arvensis var. garamantum	n.d.	0.68±0.05	17.45±1.52	0.13±0.01	3.73±0.08	3.85±0.59	0.69±0.05
M. nitens	0.14±0.05	0.72±0.19	8.30±2.44	0.35±0.11	5.79±0.53	2.05±0.24	0.35±0.06
C3							
M. moricandioides	n.d.	0.87±0.21	26.89±1.55	0.01±0.001	3.55±0.31	6.39±0.80	0.40±0.02
Phaseolus vulgaris	n.d.	0.67±0.05	5.92±2.08	0.23±0.02	2.43±0.20	54.46±4.36	0.82±0.10
Hordeum vulgare	0.02±0.01	0.86±0.13	24.91±1.28	0.39±0.04	5.09±0.85	7.98±0.44	1.69±0.19
Glycine max	0.34±0.22	0.73±0.15	4.60±0.68	0.36±0.13	1.71±0.15	4.68±0.31	0.32±0.13
Spinacia oleracea	0.01±0.002	0.42±0.01	1.94±0.54	0.13±0.03	3.61±0.15	0.23±0.08	0.09±0.02
C4							
Atriplex spongiosa	0.04±0.02	0.85±0.08	8.09±2.84	0.18±0.15	3.84±1.26	2.94±1.32	0.38±0.08
Panicum miliaceum	0.02±0.01	0.39±0.07	2.78±0.22	0.23±0.06	2.39±0.15	2.08±0.25	2.33±0.57
Zea mays	n.d.	0.30±0.11	6.12±0.37	0.35±0.003	1.30±0.25	1.06±0.05	0.28±0.03
CAM							
Kalanchoe daigremontiana	n.d.	2.10±0.54	420.1±84.1	2.35±0.36	9.21±4.08	29.10±4.23	199.7±22.1

The photosynthetic photon flux density (PPFD) was 400 μmol quanta m^{-2}s^{-1}. Leaves were frozen in liquid N$_2$ and extracted for metabolite analysis as described by Leegood & von Caemmerer (1988). Metabolites were assayed as follows: succinate (Beutler, 1985), glutamate (Lund, 1985), 2-oxoglutarate (Burlina, 1985) and other organic acids were measured according to methods described in Lowry & Passoneau (1972). Data are the means±SE of between three and six separate samples. n.d., not detectable.

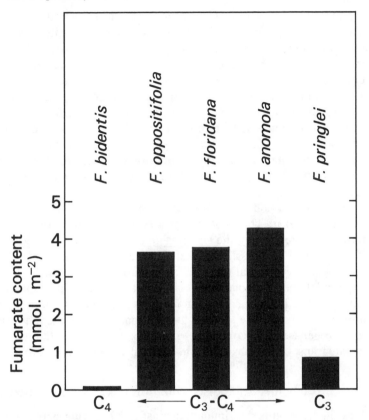

Fig. 3. Contents of fumarate in leaves of species of *Flaveria*. Leaves were freeze-clamped *in situ* on greenhouse-grown plants in full sunlight, extracted in $HClO_4$ (Leegood & von Caemmerer, 1989) and fumarate measured using the method of Lowry & Passoneau (1972). Data show means for *F. floridana* (19 samples), *F. pringlei* (7), *F. bidentis* (10), *F. oppositifolia* (3), and *F. anomola* (1).

leaves. Further comparison of a number of C3, C4 and C3–C4 intermediate *Flaveria* species showed that these elevated amounts of fumarate were confined to the C3–C4 intermediates *F. oppositifolia*, *F. floridana* and *F. anomola* (Fig. 3).

Although fumarate is a key respiratory intermediate and will be present in all plants, the data show that it is often present at, or close to, the limits of detection. As a generalisation, we can conclude that fumarate accumulation is uncommon in the leaves of plants, although Bennet-Clark (1933) noted that large quantities of fumarate had been recorded in the fumitories, *Fumaria* (from which the acid is named) and *Corydalis* and in a poppy, *Glaucium*, in which it is present in amounts equivalent to twice the amount of malate. Darnley Gibbs (1974) also noted its presence in the leaves of *Phaseolus coccineus*, *Helianthus annuus* and of a thistle, *Carduus marianus*, but without specifying the amounts.

The amount of fumarate in the leaves of C3–C4 intermediates is so high that it must be largely located in the vacuoles of the cells. For example, a content of 10 μmol mg^{-1} chlorophyll would be equivalent to a concentration of 0.4 M if confined to the cytosol with a volume of 24 μl mg^{-1} chlorophyll. The concentration would be even higher if fumarate were confined to the mitochondria and control of its oxidation would present a regulatory problem.

The accumulation of fumarate is enigmatic, in that no obvious role can be ascribed to it. The existence of a large fumarate pool permits a different interpretation of the labelling data of Monson *et al.* (1986) because it allows for the possibility that fumarate is not an end product of photosynthesis but that the labelled fumarate is simply diluted out into a large unlabelled vacuolar pool. The actual flux through fumarate could, nevertheless, be very small. Fumarate is presumably synthesised in the mitochondria because of the localisation of fumarase and succinate dehydrogenase in this organelle. It is also worth noting that the vacuolar malate transporter transports fumarate efficiently (White & Smith, 1989).

Carbon isotope discrimination in C3–C4 intermediates

Most higher plants discriminate against $^{13}CO_2$ during photosynthesis. It has been shown that the differential diffusivity of $^{13}CO_2$ and $^{12}CO_2$ and the fractionations of the carboxylating enzymes are the major components contributing to the overall fractionation (for review see O'Leary, 1988; Farquhar *et al.*, 1989).

The large carbon isotope discrimination in C3 plants compared with C4 plants reflects the intrinsically large fractionation of Rubisco. Variation in discrimination amongst C3 plants is, however, caused mainly by variation in stomatal conductance (Farquhar *et al.*, 1989). In C4 species, carbon isotope discrimination is much lower because of the lower effective discrimination of PEP carboxylase in the mesophyll. Subsequent refixation of CO_2 by Rubisco, released in the bundle sheath, allows for some discrimination by Rubisco to occur, as the bundle sheath is not completely gas-tight. In fact, the greater the portion of CO_2 supplied to the bundle sheath that subsequently leaks out, the more C3-like carbon isotope discrimination becomes (Farquhar, 1983). Carbon isotope discrimination values of C3–C4 intermediate species have, in most instances, been C3-

like (e.g. Monson *et al.*, 1988). The exceptions to this are the C4-like *Flaveria brownii* (Monson *et al.*, 1988) and F_1 hybrids made between either this species or the true C4 species *F. trinervia* and a range of C3–C4 intermediate *Flaveria* species (Brown *et al.*, 1986; Araus *et al.*, 1991).

Plants with a C3–C4 intermediate pathway also fix some, albeit a small proportion, of their carbon in the bundle sheath cells of the leaf. As discussed above, in all of the C3–C4 intermediates studied it is presumed that most of the photorespiratory CO_2, whether a consequence of mesophyll or bundle sheath Rubisco oxygenation, is released in the bundle sheath. The quantitative feasibility of localised release of photorespiratory CO_2, accounting for the observed gas exchange properties of C3–C4 intermediate species which do not have any C4-like biochemistry (e.g. *Moricandia* and *Panicum* spp.), was demonstrated by the model of von Caemmerer (1989). As a result of photorespiratory refixation in the bundle sheath, some carbon is fixed twice by Rubisco, once in the mesophyll and once again in the bundle sheath, allowing double discrimination by Rubisco to occur. Thus carbon isotope discrimination in this type of C3–C4 intermediate can be greater than is observed in C3 species, depending on the leakiness of the bundle sheath and the fraction of carbon fixed in the bundle sheath. At ambient partial pressures of CO_2 the rate of photorespiration is low, which explains the observed C3-like carbon isotope values (Edwards & Ku, 1987; von Caemmerer, 1989). At low partial pressures of CO_2, where the rate of photorespiration is greater, a carbon isotope discrimination greater than in C3-species is expected. This was experimentally verified with measurements of short-term carbon isotope discrimination during gas exchange (von Caemmerer & Hubick, 1989).

The C3-like carbon isotope value of leaf dry matter has been an enigma for species where substantial CO_2 fixation into C4 acids has been demonstrated (Edwards & Ku, 1987; Monson *et al.*, 1988). Models of C3–C4 intermediate photosynthesis that include a C4-cycle activity predict that carbon isotope becomes more C4-like as the portion of CO_2 fixed via the C4 pathway increases (Peisker, 1985; Monson *et al.*, 1988). Monson *et al.* (1988) explained the C3-like values they reported for C3–C4 intermediate *Flaveria* species by postulating a high rate of CO_2 leakage from the bundle sheath. This provides a feasible explanation and could be the result of insufficient Rubisco in the bundle sheath to assimilate the CO_2 that is released in these cells; it need not imply an anatomically induced reduction in the diffusion resistance of the bundle sheath. However, it is not clear to what extent an integrated C4-cycle activity occurs in the intermediate *Flaveria* species studied by them.

Other factors which require further research may also affect the carbon isotope discrimination. For example, insufficient carbonic anhydrase in the mesophyll cytoplasm could mean that, unlike in C4 species, bicarbonate and CO_2 are not at equilibrium.

Some *Flaveria* species show both CO_2 fixation into C4 acids and refixation of photorespiratory CO_2 (Monson et al., 1986; Hylton et al., 1988). In the following paragraphs, we formulate a simple equation which describes carbon isotope discrimination for these C3–C4 intermediate species and encompasses all these possibilities.

Following Farquhar & Richards (1984), carbon isotope discrimination is defined here as $\Delta = R_a/R_p - 1$, where R_a and R_p are the molar abundance ratios of $^{13}CO_2$ to $^{12}CO_2$ in the air and the photosynthetic product, respectively. The advantage of this definition is that the values are independent of the carbon isotope composition of the air, although the latter value needs to be known. (A simple conversion from Δ to the more conventional measure of carbon isotope composition, δ with respect to the standard Pee Dee belemnite (PDB) is $\delta = -\Delta‰ - 8‰$, where 8‰ is a common value for the carbon isotope composition of air with respect to PDB.) For a thorough discussion see Farquhar et al. (1989).

The carbon isotope discrimination in C3–C4 intermediates can then be given as

$$\Delta = a + \{b_3 - a + A_s/A(\varphi b_3 + V_p/S(b_4 - b_3))\} p_i/p_a \qquad (1)$$

where a (4.4‰), b_3 (30‰) and b_4 (−5.7‰) are the discriminations attributable to diffusion in air, by Rubisco, and the effective discrimination of PEP carboxylation (Farquhar et al., 1989). The symbols p_i and p_a stand for the intercellular and ambient partial pressures of CO_2 respectively. The ratio A_s/A denotes the fraction of net CO_2 fixation occurring in the bundle sheath, V_p denotes the rate of PEP carboxylation and S stands for the total rate of CO_2 supply to the bundle sheath, which is the sum of V_p and the photorespiratory CO_2 release. The portion of S which subsequently leaks out of the bundle sheath is denoted by φ.

When no CO_2 is fixed in the bundle sheath, ($A_s = 0$) the equation reduces to the well-known equation describing carbon isotope discrimination in C3 species. When no CO_2 fixation occurs via the C4 pathway ($V_p = 0$) the equation describes the carbon isotope discrimination for intermediate species which recycle photorespiratory CO_2. This case was given in Farquhar et al. (1989) and is similar to the equation given by von Caemmerer (1989). When $S = V_p$, the equation is identical to that given by Monson et al. (1988) and Peisker (1985), where C3–C4 intermediacy is solely attributed to a limited C4 cycle activity.

In Table 2, sample calculations are presented for the various possibilities. They show that at a similar p_i/p_a, refixation of photorespiratory CO_2

Table 2. *Carbon isotope discrimination values for C3, C4–C4 and C4 photosynthetic types calculated from Eqn 1 under different conditions*

Photosynthetic type	V_p/A	V_p/S	φ	A_s/A	p_i/p_a	$\Delta(\%)$
C3	0	0	0	0	0.75	23
C3–C4 (with photorespiratory recycling)	0	0	0.30	0.10	0.75	24
C3–C4 (with a C4 cycle)	0.40	1.00	0.30	0.28	0.75	18
	0.40	1.00	0.60	0.16	0.75	20
C3–C4 (with a C4 cycle and photorespiratory recycling)	0.40	0.85	0.30	0.33	0.75	18
	0.40	0.85	0.60	0.18	0.75	22
C4	1.42	1.00	0.30	1.00	0.75	3
	2.50	1.00	0.60	1.00	0.75	10
C4	1.42	1.00	0.30	1.00	0.40	3
	2.50	1.00	0.60	1.00	0.40	7

leads to marginally higher Δ values whether on its own or in combination with a C4 cycle.

Equation 1 describes carbon isotope discrimination of CO_2 assimilation, which can best be tested with shortened measurements of carbon isotope discrimination. Carbon isotope composition of dry matter reflects long-term carbon acquisition and respiration. It is currently not known to what degree respiration fractionates, but it may influence the Δ value of dry matter and obscure small differences in discrimination (O'Leary, 1988).

Implications for metabolite transport and metabolic regulation

The spatial separation of the production of glycine in the mesophyll and its subsequent decarboxylation in the bundle sheath has many implications for metabolism in the leaves of C3–C4 intermediate species. In a C3 plant the photorespiratory pathway has been shown to occur within a single photosynthetic cell as a metabolic cycle involving enzymes in the chloroplasts, peroxisomes and mitochondria, and integrated exchange of a number of metabolites between these subcellular compartments (Ogren, 1984; Wallsgrove *et al.*, this volume). The photorespiratory pathway permits recovery of 75% of the carbon lost as a result of glycine decarboxylation. The details of this are described by Wallsgrove *et al.* (this volume).

Transport of photorespiratory metabolites

It is clear that separation of this integrated pathway between different cell types must involve exchange of metabolites involved in photorespiration between the bundle sheath and mesophyll cells. All of the enzymes required for the synthesis of glycine during photorespiration which have been studied (no data are available for phosphoglycolate phosphatase) are present in both mesophyll and bundle sheath cells of *Moricandia arvensis* (Rawsthorne *et al.*, 1988a, b). The lack of glycine decarboxylase in the mesophyll implies that glycine is the metabolite most likely to move between the mesophyll and bundle sheath cells (Rawsthorne *et al.*, 1988a). Serine, or a product of its metabolism, must then return to the mesophyll as discussed above. Some evidence suggests that this movement of metabolites could occur down concentration gradients between the two cell types in a manner analogous to that in C4 plants (Hatch & Osmond 1976). The density of plasmodesmata on cell faces between bundle sheath and mesophyll cells of C3–C4 *Panicum* species is greater than that of the C3 *Panicum* species and is comparable to that of the C4 species *P. prionitis* (Brown *et al.*, 1983). This would facilitate the exchange of metabolites between the two cell types. However, in C4 species the arrangement of bundle sheath and mesophyll cells is such that metabolism within the C4 cycle is confined to adjacent cells (Leegood & Osmond, 1990). In C3–C4 species mesophyll cells may be spaced in such a way that there are several cell distances between outlying mesophyll cells and a bundle sheath cell (Apel & Ohle, 1979; Holaday *et al.*, 1981, 1984; Brown & Hattersley, 1989). Transport distances will therefore be much longer in the intermediates than in C4 species.

Maintenance of glycine diffusion by a downhill concentration gradient between the mesophyll and the bundle sheath is feasible in C3–C4 species given the potential of the bundle sheath cell to decarboxylate glycine at substantial rates. Immunolabelling for glycine decarboxylase is dense on bundle sheath mitochondria and the number and size of mitochondria within bundle sheath cells is much greater than that in mesophyll cells of the intermediate species or any photosynthetic cell type in C3 species (Hylton *et al.*, 1988; Brown & Hattersley, 1989). In addition, rates of glycine decarboxylation per unit chlorophyll by protoplast fractions enriched in *M. arvensis* bundle sheath cells were on average 2.5-fold greater than those reported for protoplasts from pea (C3) leaves (Walton & Woolhouse, 1986; Rawsthorne *et al.*, 1988b).

No data are available on the size of metabolite pools in specific leaf cell types of C3–C4 intermediates but amounts of glycine and serine are two-fold greater in whole leaves of *M. arvensis* than in leaves of the C3 species *M. moricandioides* (Rawsthorne & Hylton 1991). These pool sizes in leaves of *M. arvensis* at a high light intensity are about four-fold greater on a leaf area basis than those of metabolites involved in cell-to-cell diffusion in C4 photosynthesis (pyruvate, malate and aspartate) either measured experimentally or calculated from theoretical parameters (Leegood & von Caemmerer, 1988). The glycine pools decrease in both *Moricandia* species during the immediate post-illumination CO_2 burst (Rawsthorne & Hylton, 1991), consistent with the interpretation that this burst represents the continued metabolism of a pool of photorespiratory intermediates (Doehlert *et al.*, 1979). The kinetics of the dark burst are significantly slower in the intermediate than in the C3 species, as is the rate of decrease in the size of the glycine pool (Rawsthorne & Hylton, 1991). A larger glycine pool size in whole leaves of *M. arvensis* compared with that in C3 species was inferred previously from slower turnover of label through the glycine pool during $^{14}CO_2$ pulse-chase experiments (Holaday & Chollet, 1983). As discussed above, relatively large percentages of ^{14}C label appear in glycine and serine during $^{14}CO_2$ pulse-chase experiments with C3–C4 *Flaveria* species (Monson *et al.*, 1986), suggesting slower metabolism of the label owing to its dilution in a large pool. These observations are consistent with the ability to maintain a concentration gradient of glycine and with a longer diffusional path than in a C3 plant between the source of photorespiratory glycine and the site of its decarboxylation. As stated above, the concentration of serine is twice as high in leaves of *M. arvensis* as in a related C3 species and the same arguments used for glycine movement above could be applied to the movement of serine from the bundle sheath to the mesophyll.

If the compound returning to the mesophyll is serine then half of the nitrogen removed from the mesophyll cell as glycine will be returned (2 mol glycine form 1 mol serine in the mitochondria). The remainder of the nitrogen will be released as ammonia by glycine decarboxylase activity. The confinement of all the photorespiratory ammonia release in the leaf to the bundle sheath cells of a C3–C4 species implies that all of it may be reassimilated by the glutamine synthetase/glutamate synthase (GS/ GOGAT) in this cell type. The maximum catalytic activity of GS in bundle sheaths of *M. arvensis* is twice that of the maximum capacity (oxaloacetic acid-stimulated rate) for glycine decarboxylation in the same cells and GS would therefore be capable of assimilating the ammonia released (Rawsthorne *et al.*, 1988*b*). Of course, this confinement of the ammonia reassimilation to the bundle sheath cells would impose a greater energetic demand in the bundle sheath for redox equivalents and ATP for the GS/GOGAT system than in photorespiring cells of a C3 species. Once the ammonia is assimilated there is likely to be a requirement for the return of this nitrogen to the mesophyll to avoid imbalances in the transamination reactions of the photorespiratory pathway and between the cell types in general. How this nitrogen is recycled is unknown at present. In *M. arvensis* the activity of glutamate-dependent glyoxylate aminotransferases in bundle sheath cell-enriched protoplast fractions is much greater than the serine-dependent activity (Rawsthorne *et al.*, 1988*b*) whereas both activities are comparable in mesophyll-enriched protoplast fractions from *M. arvensis* (Rawsthorne *et al.*, 1988*b*) and in whole leaves of pea (Walton & Woolhouse, 1986). These data suggest that glutamate rather than serine may be utilised preferentially for photorespiratory transamination reactions in bundle sheath cells.

An intercellular PGA/triose-P shuttle in C3–C4 intermediates?

A shuttle system involving export of 3-phosphoglycerate (PGA) from bundle sheath chloroplasts to the mesophyll chloroplasts, and the return of dihydroxyacetone phosphate (triose-P) from the mesophyll to the bundle sheath chloroplasts, appears to be characteristic of all the subgroups of C4 plants (Hatch & Osmond 1976; Edwards, this volume). The operation of this shuttle can be viewed as a mechanism that reduces the concentration of O_2 in the bundle sheath because it reduces the demand for the generation of NADPH, and hence O_2 evolution by non-cyclic electron transport, in the bundle sheath. It might be regarded as the simplest means by which plants with a bundle sheath could alter the balance of carboxylation and oxygenation of Rubisco in that compartment. In C4 plants, apart from the capacity to reduce PGA to triose-P in the mesophyll cells (requiring the presence of chloroplastic NADP-glyceraldehyde-3-P dehydrogenase and glycerate-3-P kinase), shuttling of PGA is accompanied by exceptionally large pools of triose-P which drive intercellular transport back from the mesophyll to the bundle sheath cells (Hatch & Osmond, 1976; Leegood, 1985; Leegood & von Caemmerer, 1988, 1989).

The presence of such a shuttle is, therefore, revealed by the occurrence of large pools of triose-P. (It should be noted that pools of PGA are often comparable in the leaves of C3 and C4 plants, so that large pools of PGA are not a diagnostic feature.) Thus we have detected (R. C. Leegood and S. von Caemmerer, unpublished data) large pools of triose-P in leaves of *F. brownii* (C4-like: 416 nmol mg^{-1} chlorophyll) and in *F. trinervia* (C4:

Table 3. *Contents of triose-P in leaves during steady-state photosynthesis in air* (p_a 350 μbar)

Plant	Content of triose-P (nmol mg^{-1} chlorophyll)
Flaveria pringlei (C3)	45
F. bidentis (C4)	564
F. floridana (C3–C4)	67
Moricandia arvensis (C3–C4)	190

The PPFD was 1600 μmol quanta m^{-2} s^{-1}. Leaves were freeze-clamped and extracted for metabolite analysis as described by Leegood & von Caemmerer (1988).
From R.C. Leegood and S. von Caemmerer, unpublished data.

463 nmol mg^{-1} chlorophyll). In a C3 plant these values would more typically be about 50 nmol mg^{-1} chlorophyll (Table 3; and see Badger *et al.*, 1984). Table 3 compares amounts of triose-P in illuminated leaves of *F. floridana* (C3–C4), *F. pringlei* (C3), *F. bidentis* (C4) and *M. arvensis* (C3) in air. While in *F. floridana* the pool of triose-P was small and C3-like, in *M. arvensis* the triose-P pool at ambient CO_2 was four times the usual content of C3 plants. The high content of triose-P in the leaves of *M. arvensis* suggests that a PGA/triose-P shuttle between the bundle sheath and the mesophyll could be operational in the leaves of this plant.

Energetic implications of the C3–C4 mechanism

The light dependence of Γ in all the C3–C4 species in which this phenomenon has been examined provides evidence that the mechanism of C3–C4 photosynthesis is limited by the supply of ATP and NADPH at low CO_2 and is not limited by the rate of primary carboxylation by Rubisco or PEP carboxylase.

Both *F. floridana* and *M. arvensis* exhibit a number of distinctive metabolic features not observed in any C3 or C4 plants previously studied. In *F. floridana*, for example, there was a rise in the pools of PGA and pyruvate at CO_2 concentrations around Γ (Table 4). Reduction of PGA depends upon the provision of ATP and NADPH. The sharp fall in the triose-P/PGA ratio therefore indicates a lack of ATP and NADPH (Heber *et al.*, 1986). Similarly, if a C4 cycle operates, the conversion of pyruvate to PEP, catalysed by pyruvate, P_i dikinase, depends upon the provision of ATP. The rise in pyruvate and the fall in the triose-P/PGA

Table 4. *Contents of metabolites in leaves during steady-state photosyntheses in air ($_a$ 350 μbar) and close to the CO_2 compensation point (Γ)*

Plant		Content of metabolite (μmol m^{-2})		Triose-P/PGA
		RuBP	pyruvate	
Flaveria pringlei (C3)	air	42	29	0.18
	Γ	49	32	0.17
F. bidentis (C4)	air	47	335	0.41
	Γ	127	122	0.47
F. floridana (C3–C4)	air	56	82	0.19
	Γ	122	222	0.13
Moricandia arvensis (C3–C4)	air	153	85	0.41
	Γ	37	49	0.13

The PPFD was 1600 μmol quanta m^{-2} s^{-1}. Leaves were freeze-clamped and extracted for metabolite analysis as described by Leegood & von Caemmerer (1988). The mean chlorophyll contents of the leaves of each plant were: *Flaveria pringlei*, 311 mg m^{-2}; *F. floridana*, 371 mg m^{-2}; *F. bidentis*, 346 mg m^{-2} and *Moricandia arvensis*, 269 mg m^{-2}.
From R.C. Leegood and S. von Caemmerer, unpublished data.

ratio provide metabolic evidence for a decrease in energisation at low CO_2.

As in *F. floridana*, the triose-P/PGA ratio in leaves of *M. arvensis* declined drastically as P_i was lowered to Γ (Table 4). At Γ there were also large falls in the pools of RuBP (Table 4) and in the contents of triose-P and glycine (data not shown). This decline in the RuBP pool as the concentration of CO_2 was decreased contrasts with all other plants studied and we can conclude that RuBP regeneration limits photosynthesis at the lowest CO_2 concentrations in *M. arvensis*.

These observations show clearly that the system is de-energised at low CO_2. While quantum yields for CO_2 are marginally lower in intermediates such as *F. floridana* than in C3 species (Monson *et al.*, 1986), the energy requirement for operation of the C3–C4 mechanisms is likely to arise through inefficient compartmentation and coordination. Although the *Flaveria* species theoretically have an ATP demand for a C4 pump, in the *Moricandia* species there is no obvious energy-requiring step, unless transport of metabolites is energy-dependent. Even in the Flaverias, neither PEP carboxylase nor Rubisco controls photosynthesis at low CO_2, which leads us to the conclusion that the C3 and C4 cycles are poorly coordinated.

It has been shown that oxidative phosphorylation occurs in the mitochondria of C3 mesophyll cells in the light and that this process is dependent upon the production of NADH during the oxidation of photorespiratory glycine (Krömer *et al.*, 1988; Gardeström & Wigge, 1988). Furthermore, it has been proposed that transfer of redox equivalents occurs between the chloroplast, mitochondrion and cytosol in order that the supply of ATP by the mitochondria in the light is linked to the various demands imposed during photosynthetic and photorespiratory metabolism (Heldt *et al.*, 1990). For example, redox equivalents generated in the mitochondrion through glycine oxidation could contribute to the NADH required for reduction of hydroxy-pyruvate in the peroxisome via transmembrane shuttles or feed directly into oxidative phosphorylation. Clearly, the exchange of redox equivalents and the source of redox power to drive ATP synthesis via oxidative phosphorylation in the light in mesophyll and bundle sheath cells of *M. arvensis* and other species which shuttle glycine must be very different from that of a C3 plant. These cell types will have very different capacities to produce NADH from glycine oxidation and therefore different capacities to produce ATP, and these processes will in turn be influenced by the flux through the photorespiratory pathway. The mesophyll cells are therefore likely to have a greater demand than the bundle sheath cells for reducing power from other sources unless appropriate intercellular NADH shuttles exist. How the C3–C4 species maintain inter- and intracellular redox balance remains to be determined.

Acknowledgements

We wish to thank Drs Alison Smith and Simon Turner, and Mr Colin Morgan for their helpful comments on the manuscript. Research collaboration between R. C. Leegood and S. von Caemmerer was supported by the Royal Society.

References

Apel, P. & Ohle, H. (1979). CO_2-compensation point and leaf anatomy in species of the genus *Moricandia* DC. (Cruciferae). *Biochemie und Physiologie der Pflanzen* **174**, 68–75.

Apel, P. & Peisker, M. (1988). Influence of water stress on photosynthetic gas exchange in the C_3–C_4 intermediate species *Flaveria floridana*. *Biochemie und Physiologie der Pflanzen* **183**, 439–42.

134   S. RAWSTHORNE *ET AL.*

Apel, P., Tichá, I. & Peisker, M. (1978). CO_2 compensation concentrations in leaves of *Moricandia arvensis* (L.) DC. at different insertion levels and O_2 concentrations. *Biochemie und Physiologie der Pflanzen* **172**, 547–52.

Araus, J.L., Brown, R.H., Byrd, G.T. & Serret, M.D. (1991). Comparative effects of growth irradiance on photosynthesis and leaf anatomy of *Flaveria brownii* (C_4-like), *Flaveria linearis* (C_3–C_4) and their F_1 hybrid. *Planta* **183**, 497–504.

Artus, N.N., Somerville, S.C. & Somerville, C.R. (1986). The biochemistry and cell biology of photorespiration. *CRC Critical Reviews in Plant Science* **4**, 121–47.

Badger, M.R., Sharkey, T.D. & von Caemmerer, S. (1984). The relationship between steady-state gas exchange of bean leaves and the levels of carbon-reduction-cycle intermediates. *Planta* **160**, 305–13.

Bauwe, H. (1984). Photosynthetic enzyme activities and immunofluorescence studies on the localization of ribulose-1,5-bisphosphate carboxylase/oxygenase in leaves of C_3, C_4, and C_3–C_4 intermediate species of *Flaveria* (*Asteraceae*). *Biochemie und Physiologie der Pflanzen* **179**, 253–68.

Bennet-Clark, T.A. (1933). The role of organic acids in plant metabolism. *New Phytologist* **32**, 32–71.

Beutler, H.-O. (1985). Succinate. In *Methods of Enzymatic Analysis*, Vol. 7, ed. H.U. Bergmeyer, pp. 25–33. Weinheim: VCH Verlagsgesellschaft.

Bourguignon, J., Neuburger, M. & Douce, R. (1988). Resolution and characterization of the glycine-cleavage reaction in pea leaf mitochondria. *Biochemical Journal* **255**, 169–78.

Bouton, J.H., Brown, R.H., Evans, P.T. & Jernstedt, J.A. (1986). Photosynthesis, leaf anatomy and morphology of progeny from hybrids between C_3 and C_3/C_4 *Panicum* species. *Plant Physiology* **80**, 487–92.

Brown, R.H. Bassett, C.L., Cameron, R.G., Evans, P.T., Bouton, J.H., Black, C.C., Sternberg, L.O. & DeNiro, M.J. (1986). Photosynthesis of F_1 hybrids between C_4 and C_3–C_4 species of *Flaveria*. *Plant Physiology* **82**, 211–17.

Brown, R.H., Bouton, J.H., Evans, P.T., Malter, H.E. & Rigsby, L. (1984). Photosynthesis, morphology, leaf anatomy and cytogenetics of hybrids between C_3 and C_3/C_4 *Panicum* species. *Plant Physiology* **77**, 653–8.

Brown, R.H., Bouton, J.H., Rigsby, L. & Rigler, M. (1983). Photosynthesis of grass species differing in carbon dioxide fixation pathways. VIII. Ultrastructural characteristics of *Panicum* species in the *Laxa* group. *Plant Physiology* **71**, 425–31.

Brown, R.H. & Hattersley, P.W. (1989). Leaf anatomy of C_3–C_4 species as related to evolution of C_4 photosynthesis. *Plant Physiology* **91**, 1543–50.

Brown, R.H. & Morgan, J.A. (1980). Photosynthesis of grass species differing in carbon dioxide fixation pathways. VI. Differential effects of temperature and light intensity on photorespiration in C_3, C_4, and intermediate species. *Plant Physiology* **66**, 541–4.

Burlina, A. (1985). 2-oxoglutarate. In *Methods of Enzymatic Analysis*, Vol. 7, ed. H.U. Bergmeyer, pp. 20–4. Weinheim: VCH Verlagsgesellschaft.

Byrd, G.T. & Brown, R.H. (1989). Environmental effects on photorespiration of C_3–C_4 species. I. Influence of CO_2 and O_2 during growth on photorespiratory characteristics and leaf development. *Plant Physiology* **90**, 1022–8.

Cheng, S.-H., Moore, B.d., Edwards, G.E. & Ku, M.S.B. (1988). Photosynthesis in *Flaveria brownii*, a C_4-like species. Leaf anatomy, characteristics of CO_2 exchange, compartmentation of photosynthetic enzymes, and metabolism of $^{14}CO_2$. *Plant Physiology* **87**, 867–73.

Cheng, S.-H., Moore, B.d., Wu, J., Edwards, G.E. & Ku, M.S.B. (1989). Photosynthetic plasticity in *Flaveria brownii*. Growth irradiance and the expression of C_4 photosynthesis. *Plant Physiology* **89**, 1129–35.

Darnley Gibbs, R. (1974). *Chemotaxonomy of Higher Plants*. Montreal: McGill–Queen's University Press.

Doehlert, D.C., Ku, M.S.B. & Edwards, G.E. (1979). Dependence of the post-illumination burst of CO_2 on temperature, light, CO_2, and O_2 concentration in wheat (*Triticum aestivum*). *Physiologia Plantarum* **46**, 299–306.

Edwards, G.E. & Ku, M.S.B. (1987). Biochemistry of C_3–C_4 intermediates. In *The Biochemistry of Plants*, Vol. 10, ed. M.D. Hatch & N.K. Boardman, pp. 275–325. London: Academic Press.

Edwards, G.E., Ku, M.S.B. & Hatch, M.D. (1982). Photosynthesis in *Panicum milioides*, a species with reduced photorespiration. *Plant and Cell Physiology* **23**, 1185–95.

Farquhar, G.D. (1983). On the nature of carbon-isotope discrimination in C_4 species. *Australian Journal of Plant Physiology* **10**, 205–26.

Farquhar, G.D., Ehleringer, J.R. & Hubick, K.T. (1989). Carbon isotope discrimination and photosynthesis. *Annual Review of Plant Physiology and Plant Molecular Biology* **40**, 503–37.

Farquhar, G.D. & Richards, R.A. (1984). Isotopic composition of plant carbon correlates with water-use efficiency of wheat genotypes. *Australian Journal of Plant Physiology* **11**, 539–52.

Fladung, M. & Hesselbach, J. (1987). Developmental studies on photosynthetic parameters in C_3, C_3–C_4 and C_4 plants of *Panicum*. *Journal of Plant Physiology* **130**, 461–70.

Gardeström, P. & Wigge, B. (1988). Influence of photorespiration on ATP/ADP ratios in the chloroplasts, mitochondria, and cytosol, studied by rapid fractionation of barley (*Hordeum vulgare*) protoplasts. *Plant Physiology* **88**, 69–76.

Hatch, M.D. & Osmond, C.B. (1976). Compartmentation and transport in C_4 photosynthesis. In *Transport in Plants III. Encyclopedia of Plant Physiology*, New Series Vol. 3, ed. C.R. Stocking & U. Heber, pp. 114–85. Berlin: Springer Verlag.

Hattersley, P.W., Wong, S.-C., Perry, S. & Roksadic, Z. (1986). Comparative ultrastructure and gas exchange characteristics of the C_3–C_4 intermediate *Neurachne minor* S.T. Blake (Poaceae). *Plant, Cell and Environment* **9**, 217–33.

Heber, U., Neimanis, S., Dietz, K.-J. & Viil, J. (1986). Assimilatory power as a driving force in photosynthesis. *Biochimica et Biophysica Acta* **852**, 144–55.

Heldt, H.W., Heineke, D., Heupel, R., Krömer, S. & Riens, B. (1990). Transfer of redox equivalents between subcellular compartments of a leaf cell. In *Proceedings of VIIIth International Conference on Photosynthesis*, ed. M. Balttscheffsky, IV:15.1–15.7. London: Academic Press.

Holaday, A.S. & Chollet, R. (1983). Photosynthetic/photorespiratory carbon metabolism in the C_3–C_4 intermediate species, *Moricandia arvensis* and *Panicum milioides*. *Plant Physiology* **73**, 740–5.

Holaday, A.S., Harrison, A.T. & Chollet, R. (1982). Photosynthetic/photorespiratory CO_2 exchange characteristics of the C_3–C_4 intermediate species, *Moricandia arvensis*. *Plant Science Letters* **27**, 181–9.

Holaday, A.S., Lee, K.W. & Chollet, R. (1984). C_3–C_4 intermediate species in the genus *Flaveria*: leaf anatomy, ultrastructure, and the effect of O_2 on the CO_2 compensation concentration. *Planta* **160**, 25–32.

Holaday, A.S., Shieh, Y.-J., Lee, K.W. & Chollet, R. (1981). Anatomical, ultrastructural and enzymic studies of leaves of *Moricandia arvensis*, a C_3–C_4 intermediate species. *Biochimica et Biophysica Acta* **637**, 334–41.

Hunt, S., Smith, A.M. & Woolhouse, H.W. (1987). Evidence for a light-dependent system for reassimilation of photorespiratory CO_2, which does not include a C_4 cycle, in the C_3–C_4 intermediate species *Moricandia arvensis*. *Planta* **171**, 227–34.

Hylton, C.M., Rawsthorne, S., Smith, A.M., Jones, D.A. & Woolhouse, H.W. (1988). Glycine decarboxylase is confined to the bundle-sheath cells of leaves of C_3–C_4 intermediate species. *Planta* **175**, 452–9.

Krömer, S., Stitt, M. & Heldt, H.W. (1988). Mitochondrial oxidative phosphorylation participating in photosynthetic metabolism of a leaf cell. *FEBS Letters* **226**, 352–6.

Ku, M.S.B., Monson, R.K., Littlejohn, R.O., Nakamoto, H., Fisher, D.B. & Edwards, G.E. (1983). Photosynthetic characteristics of C_3–C_4 intermediate *Flaveria* species. I. Leaf anatomy, photosynthetic responses to O_2 and CO_2, and activities of key enzymes in the C_3 and C_4 pathways. *Plant Physiology* **71**, 944–8.

Leegood, R.C. (1985). The intercellular compartmentation of metabolites in leaves of *Zea mays*. *Planta* **164**, 163–71.

Leegood, R.C. & Osmond, C.B. (1990). The flux of metabolites in C_4 and CAM plants. In *Plant Physiology, Biochemistry, and Molecular Biology*, ed. D.T. Dennis and D.H. Turpin, pp. 274–98. London: Longman.

Leegood, R.C. & von Caemmerer, S. (1988). The relationship between contents of photosynthetic metabolites and the rate of photosynthetic carbon assimilation in leaves of *Amaranthus edulis* L. *Planta* **174**, 253–62.

Leegood, R.C. & von Caemmerer, S. (1989). Some relationships between contents of photosynthetic metabolites and the rate of photosynthetic carbon assimilation in leaves of *Zea mays* L. *Planta* **178**, 258–66.

Lowry, O.H. & Passoneau, J.V. (1972). *A Flexible System of Enzymatic Analysis*. New York: Academic Press.

Lund, P. (1985). Glutamine and glutamate. In *Methods of Enzymatic Analysis*, Vol. 7, ed. H.U. Bergmeyer, pp. 357–63. Weinheim: VCH Verlagsgesellschaft.

MacLennan, D.H., Beevers, H. & Harley, J. (1963). Compartmentation of acids in plant tissues. *Biochemical Journal* **89**, 316–27.

Monson, R.K. (1989). The relative contributions of reduced photorespiration, and improved water- and nitrogen-use efficiencies, to the advantages of C_3–C_4 intermediate photosynthesis in *Flaveria*. *Oecologia* **80**, 215–21.

Monson, R.K. & Moore, B.d. (1989). On the significance of C_3–C_4 intermediate photosynthesis to the evolution of C_4 photosynthesis. *Plant, Cell and Environment* **12**, 689–99.

Monson, R.K., Moore, B.d., Ku, M.S.B. & Edwards, G.E. (1986). Co-function of C_3- and C_4-photosynthetic pathways in C_3, C_4, and C_3–C_4 intermediate *Flaveria* species. *Planta* **168**, 493–502.

Monson, R.K., Teeri, J.A., Ku, M.S.B., Gurevitch, J., Mets, L.J. & Dudley, S. (1988). Carbon-isotope discrimination by leaves of *Flaveria* species exhibiting different amounts of C_3- and C_4-cycle cofunction. *Planta* **174**, 145–51.

Moore, B.d., Cheng, S.-H. & Edwards, G.E. (1986). The influence of leaf development on the expression of C_4 metabolism in *Flaveria trinervia*, a C_4 dicot. *Plant and Cell Physiology* **27**, 1159–67.

Moore, B.d., Francheschi, V.R., Cheng, S.-H., Wu, J. & Ku, M.S.B. (1987). Photosynthetic characteristics of the C_3–C_4 intermediate species *Parthenium hysterophorous*. *Plant Physiology* **85**, 984–9.

Moore, B.d., Ku, M.S.B. & Edwards, G.E. (1989). Expression of C_4-like photosynthesis in several species of *Flaveria*. *Plant, Cell and Environment* **12**, 541–9.

Moore, B.d., Monson, R.K., Ku, M.S.B. & Edwards, G.E. (1988). Activities of principal photosynthetic and photorespiratory enzymes

138   S. RAWSTHORNE *ET AL.*

in leaf mesophyll and bundle sheath protoplasts from the C_3–C_4 intermediate *Flaveria ramosissima*. *Plant and Cell Physiology* **29**, 999–1006.

Ogren, W.L. (1984). Photorespiration: pathways, regulation and modification. *Annual Review of Plant Physiology* **35**, 415–42.

O'Leary, M.H. (1988). Carbon isotopes in photosynthesis fractionation techniques may reveal new aspects of carbon dynamics in plants. *Bioscience* **38**, 329–36.

Peisker, M. (1985). Modelling carbon metabolism in C_3–C_4 intermediate species. 2. Carbon-isotope discrimination. *Photosynthetica* **19**, 300–11.

Peisker, M. (1986). Models of carbon metabolism in C_3–C_4 intermediate plants as applied to the evolution of C_4 photosynthesis. *Plant, Cell and Environment* **9**, 627–35.

Peisker, M., Apel, P., Tichá, I. & Hák, R. (1988). CO_2 compensation concentration in relation to leaf insertion level in three *Flaveria* species with different photosynthetic pathways. *Photosynthetica* **2**, 1–8.

Perrot-Rechenmann, C., Chollet, R. & Gadal, P. (1984). In-situ immunofluorescent localization of phosphoenolpyruvate and ribulose 1,5-bisphosphate carboxylases in leaves of C_3, C_4, and C_3–C_4 intermediate *Panicum* species. *Planta* **161**, 266–71.

Rajendrudu, G., Prasad, J.S.R. & Rama Das, V.S. (1986). C_3–C_4 intermediate species in *Alternanthera* (Amaranthaceae). Leaf anatomy, CO_2 compensation point, net CO_2 exchange and the activities of photosynthetic enzymes. *Plant Physiology* **80**, 409–14.

Rawsthorne, S. & Hylton, C.M. (1991). The post-illumination CO_2 burst and glycine metabolism in leaves of C_3 and C_3–C_4 intermediate species of *Moricandia*. *Planta* **186**, 122–6.

Rawsthorne, S., Hylton, C.M., Smith, A.M. & Woolhouse, H.W. (1988*a*). Photorespiratory metabolism and immunogold localization of photorespiratory enzymes in leaves of C_3 and C_3–C_4 intermediate species of *Moricandia*. *Planta* **173**, 298–308.

Rawsthorne, S., Hylton, C.M., Smith, A.M. & Woolhouse, H.W. (1988*b*). Distribution of photorespiratory enzymes between bundle-sheath and mesophyll cells in leaves of the C_3–C_4 intermediate species *Moricandia arvensis* (L.) DC. *Planta* **176**, 527–32.

Reed, J.E. & Chollet, R. (1985). Immunofluorescent localization of phosphoenolpyruvate carboxylase and ribulose 1,5-bisphosphate carboxylase/oxygenase proteins in leaves of C_3, C_4, and C_3–C_4 *Flaveria* species. *Planta* **165**, 439–45.

Sayre, R.T. & Kennedy, R.A. (1977) Ecotypic differences in the C_3 and C_4 photosynthetic activity in *Mollugo verticillata*, a C_3–C_4 intermediate. *Planta* **134**, 257–62.

Sayre, R.T., Kennedy, R.A. & Pringnitz, D.J. (1979). Photosynthetic enzyme activities and localization in *Mollugo verticillata* populations differing in the levels of C₃ and C₄ cycle operation. *Plant Physiology* **64**, 293–9.

Somerville, S.C. & Somerville, C.R. (1983). Effect of oxygen and carbon dioxide on photorespiratory flux determined from glycine accumulation in a mutant of *Arabidopsis thaliana*. *Journal of Experimental Botany* **34**, 415–24.

Ting, I.P. & Dugger, W.M. (1968). Non-autotrophic carbon dioxide metabolism in cacti. *Botanical Gazette* **129**, 9–15.

von Caemmerer, S. (1989). Biochemical models of photosynthetic CO₂-assimilation in leaves of C₃–C₄ intermediates and the associated carbon-isotope discrimination. I. A model based on a glycine shuttle between mesophyll and bundle-sheath cells. *Planta* **178**, 376–87.

von Caemmerer, S. & Hubick, K.T. (1989). Short-term carbon-isotope discrimination in C₃–C₄ intermediate species. *Planta* **178**, 475–81.

Walker, J.L. & Oliver, D.J. (1986). Glycine decarboxylase multienzyme complex. Purification and partial characterisation from pea leaf mitochondria. *Journal of Biological Chemistry* **261**, 2214–21.

Walton, N.J. & Woolhouse, H.W. (1986). Enzymes of serine and glycine metabolism in leaves and non-photosynthetic tissues of *Pisum sativum* L. *Planta* **167**, 119–28.

White, P.J. & Smith, J.A.C. (1989). Proton and anion transport at the tonoplast in crassulacean-acid-metabolism plants: specificity of the malate-influx system in *Kalanchoë daigremontiana*. *Planta* **179**, 265–74.

J.A.C. SMITH and J.H. BRYCE

Metabolite compartmentation and transport in CAM plants

Crassulacean acid metabolism (CAM) represents the third major subdivision of photosynthetic carbon assimilation types in green plants alongside the C3 and C4 pathways. As with the C4 pathway, the CAM pathway can be regarded as an ancillary biochemical mechanism, serving to provide CO_2 at elevated concentrations for fixation in the Calvin cycle of C3 photosynthesis. In contrast to the C4 pathway, however, all the key biochemical components of carbon assimilation in CAM plants are to be found within individual mesophyll cells. CAM, in essence, is a *cellular* phenomenon, and this highlights the importance of regulating carbon flow between organelles at the *subcellular* level.

The comparative biochemistry of CAM plants has been the subject of excellent reviews (Kluge & Ting, 1978; Osmond, 1978; Osmond & Holtum, 1981; Winter, 1985; Lüttge, 1987; Griffiths, 1988; Leegood & Osmond, 1990). These sources can be consulted for detailed information on metabolic pathways and enzymic characteristics. We shall focus here on the significance of subcellular compartmentation for the control of carbon flow in the CAM pathway. While describing the metabolic interconversions characteristic of each compartment, our particular aim is to show how the regulation of metabolite flux *between* organelles is also fundamental to the carbon assimilation process in CAM plants. Our understanding of these transport processes is improving, but we are only just starting to unravel details of their molecular mechanisms.

Overview of CAM physiology and taxonomy

What is a CAM plant? Several features of CAM plants distinguish them physiologically from C3 and C4 species. These can be summarised with reference to the typical gas-exchange pattern observed during the day–night cycle, as illustrated in Fig. 1. These characteristics are manifested to varying extents, depending on the species concerned and ambient environmental conditions, but they provide a useful operational definition of CAM plants.

Society for Experimental Biology Seminar Series 50: *Plant organelles*, ed. A. K. Tobin.
© Cambridge University Press 1992, pp. 141–67.

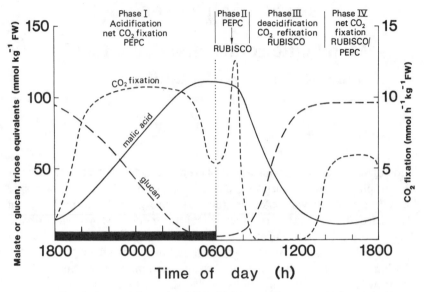

Fig. 1. Day–night cycle of CO_2 fixation and concentrations of malic acid and glucan in mesophyll cells typically observed in a well-watered constitutive CAM plant. PEPC, phosphoenolpyruvate carboxylase; RUBISCO, ribulose 1,5-bisphosphate carboxylase/oxygenase. Modified with permission from Leegood & Osmond (1990).

1. Assimilation of atmospheric CO_2 occurs predominantly at night, rather than during the day as in C3 and C4 plants. Stomata are open at night and closed for much of the daytime.

2. Malic acid accumulates during the night in the chlorenchymatous mesophyll cells as a result of dark CO_2 fixation by the enzyme phosphoenolpyruvate (PEP) carboxylase; this nocturnally accumulated malic acid is then decarboxylated in the following light period to yield CO_2 and a C3 residue.

3. Storage carbohydrate concentrations (either of insoluble glucans such as starch, or of soluble sugars) fluctuate inversely to those of malic acid, decreasing at night to provide the C3 substrate (PEP) for dark CO_2 fixation, and increasing during the day as the C3 residue generated in malate decarboxylation is conserved in gluconeogenesis.

4. Associated with these biochemical characteristics is a suite of distinctive anatomical and morphological correlates. These include a chlorenchymatous mesophyll tissue made up of

large, highly vacuolate cells with relatively thin cell walls (a non-chlorophyllous mesophyll, or 'water-storage parenchyma', is often present as well, but this tissue does not participate in the CAM rhythm). The photosynthetic leaves or stems are succulent to different degrees (showing low surface area:volume ratios, and low dry weight:fresh weight ratios), possess small stomata and low stomatal frequencies, and have relatively high water potentials (and low sap osmotic pressures) even under water-stressed conditions.

Taxonomically, these features are displayed in nearly 30 families of vascular plants, ranging from epiphytic ferns in the Polypodiaceae, through more than 20 families of dicotyledons (with large numbers of species in the Aizoaceae, Asclepiadaceae, Cactaceae, Crassulaceae and Euphorbiaceae), to 4 families of monocotyledons (Agavaceae, Bromeliaceae, Liliaceae and Orchidaceae). The essential biochemical features of CAM are even observed in some aquatic pteridophytes in the Isoetaceae, as well as in certain angiosperms with the 'isoetid' life form (Griffiths, 1988; Raven *et al.*, 1988). Given the very large numbers of CAM species occurring as tropical epiphytes (Winter, 1985; Lüttge, 1989), CAM plants may total about 20 000 species, corresponding to approximately 8% of all vascular plants. This is at least an order of magnitude greater than the number of C4 species. CAM has undoubtedly arisen many times in evolution, and may even be polyphyletic within individual genera (Smith, 1989).

These characteristics confer upon CAM plants a selective advantage in many water-limited environments at tropical and subtropical latitudes, such as semi-deserts and epiphytic microhabitats in tropical forests. Because atmospheric water vapour deficits reach high values during the daytime, stomatal opening at night can reduce by an order of magnitude the amount of water lost in transpiration per unit CO_2 assimilated (Kluge & Ting, 1978; Nobel, 1988, 1991). This, together with the properties of nocturnal PEP-carboxylase-mediated CO_2 assimilation, results in CAM plants showing high values of water-use efficiency in carbon assimilation (i.e. a high ratio of net CO_2 uptake to water loss). The biochemistry underpinning this photosynthetic mechanism is now one of the best understood metabolic examples of an ecological adaptation in plants.

Carbon flow during the day–night cycle: four phases of CAM

Before considering the importance of subcellular compartmentation for metabolism in CAM plants, we shall outline the overall flow of carbon during a typical day–night cycle. This can be conveniently discussed in terms of the four phases of gas exhange recognised by Osmond (1978), as shown in Fig. 1.

The dark period constitutes *phase I*, when CO_2 is taken up from the atmosphere and assimilated by the cytosolic enzyme PEP carboxylase, with the three-carbon substrate PEP being derived from the storage carbohydrate by glycolysis. Nocturnal CO_2 fixation results in the synthesis of malic acid (and of citric acid well as in some CAM plants: Lüttge, 1988), which accumulates in the large central vacuole of the chlorenchyma cells.

At the start of the light period, a transient increase in stomatal conductance and net CO_2 uptake may occur in *phase II*. During this time, both PEP carboxylase and ribulose bisphosphate carboxylase (Rubisco) are active in carboxylation, with the flux of carbon shifting in favour of the latter as this phase progresses (Winter, 1985). Bulk tissue concentrations of malic acid remain constant for a short time after onset of the photoperiod, but start to decline after the rate of net CO_2 uptake has passed its maximum.

As the stomata close, net efflux of malic acid from the vacuole reaches its maximum rate in *phase III*. Malate is decarboxylated in the cytoplasm, either by a malic enzyme or by PEP carboxykinase, depending on the plant concerned (this has important ramifications for the intracellular path of carbon flow in phase III). With the stomata closed, CO_2 concentrations may reach values of over 4% (v/v) in the intercellular spaces (Cockburn et al., 1979; Spalding et al., 1979). The carboxylation activity of Rubisco for refixation of this CO_2 is thereby maximised and its photorespiratory oxygenase activity suppressed.

Under appropriate environmental conditions (e.g. adequate water supply, sufficient light intensity), the stomata may open again towards the end of the light period in *phase IV* when the nocturnally accumulated malic acid has been decarboxylated and the intercellular CO_2 concentration has declined. Fixation of external CO_2 occurs via Rubisco, and the O_2-sensitivity of CO_2 exchange testifies to photorespiratory activity in this phase (Osmond & Björkman, 1975; Winter, 1985; Cote et al., 1989). However, PEP carboxylase is also catalytically active in this phase, as indicated by mass spectrometric data on labelling patterns (Ritz et al., 1986; Osmond et al., 1988), even though vacuolar accumulation of malate does not start until the dark period.

The changing metabolic flow of carbon in these four phases of the CAM rhythm is associated with a complex array of kinetic controls on enzyme activity. This applies both to key enzymes of the carboxylation and decarboxylation pathways, as well as to those mediating metabolite transport between intracellular compartments. Our understanding of these control mechanisms is still incomplete, but they ultimately determine the metabolic characteristics of CAM plants that distinguish them from conventional C3 plants.

Metabolite compartmentation and transport at night

The key enzyme in the dark CO_2 fixation pathway of CAM plants, PEP carboxylase, is a cytosolic enzyme utilising HCO_3^- (derived from dissolved CO_2) as substrate. The oxaloacetate (OAA) formed is immediately reduced by NADH to malate by the enzyme malate dehydrogenase, the equilibrium position of this reaction lying heavily in favour of malate. Conventionally this reaction is considered to occur in the cytosol, but OAA might be readily taken up into mitochondria by the high-affinity uniporter described by Zoglowek *et al.* (1988) and reduced to malate in the mitochondrial matrix (Winter, 1985; Douce & Neuburger, 1990). Malate, at any rate, can exchange freely across the mitochondrial membranes, and the final step of phase I can be regarded as transport of malate from the cytosol into the vacuole (Fig. 2).

Regulation of PEP carboxylase activity

When assayed *in vitro*, PEP carboxylase activities are many times higher than those required to catalyse the observed rates of malate synthesis at night. The enzyme *in vivo*, however, is subject to control by a number of effectors, most notably feedback inhibition by malate. Winter (1982), using rapidly desalted leaf extracts, demonstrated in a series of now classic experiments that the sensitivity of PEP carboxylase activity to effectors changes dramatically during the day–night cycle. At night, the enzyme shows low sensitivity to inhibition by malate and a high affinity for PEP, whereas during the day it is very sensitive to malate and shows lowered affinity for PEP (Winter, 1982). These properties tailor the enzyme for a role in net malate synthesis at night, and also help to prevent futile cycling of the CO_2 released from malate decarboxylation in phase III. Nimmo *et al.* (1984, 1987) and Brulfert *et al.* (1986) have now shown that the mechanism underlying these changes in PEP carboxylase properties is phosphorylation of the enzyme at night, followed by dephosphorylation during the light period (see also Carter *et al.*, 1990). Indeed, the protein kinase activity responsible for phosphorylation of the PEP carboxylase protein is under circadian control, and may explain the endogenous rhythmicity gas exchange known from CAM plants (Carter *et al.*, 1991).

Phase I: carboxylation

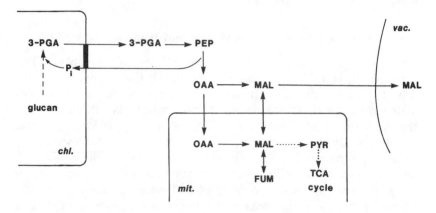

Fig. 2. Fluxes of metabolites between organelles associated with nocturnal CO_2 fixation during phase I of the CAM rhythm. *chl.*, chloroplast; *mit.*, mitochondrion; *vac.*, vacuole; the ground phase is cytosol. Only a number of key reactions central to the glycolytic flux of carbon to the level of malate are illustrated; cofactors and charges on the metabolites have been omitted. 3-PGA, 3-phosphoglycerate; PEP, phosphoenolpyruvate; OAA, oxalacetate; MAL, malate; FUM, fumarate; PYR, pyruvate; TCA cycle, tricarboxylic acid cycle. Exchange of 3-PGA and inorganic orthophosphate (P_i) across the chloroplast envelope occurs via the phosphate translocator, indicated by the rectangular box.

Carbon supply in glycolysis

The three-carbon substrate for malate synthesis, PEP, is derived from the glycolytic breakdown of storage carbohydrate. The nature of this carbohydrate has significant implications both for metabolite compartmentation and for the energetics of malate accumulation in phase I. Malic enzyme (ME) CAM plants accumulate starch (and possibly other low-molecular-weight glucans) in the light period. This storage carbohydrate is oxidised in the chloroplast to the level of 3-phosphoglycerate, which is then exported by the phosphate translocator to the cytosol. Energetically, this phosphorolytic pathway yields 0.5 ATP per malate anion synthesised (Osmond, 1978). However, malic acid accumulation in the vacuole requires 1 ATP per malate transported across the tonoplast (this being driven by an H^+-ATPase that pumps 2 H^+ per ATP hydrolysed, with 1 malate^{2-} ion following electrophoretically to maintain charge balance:

Smith *et al.*, 1984; Lüttge, 1987). The remaining 0.5 ATP per malate is supplied by oxidative phosphorylation, which appears to proceed at rates just sufficient to meet this requirement (Lüttge *et al.*, 1981; Lüttge & Smith, 1988).

In contrast to ME-CAM plants, those species using PEP carboxykinase as the decarboxylating enzyme – such as *Aechmea*, *Ananas*, *Bromelia* (all Bromeliaceae), *Aloë* (Liliaceae) and *Clusia* (Clusiaceae) – store much of their carbohydrate as soluble hexose (Kenyon *et al.*, 1985; Medina *et al.*, 1986; Ball *et al.*, 1991). Because of the additional ATP requirement of the hexokinase reaction, which produces hexose phosphate from free hexose, glycolysis to the level of malic acid results in no net production of ATP in these plants. How, then, is the ATP requirement for malic acid accumulation in the vacuole met? The answer may lie with the high activities of pyrophosphate:fructose 6-phosphate 1-phosphotransferase (PFP) found in these tissues (Fahrendorf *et al.*, 1987; Carnal & Black, 1989). This enzyme would use cytosolic PP_i rather than ATP in the synthesis of fructose 1,6-bisphosphate, thus reducing the total ATP required for malate synthesis. Such a scheme assumes that PFP acts in a glycolytic direction in these tissues and that the activity of the ATP-dependent phosphofructokinase is genuinely insufficient to support glycolysis (Carnal & Black, 1989), two notions that need further study (Kruger, 1990). Also, such species may have high rates of respiration relative to malate accumulation, which would increase the supply of ATP through oxidative phosphorylation (Lüttge & Ball, 1987). On volumetric grounds, the soluble sugars are probably stored mainly in the vacuole (Kenyon *et al.*, 1985), but neither the mechanism nor the energetic cost of sugar transport across the tonoplast has been investigated.

Mitochondrial involvement

Malate accumulation in the vacuole during phase I must closely match the rate of malate synthesis from PEP. However, it is now clear that a substantial fraction of the malate synthesised at night passes first through the mitochondria before being transported across the tonoplast into the vacuole (Fig. 2). Isotopic CO_2 initially labels malate in the C-4 position, but mass spectrometry shows that label randomisation gradually occurs between the C-4 and C-1 positions (Cockburn & McAulay, 1975; Osmond *et al.*, 1988). The only mechanism known to account for this labelling pattern is randomisation after carboxylation by the mitochondrial enzyme fumarase. Mass spectrometry also shows that only singly labelled malate molecules are found, ruling out the possibility of a double carboxylation involving Rubisco followed by PEP carboxylase (Cockburn

& McAulay, 1975). Rubisco appears to be inactivated in phase I, along with other enzymes of the photosynthetic carbon reduction cycle (Winter, 1985).

The factors controlling the extent of label randomisation between malate and fumarate *in vivo* are not yet properly understood. Depending on the species concerned, between 30% and 100% of the total malate pool may be exposed to fumarase in the course of the dark period (Osmond *et al.*, 1988). However, Kalt *et al.* (1990) have shown in an elegant experiment that the extent of label randomisation by fumarase is inversely related to the rate of malate synthesis. In the course of phase I, rates of net CO_2 uptake and malate synthesis tend to be highest in the middle of the night, whereas concurrent rates of label randomisation show the opposite trend (Fig. 3). Kalt *et al.* (1990) thus believe that malate uptake into the mitochondria is at its minimum when the flux of malate into the vacuole is highest. Conversely, when rates of malate synthesis and vacuolar accumulation are low, this may increase the 'residence time' of malate in the cytoplasm and allow greater exchange with the mitochondrial pool. This interpretation is supported by the effect of increasing the night temperature from the optimum of about 15 °C to 25 °C (Fig. 3): this tends to decrease the rate of net CO_2 fixation and vacuolar malate accumulation while increasing the flux of malate through the mitochondria (Ritz *et al.*, 1987; Kalt *et al.*, 1990). Respired CO_2 released from the mitochondria can be assimilated via PEP carboxylase into malate, and in many CAM plants this may account for a substantial proportion of the vacuolar malate accumulated at night, especially at high ambient temperatures (Winter *et al.*, 1986; Lüttge & Ball, 1987; Griffiths, 1988).

Malate uptake into the vacuole

Apart from its role in storage of the carbon assimilated by PEP carboxylase, malate accumulation at night in the vacuole is imperative for cytoplasmic homeostasis. First, synthesis of malate from a neutral carbohydrate precursor is accompanied by the stoichiometric 'production' of 2 H^+ per malate^{2-} anion (Smith & Raven, 1979). These H^+ must be quantitatively removed from the cytoplasm to prevent a catastrophic decrease in cytoplasmic pH: on both volumetric and biochemical grounds, the total buffering capacity of the cytoplasm is limited (Kurkdjian & Guern, 1989). Second, malate accumulation in the cytosol would feedback-inhibit PEP carboxylase activity, even in the enzyme's phosphorylated (night) state. Given the limited volume of the apoplast, which accounts for less than 2% of total cell volume in such tissues (Steudle *et*

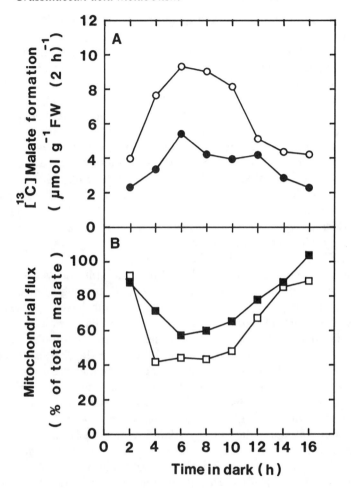

Fig. 3. Rate of [^{13}C] malate synthesis and flux of [^{13}C]malate through the mitochondria during a 16-h dark period in phyllodia of *Kalanchoë tubiflora*. *A*, Replicate plants maintained at either 15 °C (○) or 25 °C (●) were pulse-labelled with $^{13}CO_2$ for 2 h at different times during the dark period. At the end of each labelling period, the concentration of [^{13}C]malate in tissue extracts was determined by nuclear magnetic resonance spectroscopy. *B*, Flux of recently formed [^{13}C]malate through the mitochondria during the dark period at either 15 °C (□) or 25 °C (■), estimated from the extent of ^{13}C-label randomisation between the C-4 and C-1 atoms of [^{13}C]malate synthesised during the same 2-h labelling intervals. Modified with permission from Kalt *et al.* (1990).

al., 1980), the only sink of any quantitative significance for malic acid storage is the vacuole.

Malate is an important vacuolar anion in many types of plants, but it is usually largely charge-balanced by inorganic cations such as K^+ and Ca^{2+} (Osmond, 1976; Smith, 1987). During the CAM cycle, however, malate accumulation is charge-balanced exactly by two titratable H^+ per malate: in effect, free malic acid accumulates in the vacuole (Lüttge *et al.*, 1982). The vacuolar sap is strongly buffered on account of its carboxylate content. Nonetheless, in the leaf-succulent species of *Kalanchoë*, nocturnal accumulation of 150 mol m^{-3} malate (and 300 mol m^{-3} H^+) can cause sap pH to decrease to values of 3.3 at dawn (Lüttge *et al.*, 1981; Lüttge & Smith, 1984). The greatest nocturnal increase in titratable acidity so far observed in a CAM plant is 998 mol H^+ m^{-3} in the strangling fig *Clusia rosea* (Franco *et al.*, 1990). In this genus, sap pH can be as low as 2.85 at dawn, and citric acid typically makes a large contribution to nocturnal acid accumulation (Franco *et al.*, 1990; Ball *et al.*, 1991).

Transport of malic acid (as 2 H^+ + 1 malate^{2-}) into the vacuole across the tonoplast requires energy, since the H^+ are moving against a steep electrochemical potential gradient. Lüttge's group have suggested that this process is energised by the tonoplast H^+-ATPase, which can transport 2 H^+ per ATP hydrolysed, with malate^{2-} anions following electrophoretically to provide charge balance (Lüttge & Ball, 1979; Lüttge *et al.*, 1982). This hypothesis is supported by the finding that the tonoplast H^+-ATPase activity is sufficient to drive such a mechanism (Smith *et al.*, 1984), that malate can dissipate the inside-positive membrane potential generated by the H^+-ATPase (Jochem & Lüttge, 1987; White & Smith, 1989), and that malate uptake by isolated vacuoles is stimulated by ATP (Nishida & Tominaga, 1987). As in other plants, the tonoplast of CAM plants also possesses a second H^+-translocating enzyme, an inorganic pyrophosphatase (H^+-PP$_i$ase), but the relative contribution of these two proton pumps to energisation of the membrane is not yet known (Marquardt & Lüttge, 1987; White *et al.*, 1990; see also Sanders *et al.*, this volume). Bremberger *et al.* (1988) have made the interesting observation that the activity of the tonoplast H^+-ATPase increases strongly on CAM induction in *Mesembryanthemum crystallinum*, whereas that of the H^+-PP$_i$ase declines. This suggests that the H^+-ATPase might play the more important physiological role in energising malic acid accumulation.

The molecular pathway for malate^{2-} transport across the tonoplast has been more difficult to elucidate. Until recently, it was presumed to represent some form of 'carrier' or permease (Buser-Suter *et al.*, 1982; Lüttge, 1987; Nishida & Tominaga, 1987). Biochemically, the malate transport system is known to have a relatively high K_m (in the millimolar rather than micromolar range), to show a high selectivity for four-carbon, *trans* 1,4-dicarboxylates as substrates, and to be sensitive to a number of inhibitors of rather low specificity (Buser-Suter *et al.*, 1982; Jochem & Lüttge, 1987; Nishida & Tominaga, 1987; White & Smith, 1989). Recent experiments with isolated vacuoles of *Kalanchoë daigremontiana* using the patch-clamp technique have shown that malate^{2-} is in fact transported through a voltage-regulated ion channel. Large currents associated with maleate^{2-} movement from the cytosolic side of the membrane into the vacuole are observed when the membrane potential is clamped at inside-positive values (Fig. 4*A*). The whole-vacuole currents also display the same order with respect to carboxylates (i.e. fumarate^{2-} > L-malate^{2-} > maleate^{2-}) as seen in experiments with tonoplast membrane vesicles (White & Smith, 1989). The non-linearity of the current–voltage plot indicates that the current is strongly rectifying, favouring malate influx into the vacuole in symmetrical solutions rather than malate efflux (Fig. 4*A*). Recordings with small patches of isolated membrane reveal the presence of ion channels of relatively low unitary conductance (3.8 pS) but high open probability (Fig. 4*B*). Calculation shows that the activity of this malate-selective ion channel is sufficient to explain the transtonoplast flux of malate^{2-} ions (62 nmol m^{-2} s^{-1} averaged over a 12-h dark period) associated with the maximum rates of malate accumulation observed in *K. daigremontiana* (A. J. Pennington and J. A. C. Smith, unpublished data).

These properties of the tonoplast transport systems suggest a more detailed model for regulation of malic acid accumulation at the membrane level. An almost immediate response to the start of malic acid synthesis in phase I must be a significant acidification of the cytosol. This will stimulate the activity of the tonoplast H$^+$-ATPase (and possibly the H$^+$-PP$_i$ase), causing the transmembrane potential difference to become more inside-positive. This in turn will increase the activity of the malate-selective ion channel. Covalent modification of membrane proteins might also modulate pump and channel activity (Hager & Lanz, 1989; Graham *et al.*, 1989), but the functional importance of these mechanisms has not yet been established for the tonoplast.

The role of the vacuole remains wholly unexplored in CAM plants such as *Ananas* (pineappple), in which much of the carbohydrate reserve is in the form of soluble sugars. The major end product of photosynthesis in these species is likely to be sucrose (cf. Kruger, 1990), which may then be transported from the cytosol into the vacuole by a sucrose-H$^+$ antiport mechanism or by 'facilitated diffusion' (see Rausch, 1991; Preisser & Komor, 1991). Leaf tissue of *Ananas* contains high acid invertase activity (Chen & Black, 1990), presumably localised in the vacuole. This would

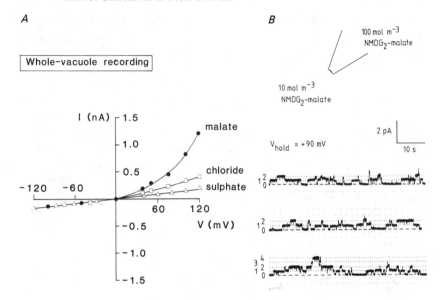

Fig. 4. Malate-selective ion channels in the vacuolar membrane of the CAM plant *Kalanchoë daigremontiana* recorded by the patch-clamp technique (Hamill *et al.*, 1981). *A*, Whole-vacuole currents (I) recorded at different voltages (V, defined with reference to the vacuole interior) in the presence of symmetrical 100 mM concentrations inside and outside the vacuole of three anions (malate, chloride, or sulphate) balanced by the impermeant cation N-methyl-D-glucamine (NMDG). Both solutions contained in addition 5 mM $MgSO_4$, 2 mM EGTA, 90 mM mannitol, and were adjusted to pH 7.5 with NMDG. *B*, Part of a continuous current recording from an isolated patch of vacuolar membrane showing up to four ion channels opening simultaneously. A malate concentration gradient was directed into the vacuole (as indicated by the patch configuration in the inset at top), and the patch held at an inside-vacuole positive potential of +90 mV. Upward deflections reflect anion movement into the vacuole. The unitary current was 0.34 pA, which corresponds to a single-channel conductance of 3.8 pS. Unpublished data of A. J. Pennington and J. A. C. Smith.

hydrolyse the imported sucrose and account for the observation of Carnal & Black (1989) that glucose and fructose accumulate in a 1:1 ratio in the course of the light period (see also Ball *et al.*, 1991). At night, glucose and fructose would be mobilised from the vacuole to provide carbon substrates for the glycolytic production of malic acid. Quantitatively, these transmembrane fluxes of sugar would be considerable. For example, if the entire carbohydrate reserve involved in the CAM cycle were made up of soluble sugar, then on a molar basis the flux of sucrose *into* the vacuole during the day (or the combined flux of glucose and fructose *out of* the vacuole at night) should be at least 25% of that of malate. Even if restricted to a limited subset of CAM plants, these sugar fluxes deserve study in their own right, especially as a model for the kinetic factors controlling net influx and efflux of solutes across the tonoplast.

Metabolite compartmentation and transport during the day

With the beginning of the light period, the direction of carbon flow switches to bring about decarboxylation of the nocturnally accumulated malate. As is often the case in metabolism, this occurs not simply by reversal of all the phase I reactions, but involves a number of enzymes and control steps unique to the light period.

During the initial part of the light period, before the stomata close completely, net CO_2 uptake may increase briefly as a result of fixation by both PEP carboxylase and Rubisco. This transitional period, phase II, has characteristics intermediate between those of phase I and phase III. Winter (1985) has discussed these in detail, and they will not be considered further here. We shall focus on the interactions between organelles in the main period of malate decarboxylation during phase III, and on the implications of CO_2 fixation in phase IV for intracellular metabolite fluxes.

Malate efflux from the vacuole

At the start of the light period, with tissue malic acid concentrations at their maximum, vacuolar malate exists largely as a mixture of the three species malate^{2-}, Hmalate^{1-}, and H$_2$malate^0. (Some of the vacuolar malate is also chelated to calcium and magnesium; although this represents less than 15% of the total malate at dawn: J. A. C. Smith and N. J. A. Cook, unpublished data.) Application of the Nernst equation shows that malate^{2-} may be distributed close to thermodynamic equilibrium across the tonoplast, but large passive driving forces exist for both Hmalate^{1-} and H$_2$malate^0 efflux out of the vacuole into the cytosol (Lüttge & Ball, 1979; Rona *et al.*, 1980). This raises two important questions, neither of which has yet been satisfactorily answered. What is the principal mechanism of malate efflux during phase III, and what prevents malate efflux occurring on a similar scale at night?

Any mechanism for malate efflux must explain the stoichiometric removal from the vacuole of 2 H$^+$ per malate. In principle, malate^{2-}

could efflux through the malate-selective ion channel described above, although the strongly-rectifying characteristics of the ion channel would suggest this is unlikely. When malic acid production ceases and reduces the 'acid load' on the cytosol, this should also reduce the activity of the tonoplast H^+ pump(s); as a result, the membrane potential will become less positive, thereby closing the malate channel (cf. Fig. 4A). Furthermore, no plausible pathway for H^+ efflux has yet been identified, since the tonoplast H^+ pumps are not thought to be freely reversible (Rea & Sanders, 1987).

At very high malic acid concentrations, when pH values in the vacuole are below about 4.0, $H_2malate^0$ is present at substantial concentrations (>10 mol m^{-3}). This species is relatively lipophilic and could thus diffuse passively out of the vacuole by a 'lipid solution' mechanism (Lüttge & Smith, 1984). Because such a leak pathway is essentially uncontrollable, accumulation of $H_2malate^0$ may set an upper limit to the maximum attainable concentrations of vacuolar malate during the CAM rhythm.

For most of the day–night cycle, $Hmalate^{1-}$ is the dominant species of malate in the vacuole. It is likely that a catalysed mechanism for $Hmalate^{1-}$ efflux exists (and conceivably for $malate^{2-}$ efflux), but we know nothing of its properties. An attractive mechanism would be an H^+-linked symport system, such as $H^+/Hmalate^{1-}$ or $2 H^+/malate^{2-}$ cotransport, since this would satisfy the required stoichiometry for malic acid efflux. As yet, however, we lack any experimental data bearing directly on the efflux mechanism, and the existence of such a transport system remains conjectural.

While the thermodynamic arguments indicate that malate efflux from the vacuole can occur passively, any efflux at night during phase I would clearly be energetically wasteful. Kinetic regulation of the efflux process to prevent malate leakage at night might conceivaby be achieved by reversible covalent modification of membrane proteins. A similar case could also be made for inactivation of the malate influx pathway during the day. But again, we do not have even a rudimentary understanding of these possible control processes in CAM plants. This will be an important area for future research.

Malate decarboxylation and carbon flow in gluconeogenesis

As in C4 plants, three different enzymes are known from CAM plants capable of decarboxylating malate: $NADP^+$-malic enzyme, NAD^+-malic enzyme, and PEP carboxykinase. CAM plants do not lend themselves to quite the same subdivision of decarboxylation types known for C4 plants

Phase III : decarboxylation

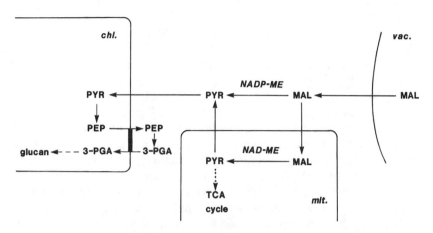

Fig. 5. Fluxes of metabolites between organelles associated with malate decarboxylation during phase III in CAM plants possessing malic enzyme (ME) activities. Stomata are closed during this phase, and the CO_2 released by NADP-ME and/or NAD-ME is reassimilated in the chloroplast by Rubisco via the Calvin cycle (not shown). Conversion of pyruvate to glucan in gluconeogenesis involves the exchange of PEP and 3-PGA across the chloroplast envelope by the phosphate translocator. Abbreviations as in Fig. 2.

(see Edwards & Krall, this volume), since many species contain substantial activities of both NADP[+]- and NAD[+]-malic enzymes. The intracellular location of these enzymes has important implications for carbon flow during phase III.

NADP[+]- and NAD[+]-malic enzymes are found in large numbers of CAM plants in families such as the Cactaceae, Crassulaceae, Mesembryanthemaceae and Orchidaceae (Winter, 1985). The relative activities of the two enzymes vary, but in many species they are probably both operative in oxidising malate to pyruvate (Fig. 5). NADP[+]-malic enzyme is believed to be cytosolic (in contrast to the enzyme from C3 plants, which is located in the chloroplast: El-Shora & ap Rees, 1991), whereas NAD[+]-malic enzyme – as in C3 plants – is confined to the mitochondria (Artus & Edwards, 1985; Winter, 1985; Wedding, 1989). Pyruvate formed in these reactions is then transported into the chloroplast for phosphorylation to PEP by the enzyme pyruvate, P_i dikinase. Thereafter,

PEP is transported back to the cytosol via the chloroplast phosphate translocator, which in CAM plants can exchange PEP for 3-phosphoglycerate at high rates (Neuhaus *et al.*, 1988). In contrast, in PEP carboxykinase species such as members of the Asclepiadaceae, Bromeliaceae and Euphorbiaceae, the sequential action of NAD^+-malate dehydrogenase and PEP carboxykinase in the cytosol results in the formation of PEP directly.

A decrease in cytoplasmic pH, as might occur with the onset of net malic acid efflux from the vacuole at the start of phase III (Marigo *et al.*, 1983; Nungesser *et al.*, 1984), will favour malate oxidation by NAD^+-malic enzyme (Day, 1980). But to conserve in gluconeogenesis the three-carbon residue resulting from malate decarboxylation, any pyruvate formed by NAD^+-malic enzyme in the mitochondria must be preferentially exported to the cytosol rather than oxidised in the tricarboxylic acid cycle. One candidate for this important control point is the pyruvate dehydrogenase complex (PDC), which oxidatively decarboxylates pyruvate in a reaction with acetyl coenzyme A and NAD^+. In C3 plants, PDC is regulated by covalent modification involving protein kinase and phosphatase activities, being inactivated by phosphorylation and activated by dephosphorylation (Randall & Miernyk, 1990). Using leaves of *Kalanchoë daigremontiana*, we have found that PDC activity can be assayed in extracts from isolated mitochondria by following the release of $^{14}CO_2$ from [1-^{14}C]pyruvate. PDC activity declines gradually with time, but is considerably lower in mitochondria prepared from plants in the light period compared with the dark period (Table 1; M. J. Cairns, J. H. Bryce and J. A. C. Smith, unpublished data). Whether the CAM-plant enzyme is regulated by phosphorylation/dephosphorylation in a similar way to the C3 enzyme remains to be investigated.

Pyruvate oxidation through the respiratory chain may occur to a limited extent in phase III (Adams *et al.*, 1986), but PDC activity will also be subject to feedback inhibition *in vivo* by NADH, acetyl coenzyme A and ATP (Winter, 1985; Budde *et al.*, 1991). However, isolated *Kalanchoë* mitochondria are able to export pyruvate even in the face of considerable PDC activity (Bryce *et al.*, 1990). Fumarase randomisation of the malate imported into the mitochondria seems not to occur, since the label distribution between C-1 and C-4 of malate synthesised in the preceding dark period does not change during phase III (Osmond *et al.*, 1988). Futile cycling of the CO_2 released in malate decarboxylation is also avoided because PEP carboxylase at this time is in its dephosphorylated, malate-sensitised form (Nimmo *et al.*, 1987).

Recovery of the pyruvate exported from the mitochondria in gluconeogenesis requires up to 6.7 ATP and 2 NADPH in the chloroplasts per

Table 1. *Activity of the pyruvate dehydrogenase complex (PDC), assayed as $^{14}CO_2$ release from [1−^{14}C]pyruvate, in extracts of isolated mitochondria from leaves of* Kalanchoë daigremontiana

Incubation period (min)	$^{14}CO_2$ released (Bq mg^{-1} protein)	
	Phase I mitochondria	Phase III mitochondria
30	51	16
60	76	27
120	98	34

Mitochondria were prepared by the method of Day (1980) from leaf tissue harvested during the last 2 h of either the dark period (phase I) or light period (phase III). The assay medium contained 80 mol m^{-3} Tes (pH 7.6), 0.5 mol m^{-3} MgCl$_2$, 0.2 mol m^{-3} thiamine pyrophosphate, 2.0 mol m^{-3} NAD$^+$, 0.12 mol m^{-3} lithium coenzyme A, 2.0 mol m^{-3} cysteine, 0.1% (v/v) Triton X-100 and 1.0 mol m^{-3} [1−^{14}C]pyruvate (16.7 GBq mol^{-1}). The reaction was contained in sealed Warburg flasks. Released $^{14}CO_2$ was absorbed by filter paper soaked in 5.0 kmol m^{-3} ethanolamine and counted by liquid scintillation spectrometry. Values are expressed per mg mitochondrial protein and are the means of two experiments. Unpublished data of M.J. Cairns, J.H. Bryce and J.A.C. Smith.

malate converted to glucan (Winter, 1985). In PEP carboxykinase plants the maximum requirement is only 5.7 ATP and 2 NADPH per malate. The exact stoichiometry of this reaction depends on the extent to which substrate oxidation by mitochondria can contribute ATP. Light-scattering measurements suggest that photophosphorylation can meet the requirement for a high ATP/NADPH ratio in the chloroplasts during deacidification (Köster & Winter, 1985). Consistent with the high energy demand of the gluconeogenic pathway, the ATP/ADP ratio in the bulk tissue tends to be lowest during the deacidification phase (Smith *et al.*, 1982).

Metabolite transport in phase IV

When decarboxylation of the nocturnally accumulated malate is complete, the stomata may open for several hours towards the end of the light period in phase IV. Because CO$_2$ fixation occurs through both Rubisco and PEP carboxylase, there will be complex patterns of intracellular

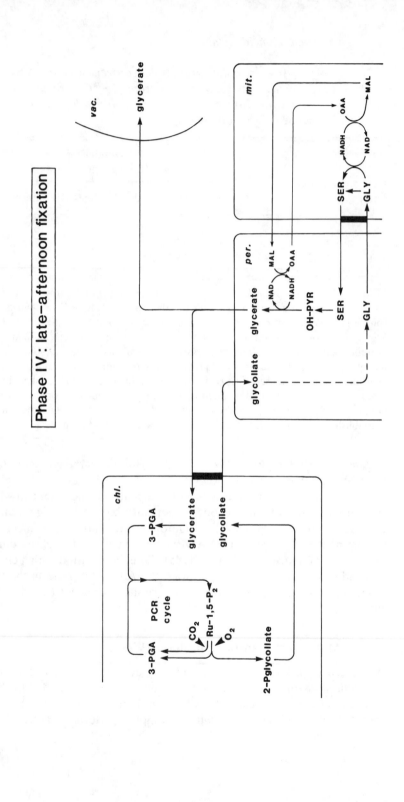

Phase IV : late-afternoon fixation

metabolite transport associated with the photorespiratory carbon oxidation (PCO) cycle (Osmond *et al.*, 1982; Whitehouse *et al.*, 1991). However, net influx of malate into the vacuole is generally not observed until the start of the dark phase (Winter, 1985).

Although PEP carboxylase is catalytically active in phase IV (Osmond *et al.*, 1988), the O_2-sensitivity and relatively high CO_2-compensation point of photosynthesis indicate that significant CO_2 fixation is also occurring via Rubisco (Osmond & Björkman, 1975). The main products of phase IV fixation may be soluble sugars that are preferentially transported in the phloem to growing sink tissues (Winter, 1985). At the same time, the photorespiratory cycle allows 75% of the carbon associated with 2-phosphoglycolate produced in the oxygenase reaction of Rubisco to be recovered as triose phosphate (Osmond *et al.*, 1982; Canvin, 1990; see also Wallsgrove *et al.*, this volume).

The involvement of specific transport systems across organelle membranes in the photorespiratory cycle is illustrated in Fig. 6. Exchange of glycolate and D-glycerate occurs across the chloroplast inner envelope (Howitz & McCarty, 1991), together with exchange of glycine and serine between the peroxisomes and mitochondria (Oliver, 1987). Both these reactions proceed with a stoichiometry of 2:1 (glycolate/glycerate, and glycine/serine), and appear to be associated with H^+ symport (or OH^- antiport) activity for the purposes of charge balance. Metabolite transport is also required to provide reducing equivalents in the peroxisomes for the reduction of hydroxypyruvate to glycerate (Fig. 6). Since the

---

Fig. 6. Metabolite exchange between organelles associated with late-afternoon stomatal opening and CO_2 fixation in phase IV of the CAM cycle. This phase is characteristic of well-watered CAM plants, and both CO_2 and O_2 are fixed directly from the atmosphere by Rubisco in reactions with ribulose 1,5-bisphosphate (Ru-1,5-P_2). As in conventional C3 plants, 3-PGA can be regarded as the net product of the photosynthetic carbon reduction (PCR) cycle: it is then exported from the chloroplast and converted to sucrose in the cytosol (not shown). Operation of the photorespiratory carbon oxidation cycle involves a 2:1 glycolate/glycerate exchange across the chloroplast envelope and a 2:1 glycine (GLY)/serine (SER) exchange across the inner mitochondrial membrane. Both exchange reactions are thought to be mediated by specific translocators, indicated by rectangular boxes. Conversion of glycine to serine in the mitochondrion releases CO_2 and NH_3, the latter being reassimilated into organic form in the chloroplast. The reducing equivalents produced on glycine decarboxylation may be transferred by a redox shuttle to the peroxisome (*per.*) and utilised in reduction of hydroxypyruvate (OH-PYR) to glycerate.

peroxisomes contain an isozyme of malate dehydrogenase, the reducing equivalents are most likely supplied in the form of malate exported from the mitochondria (Ebbighausen *et al.*, 1985; Zoglowek *et al.*, 1988; Canvin, 1990). Whether substrate-specific proteins are required at the peroxisomal membrane to catalyse transport is currently unclear, because these organelles appear to be relatively permeable to a range of metabolites (see Heldt & Flügge, this volume).

Malate exported from the mitochondria in phase IV may equilibrate with the cytosolic pool, although the mitochondria and peroxisomes are usually closely appressed. The same is true for glycerate leaving the peroxisome destined for the chloroplast. We have recently obtained evidence from high-performance liquid chromatography and ^{1}H-nuclear magnetic resonance spectroscopy for the presence of substantial amounts of D-glycerate in leaves of two species of the Crassulaceae native to the British Isles, *Hylotelephium* (=*Sedum*) *telephium* and *Umbilicus rupestris* (J. A. C. Smith and A.-L. McWilliam, unpublished data). Given the ultrastructural characteristics of leaf mesophyll cells in this family, most of this glycerate must be located in the vacuole, suggesting the existence of a glycerate transport system at the tonoplast (Fig. 6). Both these species show relatively weak dark CO_2 fixation under water-stressed conditions, and probably assimilate most of their carbon in temperate environments by daytime fixation through the C3 pathway (Daniel *et al.*, 1985; Groenhof *et al.*, 1986; Lee & Griffiths, 1987). Since the photorespiratory cycle is the only known source of glycerate in photosynthetic cells, this may represent an important pathway for net synthesis of vacuolar anions under photorespiratory conditions.

Conclusions

Control of the huge day-night changes in carbon flow occurring during the CAM cycle is exercised by subcellular compartmentation of key enzymes, regulation of enzyme activity by covalent modification, and transport of metabolites between organelles. Malate storage and remobilisation is dominated by the central vacuole in CAM tissues, but the factors regulating malate influx and efflux across the tonoplast are not yet understood. Further control is exerted at the level of gene expression, since CAM activity changes during development and is influenced by environmental conditions (Winter, 1985). The detailed studies of PEP carboxylase gene expression associated with CAM induction in *Mesembryanthemum crystallinum* provide a superb model for future investigations (Cushman *et al.*, 1989). Extension of this approach to the molecular basis of metabolite compartmentation and transport will provide new insights into the control of the CAM pathway in photosynthetic tissues.

References

Adams, W.W., III, Nishida, K. & Osmond, C.B. (1986). Quantum yields of CAM plants measured by photosynthetic O_2 exchange. *Plant Physiology* **81**, 297–300.

Artus, N.N. & Edwards, G.E. (1985). NAD-malic enzyme from plants. *FEBS Letters* **182**, 225–33.

Ball, E., Hann, J., Kluge, M., Lee, H.S.J., Lüttge, U., Orthen, B., Popp, M., Schmitt, A. & Ting, I. P. (1991). Ecophysiological compartment of the tropical CAM-tree *Clusia* in the field. II. Modes of photosynthesis in trees and seedlings. *New Phytologist* **117**, 483–91.

Bremberger, C., Haschke, H.-P. & Lüttge, U. (1988). Separation and purification of the tonoplast ATPase and pyrophosphatase from plants with constitutive and inducible Crassulacean acid metabolism. *Planta* **175**, 465–70.

Brulfert, J., Vidal, J., Le Marechal, P., Gadal, P., Queiroz, O., Kluge, M. & Krüger, I. (1986) Phosphorylation–dephosphorylation process as a probable mechanism for the diurnal regulatory changes of phosphoenolpyruvate carboxylase in CAM plants. *Biochemical and Biophysical Research Communications* **136**, 151–9.

Bryce, J.H., Nelson, J.P.F. & Smith, J.A.C. (1990). Malate oxidation by mitochondria from the CAM plant *Kalanchoë daigremontiana*. *Journal of Experimental Botany* **41**, P4-4.

Budde, R.J.A., Fang, T.K., Randall, D.D. & Miernyk, J.A. (1991). Acetyl-coenzyme A can regulate activity of the mitochondrial pyruvate dehydrogenase complex *in situ*. *Plant Physiology* **95**, 131–6.

Buser-Suter, C., Wiemken, A. & Matile, P. (1982). A malic acid permease in isolated vacuoles of a Crassulacean Acid Metabolism plant. *Plant Physiology* **69**, 456–9.

Canvin, D.T. (1990). Photorespiration and CO_2-concentrating mechanisms. In *Plant Physiology, Biochemistry and Molecular Biology*, ed. D.T. Dennis & D.H. Turpin, pp. 253–73. Harlow, Essex: Longman Scientific & Technical.

Carnal, N.W. & Black, C.C. (1989). Soluble sugars as the carbohydrate reserve for CAM in pineapple leaves. Implications for the role of pyrophosphate:6-phosphofructokinase in glycolysis. *Plant Physiology* **90**, 91–100.

Carter, P.J., Nimmo, H.G., Fewson, C.A. & Wilkins, M.B. (1990). *Bryophyllum fedtschenkoi* protein phosphatase type 2A can desphosphorylate phosphoenolpyruvate carboxylase. *FEBS Letters* **263**, 233–6.

Carter, P.J., Nimmo, H.G., Fewson, C.A. & Wilkins, M.B. (1991). Circadian rhythms in the activity of a plant protein kinase. *EMBO Journal* **10**, 2063–8.

Chen, J. & Black, C.C. (1990). Purification, characterization and immunochemistry of invertases from potato stems and pineapple leaves. *Plant Physiology* **93**, S59.

Cockburn, W. & McAulay, A. (1975). The pathway of carbon dioxide fixation in crassulacean plants. *Plant Physiology* **55**, 87–9.

Cockburn, W., Ting, I.P. & Sternberg, L.O. (1979). Relationships between stomatal behavior and internal carbon dioxide concentration in Crassulacean acid metabolism plants. *Plant Physiology* **63**, 1029–32.

Cote, F.X., Andre, M., Folliot, M., Massimino, D. & Daguenet, A. (1989). CO_2 and O_2 exchanges in the CAM plant *Ananas comosus* (L.) Merr. *Plant Physiology* **89**, 61–8.

Cushman, J.C., Meyer, G., Michalowski, C.B., Schmitt, J.M. & Bohnert, H.J. (1989). Salt stress leads to differential expression of two isogenes of phosphoenolpyruvate carboxylase during Crassulacean acid metabolism induction in the common ice plant. *The Plant Cell* **1**, 715–25.

Daniel, P.P., Woodward, F.I., Bryant, J.A. & Etherington, J.R. (1985). Nocturnal accumulation of acid in leaves of wall pennywort (*Umbilicus rupestris*) following exposure to water stress. *Annals of Botany* **55**, 217–23.

Day, D.A. (1980). Malate decarboxylation by *Kalanchoë daigremontiana* mitochondria and its role in Crassulacean acid metabolism. *Plant Physiology* **65**, 675–9.

Douce, R. & Neuburger, M. (1990). Metabolite exchange between the mitochondrion and the cytosol. In *Plant Physiology, Biochemistry and Molecular Biology*, ed. D.T. Dennis & D.H. Turpin, pp. 173–90. Harlow, Essex: Longman Scientific & Technical.

Ebbighausen, H., Jia, C. & Heldt, H.W. (1985). Oxalacetate translocator in plant mitochondria. *Biochimica et Biophysica Acta* **810**, 184–99.

El-Shora, H.M. & ap Rees, T. (1991). Intracellular location of $NADP^{+}$-linked malic enzyme in C_3 plants. *Planta* **185**, 362–7.

Fahrendorf, T., Holtum, J.A.M., Mukherjee, U. & Latzko, E. (1987). Fructose 2,6-bisphosphate, carbohydrate partitioning, and crassulacean acid metabolism. *Plant Physiology* **84**, 182–7.

Franco, A.C., Ball, E. & Lüttge, U. (1990). Patterns of gas exchange and organic acid oscillations in tropical trees of the genus *Clusia*. *Oecologia* **85**, 108–14.

Graham, B.M., Blowers, D.P., Trewavas, A.J. & Smith, J.A.C. (1989). Calcium-dependent phosphorylation of tonoplast proteins in the CAM plant *Kalanchoë daigremontiana*. In *Plant Membrane Transport: The Current Position*, ed. J. Dainty, M.I. de Michelis, E. Marré & F. Rasi-Caldogno, pp. 283–4. Amsterdam: Elsevier.

Griffiths, H. (1988). Crassulacean acid metabolism: a re-appraisal of physiological plasticity in form and function. *Advances in Botanical Research* **15**, 43–92.

Groenhof, A.C., Bryant, J.A. & Etherington, J.R. (1986). Photosynthetic changes in the inducible CAM plant *Sedum telephium* L. following the imposition of water stress. 1. General characteristics. *Annals of Botany* **57**, 689–95.

Hager, A. & Lanz, C. (1989). Essential sulfhydryl groups in the catalytic center of the tonoplast H^+-ATPase from coleoptiles of *Zea mays* L. as demonstrated by the biotin-streptavidin-peroxidase system. *Planta* **180**, 116–22.

Hamill, O.P., Marty, A., Neher, E., Sakmann, B. & Sigworth, F.J. (1981). Improved patch-clamp techniques for high-resolution current recording from cells and cell-free membrane patches. *Pflügers Archiv* **391**, 85–100.

Howitz, K.T. & McCarty, R.E. (1991). Solubilization, partial purification, and reconstitution of the glycolate/glycerate transporter from chloroplast inner envelope membranes. *Plant Physiology* **96**, 1060–9.

Jochem, P. & Lüttge, U. (1987). Proton transporting enzymes at the tonoplast of leaf cells of the CAM plant *Kalanchoë daigremontiana*. I. The ATPase. Journal of Plant Physiology **129**, 251–68.

Kalt, W., Osmond, C.B. & Siedow, J.N. (1990). Malate metabolism in the dark after $^{13}CO_2$ fixation in the Crassulacean plant *Kalanchoë tubiflora*. *Plant Physiology* **94**, 826–32.

Kenyon, W.H., Severson, R.F. & Black, C.C. (1985). Maintenance carbon cycle in Crassulacean acid metabolism plant leaves. Source and compartmentation of carbon for nocturnal malate synthesis. *Plant Physiology* **77**, 183–9.

Kluge, M. & Ting, I.P. (1978). *Crassulacean Acid Metabolism. Analysis of an Ecological Adaptation*. Berlin: Springer-Verlag.

Köster, S. & Winter, K. (1985). Light scattering as an indicator of the energy state in leaves of the crassulacean acid metabolism plant *Kalanchoë pinnata*. *Plant Physiology* **79**, 520–4.

Kruger, N.J. (1990). Carbohydrate synthesis and degradation. In *Plant Physiology, Biochemistry and Molecular Biology*, ed. D.T. Dennis & D.H. Turpin, pp. 59–76. Harlow, Essex: Longman Scientific & Technical.

Kurkdjian, A. & Guern, J. (1989). Intracellular pH; measurement and importance in cell activity. *Annual Review of Plant Physiology and Plant Molecular Biology* **40**, 271–303.

Lee, H.S.J. & Griffiths, H. (1987). Induction and repression of CAM in *Sedum telephium* L. in response to photoperiod and water stress. *Journal of Experimental Botany* **38**, 834–41.

Leegood, R.C. & Osmond, C.B. (1990). The flux of metabolites in C4 and CAM plants. In *Plant Physiology, Biochemistry and Molecular*

Biology, ed. D.T. Dennis & D.H. Turpin, pp. 274–98. Harlow, Essex: Longman Scientific & Technical.

Lüttge, U. (1987). Carbon dioxide and water demand: Crassulacean acid metabolism (CAM), a versatile ecological adaptation exemplifying the need for integration in ecophysiological work. *New Phytologist* **106**, 593–629.

Lüttge, U. (1988). Day–night changes of citric-acid levels in crassulacean acid metabolism: phenomenon and ecological significance. *Plant, Cell and Environment* **11**, 445–51.

Lüttge, U. (ed.) (1989). *Vascular Plants as Epiphytes; Evolution and Ecophysiology*. Berlin: Springer-Verlag.

Lüttge, U. & Ball, E. (1979). Electrochemical investigation of active malic acid transport at the tonoplast into the vacuoles of the CAM plant *Kalanchoë daigremontiana*. *Journal of Membrane Biology* **47**, 401–22.

Lüttge, U. & Ball, E. (1987). Dark respiration of CAM plants. *Plant Physiology and Biochemistry* **25**, 3–10.

Lüttge, U. & Smith, J.A.C. (1984). Mechanism of passive malic-acid efflux from vacuoles of the CAM plant *Kalanchoë daigremontiana*. *Journal of Membrane Biology* **81**, 149–58.

Lüttge, U. & Smith, J.A.C. (1988). CAM plants. In *Solute Transport in Plant Cells and Tissues*, ed. D.A. Baker & J.L. Hall, pp. 417–52. Harlow, Essex: Longman Scientific & Technical.

Lüttge, U., Smith, J.A.C. & Marigo, G. (1982). Membrane transport, osmoregulation, and the control of CAM. In *Crassulacean Acid Metabolism*, ed. I.P. Ting & M. Gibbs, pp. 69–91. Rockville, MD: American Society of Plant Physiologists.

Lüttge, U., Smith, J.A.C., Marigo, G. & Osmond, C.B. (1981). Energetics of malate accumulation in the vacuoles of *Kalanchoë tubiflora* cells. *FEBS Letters* **126**, 81–4.

Marigo, G., Lüttge, U. & Smith, J.A.C. (1983). Cytoplasmic pH and the control of crassulacean acid metabolism. *Zeitschrift für Pflanzenphysiologie* **109**, 405–13.

Marquardt, G. & Lüttge, U. (1987). Proton transporting enzymes at the tonoplast of leaf cells of the CAM plant *Kalanchoë daigremontiana*. II. The pyrophosphatase. *Journal of Plant Physiology* **129**, 269–86.

Medina, E., Olivares, E. & Diaz, M. (1986). Water stress and light intensity effects on growth and nocturnal acid accumulation in a terrestrial CAM bromeliad (*Bromelia humilis* Jacq.) under natural conditions. *Oecologia* **70**, 441–6.

Neuhaus, H.E., Holtum, J.A.M. & Latzko, E. (1988). Transport of phosphoenolpyruvate by chloroplasts from *Mesembryanthemum crystallinum* L. exhibiting Crassulacean acid metabolism. *Plant Physiology* **87**, 64–8.

Nimmo, G.A., Nimmo, H.G., Fewson, C.A. & Wilkins, M.B. (1984). Diurnal changes in the properties of phosphoenolpyruvate carboxy-

Crassulacean acid metabolism   165

lase in *Bryophyllum* leaves: a possible covalent modification. *FEBS Letters* **178**, 199–203.

Nimmo, G.A., Wilkins, M.B., Fewson, C.A. & Nimmo, H.G. (1987). Persistent circadian rhythms in the phosphorylation state of phosphoenolpyruvate from *Bryophyllum fedtschenkoi* leaves and its sensitivity to inhibition by malate. *Planta* **170**, 408–15.

Nishida, K. & Tominaga, O. (1987). Energy-dependent uptake of malate into vacuoles isolated from CAM plant, *Kalanchoë daigremontiana*. *Journal of Plant Physiology* **127**, 385–93.

Nobel, P.S. (1988). *Environmental Biology of Agaves and Cacti*. Cambridge: Cambridge University Press.

Nobel, P.S. (1991). Achievable productivities of certain CAM plants: basis for high values compared with C_3 and C_4 plants. *New Phytologist* **119**, 183–205.

Nungesser, D., Kluge, M., Tolle, H. & Oppelt, W. (1984). A dynamic computer model of the metabolic processes in Crassulacean acid metabolism. *Planta* **162**, 204–14.

Oliver, D.J. (1987). Glycine uptake by pea leaf mitochondria: a proposed model for the mechanism of glycine–serine exchange. In *Plant Mitochondria. Structural, Functional and Physiological Aspects*, ed. A.L. Moore & R.B. Beechey, pp. 219–26. New York: Plenum Press.

Osmond, C.B. (1976). Ion absorption and carbon metabolism in cells of higher plants. In *Encyclopedia of Plant Physiology*, New Series, Vol. 2, *Transport in Plants II*, Part A, *Cells*, ed. U. Lüttge & M.G. Pitman, pp. 347–72. Berlin: Springer-Verlag.

Osmond, C.B. (1978). Crassulacean acid metabolism: A curiosity in context. *Annual Review of Plant Physiology* **29**, 379–414.

Osmond, C.B. & Björkman, O. (1975). Pathways of CO_2 fixation in the CAM plant *Kalanchoë daigremontiana*. II. Effects of O_2 and CO_2 concentration on light and dark CO_2 fixation. *Australian Journal of Plant Physiology* **2**, 155–62.

Osmond, C.B. & Holtum, J.A.M. (1981). Crassulacean acid metabolism. In *The Biochemistry of Plants*, Vol. 8, *Photosynthesis*, ed. M. D. Hatch & N.K. Boardman, pp. 283–328. New York: Academic Press.

Osmond, C.B., Holtum, J.A.M., O'Leary, M.H., Roeske, C., Wong, O.C., Summons, R.E. & Avadhani, P.N. (1988). Regulation of malic-acid metabolism in Crassulacean-acid-metabolism plants in the dark and light: In-vivo evidence from ^{13}C-labeling patterns after [13]CO_2 fixation. *Planta* **175**, 184–92.

Osmond, C.B., Winter, K. & Ziegler, H. (1982). Functional significance of different pathways of CO_2 fixation in photosynthesis. In *Encyclopedia of Plant Physiology*, New Series, Vol. 12B, *Physiological Plant Ecology II, Water Relations and Carbon Assimilation*, ed. O.L. Lange, P.S. Nobel, C.B. Osmond & H. Ziegler, pp. 479–547. Berlin: Springer-Verlag.

Preisser, J. & Komor, E. (1991). Sucrose uptake into vacuoles of sugarcane suspension cells. *Planta* **186**, 109–14.

Randall, D.D. & Miernyk, J.A. (1990). The mitochondrial pyruvate dehydrogenase complex. In *Methods in Plant Biochemistry*, Vol. 3, *Enzymes of Primary Metabolism*, ed. P.J. Lea, pp. 175–92. London: Academic Press.

Rausch, T. (1991). The hexose transporters at the plasma membrane and the tonoplast of higher plants. *Physiologia Plantarum* **82**, 134–42.

Raven, J.A., Handley, L.L., MacFarlane, J.J., McInroy, S., McKenzie, L., Richards, J.H. & Samuelsson, G. (1988). The role of CO_2 uptake by roots and CAM in acquisition of inorganic C by plants of the isoetid life-form: a review, with new data on *Eriocaulon decangulare* L. *New Phytologist* **108**, 125–48.

Rea, P.A. & Sanders, D. (1987). Tonoplast energization: Two H^+ pumps, one membrane. *Physiologia Plantarum* **71**, 131–41.

Ritz, D., Kluge, M. & Veith, H.J. (1986). Mass-spectrometric evidence for the double-carboxylation pathway of malate synthesis by Crassulacean acid metabolism plants in the light. *Planta* **167**, 284–91.

Ritz, D., Kluge, M. & Veith, H.J. (1987). Effect of temperature and CO_2-concentration on malate accumulation during CAM: evidence by ^{13}C-mass-spectrometry. *Plant Physiology and Biochemistry* **25**, 391–9.

Rona, J.-P., Pitman, M.G., Lüttge, U. & Ball, E. (1980). Electrochemical data on compartmentation into cell wall, cytoplasm, and vacuole of leaf cells in the genus *Kalanchoë*. *Journal of Membrane Biology* **57**, 25–35.

Smith, F.A. & Raven, J.A. (1979). Intracellular pH and its regulation. *Annual Review of Plant Physiology* **30**, 289–311.

Smith, J.A.C. (1987). Vacuolar accumulation of organic acids and their anions in CAM plants. In *Plant Vacuoles: Their Importance in Solute Compartmentation in Cells and their Applications in Plant Biotechnology*, ed. B. Marin, pp. 79–87. New York: Plenum Press.

Smith, J.A.C. (1989). Epiphytic bromeliads. In *Vascular Plants as Epiphytes: Evolution and Ecophysiology*, ed. U. Lüttge, pp. 109–38. Berlin: Springer-Verlag.

Smith, J.A.C., Marigo, G., Lüttge, U. & Ball, E. (1982). Adenine-nucleotide levels during crassulacean acid metabolism and the energetics of malate accumulation in *Kalanchoë tubiflora*. *Plant Science Letters* **26**, 13–21.

Smith, J.A.C., Uribe, E. G., Ball, E., Heuer, S. & Lüttge, U. (1984). Characterization of the vacuolar ATPase activity of the crassulacean-acid-metabolism plant *Kalanchoë daigremontiana*. *European Journal of Biochemistry* **141**, 415–20.

Spalding, M.H., Stumpf, D.K., Ku, M.S.B., Burris, R.H. & Edwards, G.E. (1979). Crassulacean acid metabolism and diurnal variations of internal CO_2 and O_2 concentrations in *Sedum praealtum* DC. *Australian Journal of Plant Physiology* **6**, 557–67.

Steudle, E., Smith, J.A.C. & Lüttge, U. (1980). Water-relation parameters of individual mesophyll cells of the crassulacean acid metabolism plant *Kalanchoë daigremontiana. Plant Physiology* **66**, 1155–63.

Wedding, R.T. (1989). Malic enzymes of higher plants. Characteristics, regulation, and physiological function. *Plant Physiology* **90**, 367–71.

White, P.J., Marshall, J. & Smith, J.A.C. (1990). Substrate kinetics of the tonoplast H^+-translocating inorganic pyrophosphatase and its activation by free Mg^{2+}. *Plant Physiology* **93**, 1063–70.

White, P.J. & Smith, J.A.C. (1989). Proton and anion transport at the tonoplast in crassulacean-acid-metabolism plants: specificity of the malate-influx system in *Kalanchoë daigremontiana. Planta* **179**, 265–74.

Whitehouse, D.G., Rogers, W.J. & Tobin, A.K. (1991). Photorespiratory enzyme activities in C_3 and CAM forms of the facultative CAM plant, *Mesembryanthemum crystallinum* L. *Journal of Experimental Botany* **42**, 485–92.

Winter, K. (1982). Properties of phosphoenolpyruvate carboxylase in rapidly prepared, desalted leaf extracts of the Crassulacean acid metabolism plant *Mesembryanthemum crystallinum* L. *Planta* **154**, 298–308.

Winter, K. (1985). Crassulacean acid metabolism. In *Photosynthetic Mechanisms and the Environment*, ed. J. Barber & N.R. Baker, pp. 329–87. Amsterdam: Elsevier.

Winter, K., Schröppel-Meier, C. & Caldwell, M.M. (1986). Respiratory CO_2 as carbon source for nocturnal acid synthesis at high temperatures in three species exhibiting crassulacean acid metabolism. *Plant Physiology* **81**, 390–4.

Zoglowek, C., Krömer, S. & Heldt, H.W. (1988). Oxaloacetate and malate transport by plant mitochondria. *Plant Physiology* **87**, 109–15.

DALE SANDERS, JULIA M. DAVIES, PHILIP
A. REA, JAMES M. BROSNAN and EVA
JOHANNES

Transport of H^+, K^+ and Ca^{2+} at the vacuolar membrane of plants

The vacuole is, in terms of volume, the dominant organelle of the vast majority of mature plant cells. Many of its functions have long been recognised (Dainty, 1968; Boller & Wiemken, 1986; Matile, 1987; Raven, 1987). Thus, the vacuole provides a storage compartment for nutrients and metabolites, exhibits lysosomal characteristics as a major lytic compartment, and (by virtue of its size and relative metabolic inertness) enables the cell to achieve a large volume without compromising the high cytoplasmic surface-to-volume ratio that is necessary for efficient gas and nutrient exchange with the environment.

Increasingly, however, additional functions for the vacuole are being discovered. The presence of ion translocating phosphohydrolases at the vacuolar membrane (the tonoplast) implies that the vacuole plays an important role in cellular energetics (Rea & Sanders, 1987). Indeed the H^+-pumping pyrophosphatase at this membrane appears (in photosynthetic tissue at least) to constitute the only means of hydrolysis of cytosolic pyrophosphate (PP_i), which suggests that the organelle must be primarily responsible for the removal of PP_i produced during biosynthetic reactions (Weiner et al., 1987). Furthermore, in accord with its role as a storage organelle, the vacuole seems likely to constitute a major source of second messengers which can be released into the cytoplasm during signal transduction (Boudet & Ranjeva, 1989). Changes in the level of cytosolic free Ca^{2+}, for example, are becoming increasingly recognised as playing a crucial role in stimulus–response coupling in plant cells (Poovaiah & Reddy, 1987; Johannes et al., 1991), and recent Ca^{2+} imaging data are consistent with vacuolar Ca^{2+} release providing a significant component of elevated cytosolic Ca^{2+} (Gilroy et al., 1991).

How, then, are ions and solutes compartmented within the vacuole? A number of technical advances over the past decade have enabled detailed studies of tonoplast transport processes. Transport-competent membrane vesicles can be employed to characterise the basic mechanisms of transport (Sze, 1985), and the more recent application of the patch clamp

Society for Experimental Biology Seminar Series 50: *Plant organelles*, ed. A. K. Tobin.
© Cambridge University Press 1992, pp. 169–88.

technique (Hamill *et al.*, 1981) has both confirmed the existence of electrogenic pumps (Hedrich *et al.*, 1989) and succeeded in identifying novel channels which catalyse the passive transport of ions across the tonoplast (Hedrich & Neher, 1987; Hedrich & Kurkdjian, 1988).

The importance of maintaining a permeability barrier at the tonoplast is clearly exemplified by consideration of the compartmentation of two key ions: H^+ and Ca^{2+}. Both are sequestered in the vacuole at activities which would result in immediate cell death were the permeability barrier to fail. Typically, the vacuolar H^+ activity is in excess of 3 μM (i.e. pH <5.5), compared with a cytosolic activity 100-fold lower (pH 7.5) (Bertl *et al.*, 1984). Compartmentation of Ca^{2+} is even more impressive, since vacuolar free Ca^{2+} is maintained at millimolar levels with cytosolic free Ca^{2+} resting at a value four orders of magnitude lower (Miller & Sanders, 1987). It is equally clear that tonoplast fluxes of H^+ and Ca^{2+} must be under acute control: the ions are not merely locked inside the vacuole once they have been accumulated because there exist defined and selective pathways for their release.

Figure 1 summarises our current knowledge of the transport pathways for uptake and release of H^+ and Ca^{2+}. The primary mechanism for H^+ accumulation in most cells is via a H^+-ATPase. With respect to its substrate and inhibitor specificities, as well as to its reaction mechanism and polypeptide composition, this enzyme is typical of the V-type H^+-ATPases that are widely distributed among the endomembranes of eukaryotic cells (Rea & Poole, 1992). Thus, the vacuolar enzyme

1. displays a selectivity for ATP over GTP which is weak in comparison with the plant plasma membrane H^+-ATPase;
2. is potently and selectively inhibited by the antibiotic Bafilomycin A_1;
3. does not form a covalent enzyme–phosphate intermediate as part of its hydrolytic mechanism (again in contrast to the H^+-ATPase of plasma membrane);
4. comprises at least six subunit types of 100, 70, 60, 44, 33, and 16 kDa, with the largest and smallest intimately associated with the membrane and the other subunits more peripherally bound (Parry *et al.*, 1989).

The inorganic pyrophosphatase (PPase) has also been demonstrated to pump H^+ into purified tonoplast vesicles where, like the H^+-ATPase, it sets up an inside-positive trans-tonoplast electrical potential (Wang *et al.*, 1986) as a result of its current-passing capabilities (Hedrich *et al.*, 1989). The enzyme has been purified to homogeneity and appears to comprise a single polypeptide type, of molecular mass 64 kDa (Britten *et al.*, 1989).

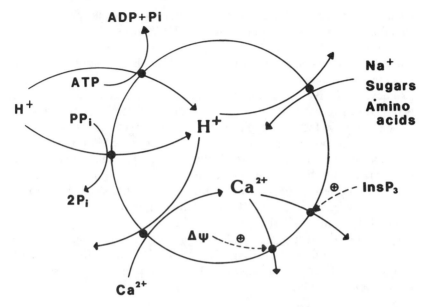

Fig. 1. Transport of H^+ and Ca^{2+} into and out of the vacuole. Transport systems on the left energise accumulation of the ions; the systems are, respectively, for H^+, an ATPase and a PPase, and for Ca^{2+}, a H^+/Ca^{2+} exchanger powered by the protonmotive force. Transport systems on the right hand side show the mechanisms of vacuolar release of H^+ and Ca^{2+}. The H^+ gradient is likely to be dissipated primarily by exchange systems for ions and solutes analogous to the H^+/Ca^{2+} exchanger. For Ca^{2+}, an inositol (1,4,5)-trisphosphate-gated channel is one established pathway for exit. A novel voltage-dependent Ca^{2+} channel is also described later in this chapter.

Passive, return flow of H^+ out of the vacuole is used to drive uptake of a number of solutes via so-called antiport systems. These solutes include ions such as Na^+ (Blumwald & Poole, 1985) as well as amino acids (Homeyer & Schultz, 1988; Homeyer *et al.*, 1989) and possibly sugars (Briskin *et al.*, 1985). Calcium also enters the vacuole in exchange for H^+ (Schumaker & Sze, 1985). The large transmembrane Ca^{2+} gradient can be generated from the smaller H^+ gradient because the stoichiometry of exchange is at least $3\,H^+:Ca^{2+}$ (Blackford *et al.*, 1990).

The sole pathway for vacuolar Ca^{2+} release which has been documented to date is via a ligand-gated channel. This channel is specifically opened by inositol (1,4,5)-trisphosphate ($InsP_3$) (Schumaker & Sze, 1987) which, by analogy with animal cells (Berridge & Irvine, 1989), is envisaged to be released by receptor-mediated hydrolysis of the plasma membrane trace lipid phosphatidylinositol (4,5)-bisphosphate during stimulus–response coupling in plants (Rincón & Boss, 1990). An intriguing feature of the plant $InsP_3$-gated Ca^{2+} channel concerns its intracellular location. Thus, while both biochemical and biophysical studies on intact vacuoles have unambiguously assigned this channel to the tonoplast (Ranjeva *et al.*, 1988; Alexandre *et al.*, 1990), its mammalian counterpart is thought to reside in the endoplasmic reticulum (Streb *et al.*, 1984).

This chapter focuses specifically on several key issues relating to tonoplast transport of H^+ and Ca^{2+}. In the first section, we examine the question of just why two independently powered H^+ pumps reside at the same membrane by examining the mechanism and control of the H^+-PPase. In the second section, we address two questions pertaining to the release of vacuolar Ca^{2+}: (i) do the plant and animal $InsP_3$-gated Ca^{2+} channels bear any similarities in view of their disparate cellular locations? and (ii) does the $InsP_3$-gated pathway constitute the sole mechanism for vacuolar Ca^{2+} release? All experiments have been performed on the storage root of beet (*Beta vulgaris* L.), which represents a rich source of the tonoplast membrane.

Role of the H^+-PPase

Physiological poise

Some years ago, we considered the possibility that the H^+-PPase might function *in vivo* as a PP_i synthase by catalysing an *outward* flow of H^+ from the vacuole (Rea & Sanders, 1987). The physiological poise of the enzyme was shown to depend critically on the stoichiometric ratio of H^+ translocated per PP_i hydrolysed, with $H^+:PP_i$ ratios >1 signifying synthetic activity under the influence of the protonmotive force generated by the H^+-ATPase. More recent work with intact vacuoles has demonstrated, however, that the H^+-PPase is capable of augmenting the ATP-generated protonmotive force in the presence of substrate and product concentrations that are likely to reflect *in vivo* conditions (Johannes & Felle, 1990).

With the H^+-PPase firmly assigned the physiological function of a PP_i hydrolase, it is pertinent to ask what role this auxiliary H^+ pump plays at the tonoplast. One simple answer is that the enzyme salvages free energy (Rea & Sanders, 1987). Thus, while a soluble PPase would merely thermally dissipate the free energy available from hydrolysis of the phosphoanhydride bond (which in cellular conditions amounts to around 15 kJ mol^{-1}), a biologically useful output is achieved if some of this free energy is conserved as a protonmotive force to be utilised in subsequent solute translocation.

K^+ energetics at the tonoplast

A supplementary role for the H^+-PPase is, however, also suggested by observations that the enzyme is strongly activated by potassium ions. Thus, PP_i-dependent H^+ pumping in tonoplast vesicles is abolished in the absence of K^+, and hydrolytic activity is also strongly suppressed in similar conditions (Rea & Poole, 1985; Wang *et al.*, 1986). Is it possible, then, that the H^+-PPase translocates K^+ as well as H^+? In order to evaluate the requirement for an energised mechanism for K^+ transport into the vacuole, it is first necessary to digress briefly on the nature of K^+ gradients across the tonoplast.

In nutrient-replete conditions, K^+ is accumulated against its electrochemical gradient into vacuoles. The overall electrochemical potential difference for K^+ across the tonoplast ($\Delta\bar{\mu}_{K^+}$) can be calculated from the Nernst equation as

$$\Delta\bar{\mu}_{K^+} = F\Delta\psi + RT\ln([K^+]_v/[K^+]_c), \tag{1}$$

where $\Delta\psi$ is the trans-tonoplast electrical potential (vacuole with respect to cytoplasm), the square brackets signify ionic activities, the subscripts v and c represent the vacuolar and cytoplasmic compartments, respectively, and F, R and T have their usual meanings. The value of $\Delta\psi$, generated by the activity of the two phosphohydrolases, resides between +20 and +50 mV (Spanswick & Williams, 1964; Bates *et al.*, 1982), while $[K^+]_c$ is about 100 mM and $[K^+]_v$ can reach 200 mM (Leigh & Wyn Jones, 1984). Thus the overall electrochemical gradient against which K^+ must be accumulated is +3 and +7 kJ mol^{-1}.

Experimental analysis of K^+ effects on the H^+-PPase

It is not possible to assay tonoplast vesicles for PP_i-dependent K^+ transport because of their high K^+ permeability. Intact vacuoles, on the other hand, offer a less K^+-permeable system in which the properties of the H^+-PPase can be explored with electrophysiological ('whole-vacuole' patch clamp) techniques: PP_i-driven cation transport into the vacuole is detected as an inward current. This approach offers the further advantage that the orientation of the membrane is established and enables simple analysis of the variation in pump current ($\equiv$ enzyme activity) as a function of one of the driving forces – the membrane potential.

If the H^+-PPase is responsible for pumping K^+ into the vacuole, then two predictions emerge which can be tested. First, the site at which K^+ stimulates hydrolytic activity should be located on the cytosolic side of the enzyme (i.e. the side from which K^+ is pumped). Second, variation in the trans-tonoplast K^+ concentration difference should influence the thermodynamic properties of the enzyme in a defined manner.

The vectorial characteristics of the H^+-PPase with respect to K^+ are clearly consistent with K^+ translocation into the vacuole. Thus, PP_i-dependent membrane currents of the order 0.3 μA cm^{-2} are apparent in patch-clamped intact vacuoles in control conditions with 50 mM K^+ in both the bathing medium (cytoplasmic side) and the recording pipette (Davies *et al.*, 1991). (Exchange of the medium in the patch pipette with that in the vacuole is apparent during the first minute or two after electrical access is gained to the interior of the vacuole either from the disappearance of pigment (red beet) or from a marked decrease in refractive index (sugar beet).) In the absence of K^+ in the bathing medium, no PP_i-dependent current is detected, regardless of the K^+ status within the vacuole. After supplying K^+ to the medium surrounding these same vacuoles, the PP_i-dependent current is restored.

However, this kind of kinetic effect of K^+ cannot *a priori* constitute unambiguous evidence for K^+ transport by the H^+-PPase, since stimulation by cytoplasmic K^+ could equally well result from binding to a modifier (allosteric) site as to a bona fide transport site. A more refined investigation results from a consideration of the thermodynamic properties of the enzyme. As a prelude to a description of the experimental strategy behind such thermodynamic investigations, we briefly consider the principles which enable information on energetics to be derived from current–voltage analysis.

Figure 2 draws on general theoretical considerations concerning the properties of ion pumps (Hansen *et al.*, 1981) to illustrate how the current-carrying behaviour of the PPase might be expected to vary with membrane voltage. The region of negative (inward) current (below the voltage axis) represents hydrolytic pumping of positive charge into the vacuole. As the opposing inside-positive voltage becomes more negative, the transport current reaches a saturation point where activity of the enzyme becomes rate-limited by factors such as ligand binding. Above the voltage axis, the transport system behaves as a PP_i synthase, since the more positive voltage forces the enzyme to reverse direction and pass outward current. For the purposes of the present discussion, the point at which the so-called current–voltage relationship crosses the voltage axis is crucial. This voltage is known as the reversal potential (E_{rev}) for the system: it is the point at which the transport system is poised at equilibrium because the electrical driving force on it is exactly offset by

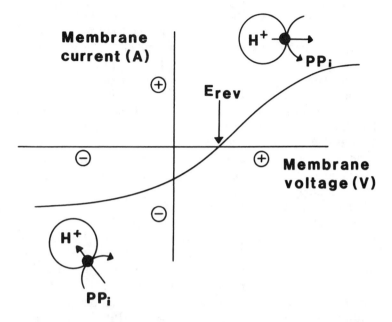

Fig. 2. Idealised representation of a current–voltage relationship for the H$^+$-PPase. The net directions of transport (negative or positive of the voltage axis) are shown, together with the reversal potential (E_{rev}) for the transport system. For further description, see text.

the other driving forces (which comprise PP$_i$ hydrolysis and transmembrane ion gradients).

For the hypothetical condition in which the H$^+$-PPase translocates nH$^+$ and mK$^+$ into the vacuole per PP$_i$ hydrolysed, we can write the overall free energy relationship for the enzyme as

$$\Delta G' = (n+m)F\Delta\psi + RT\ln\left(\frac{[P_i]^2[H^+]_v^n[K^+]_v^m}{K_{PP_i}[PP_i][H^+]_c^n[K^+]_c^m}\right) \qquad (2)$$

where K_{PP_i} is the equilibrium constant for PP$_i$ hydrolysis. For the special case in which the whole catalytic reaction is at equilibrium, $\Delta G' = 0$ and $\Delta\psi = E_{rev}$. Equation 2 can then be rearranged to yield

$$E_{rev} = \{RT/(n+m)F\}\ln\left(\frac{K_{PP_i}[PP_i][H^+]_c^n[K^+]_c^m}{[P_i]^2[H^+]_v^n[K^+]_v^m}\right) \qquad (3)$$

Equation 3 now makes specific predictions about the behaviour of E_{rev} as a function of the ionic gradients. In the simplest case, in which K^+ is not transported by the enzyme ($m = 0$) and only H^+ is translocated, Equation 3 predicts that E_{rev} will be insensitive to $[K^+]$ on either side of the membrane. (This does not, of course, preclude a *kinetic* effect of K^+ in the regions of the current–voltage relationship in which a defined current is observed.) Considering the case in which a single H^+ is translocated with a single K^+, the predicted change in E_{rev} is +30 mV for a 10-fold decrease in the K^+ concentration gradient. This conclusion is valid regardless of the actual reaction mechanism of the enzyme, and is independent of how the change in K^+ concentration gradient is imposed experimentally: a 10-fold decrease in $[K^+]_v$ *or* a 10-fold increase in $[K^+]_c$ will have exactly the same effect on E_{rev}.

Our strategy in testing these predictions has been as follows. First, the value of E_{rev} has to be placed within the range of membrane potentials experimentally accessible with the patch clamp technique. In practice, this means a value around 50 mV either side of zero, which can be set by judicious consideration of the prevailing pH gradient as well as of the PP_i and P_i concentrations. The appropriate conditions turn out to be a cytosolic pH of 8.0, intravacuolar pH = 6.0, $[PP_i] = 100\ \mu$M and $[P_i] = 10$ mM. Next, the current–voltage relationship of the tonoplast is determined in the absence of enzymatic activity, i.e. without PP_i and P_i in the bathing medium. The difference between this current–voltage relationship and that determined in the presence of PP_i and P_i then yields the current–voltage relationship of the electrogenic pump. A second current–voltage difference relationship is then measured in the presence of a new $[K^+]$ in the bathing medium.

The results show shifts in E_{rev} which are in close accord with the notion that the enzyme translocates K^+ (J. M. Davies *et al.*, unpublished data). For $[K^+]_v = 100$ mM and $[K^+]_c = 30$ mM, $E_{rev} = -18 \pm 1$ mV ($n = 6$), and on shifting $[K^+]_c$ to 100 mM, $E_{rev} = +2 \pm 1$ mV. The mean K^+-induced change in E_{rev} is +20 mV, which compares with a value of +16 mV predicted for this change in $[K^+]_c$ from Equation 3 if $n = m = 1$. A similar story emerges with $[K^+]_v$ fixed at 30 mM, rather than 100 mM: there the increase in $[K^+]_c$ from 30 to 100 mM results in a mean change of E_{rev} from -1 ± 1 mV to $+15 \pm 1$ mV ($n = 5$), a shift of +16 mV.

The results of both the kinetic and thermodynamic analyses of PPase activity therefore suggest that the enzyme pumps one H^+ and one K^+ into the vacuole at the expense of each PP_i molecule hydrolysed. This finding therefore not only provides a potential *raison d'être* for the presence of two primary H^+-pumping phosphohydrolases at the tonoplast; it also suggests that the principal mechanism for K^+ accumulation in plant vacuoles is the hydrolysis of PP_i.

Control of PPase activity by Ca^{2+}

The dominant role of the metabolically coupled pumps in ion transloca- tion suggests that they should be subject to homeostatic control, yet hitherto the regulatory mechanisms acting on the transport systems at the tonoplast have remained obscure. However, the effect of the Ca^{2+} chel- ator ethyleneglycol-bis-(animoethylether)-N, N, N', N'-tetraacetic acid (EGTA) provides a preliminary indication that cytosolic free Ca^{2+} might play a role in selective inhibition of tonoplast PPase activity. Thus, we have observed that the inclusion of EGTA in the assay medium results in a small (30%) but significant stimulation of PPase activity in membrane preparations, suggesting that low levels of Ca^{2+} contaminating the assay media might inhibit the enzyme. By contrast, neither EGTA nor Ca^{2+} at 100 μM has an effect on the activity of the tonoplast H^+-ATPase.

The effects of Ca^{2+} on PPase activity were investigated in more detail (P. A. Rea *et al.*, unpublished data), and some results are shown in Fig. 3. All reaction media were strongly buffered with 5 mM EGTA, and the activities of free Ca^{2+} and Mg^{2+} were set at defined values calculated from their respective dissociation constants with PP_i and EGTA. Two features of the results are noteworthy. First, regardless of the free $[Mg^{2+}]$, signifi- cant inhibition of PPase activity is observed in the sub-micromolar range of $[Ca^{2+}]$. Indeed, the $K_{0.5}$ for free $[Ca^{2+}]$ (which we define as the activity required to elicit half the maximal inhibition) lies between 2 and 4 μM, and is independent of the prevailing free $[Mg^{2+}]$. Secondly, and by con- trast, the maximal inhibition (I_{max}) itself falls significantly as free $[Mg^{2+}]$ is raised. Thus, the least squares estimates of I_{max} for the ascending series of free $[Mg^{2+}]$ at 0.25, 0.5, 1.5 and 3.0 mM are, respectively, 94, 87, 66 and 55%. Ca^{2+}-induced inhibition of the PPase is therefore non-competitive with Mg^{2+}, which implies the existence of a discrete inhibitory site for Ca^{2+} binding to the enzyme.

Measurements of cytosolic free Ca^{2+} in the Charophyte alga *Nitellopsis* have demonstrated a marked decline in the dark resting level during the onset of photosynthesis from 350 to 50 nM (Miller & Sanders, 1987). Is it possible that this change in free Ca^{2+} is responsible for activating the PPase in the light, with a concomitant stimulation of ion transport and PP_i turnover? On the basis of the present data, the $K_{0.5}$ for Ca^{2+} appears to be somewhat too high to yield a major change in PPase activity for Ca^{2+} changes over this range: at 0.5 mM Mg^{2+}, for example, the enzyme would

178   D. SANDERS *ET AL.*

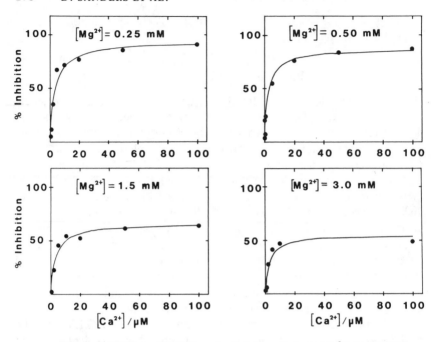

Fig. 3. Inhibition of the PPase hydrolytic activity by Ca^{2+}. Total [PPi] = 250 μM. Absolute hydrolytic rates ($\equiv$ 0% inhibition) were 5.1, 7.1, 7.7 and 6.7 μmol PPi mg^{-1} h^{-1} at 0.25, 0.5, 1.5 and 3.0 mM free Mg^{2+}, respectively. The solid lines are fits of the data to rectangular hyperbolic functions, with half saturation constants = 3.4 $\pm$ 0.6, 2.54 $\pm$ 0.3, 3.2 $\pm$ 0.7 and 2.7 $\pm$ 0.9 μM Ca^{2+} for the ascending series of [Mg^{2+}], with maximal inhibition values of 96, 87, 66 and 55%, respectively.

be expected to retain 90% of its full activity even in darkness. Nevertheless, elevation of cytosolic free Ca^{2+} to supra-micromolar levels, such as occurs during some types of signal transduction (e.g. stomatal closure: Gilroy *et al.*, 1991), would result in a significant reduction of PPase activity. At 2 μM Ca^{2+} and 0.5 mM Mg^{2+}, catalytic activity would amount to only one-third of the uninhibited rate. It seems likely, then, that if these results from tonoplast of beet storage root can be extrapolated to other tissues the inhibitory effects of Ca^{2+} on PPase activity will emerge to have physiological significance. In the particular case of stomatal guard cells, the function of the rise in cytosolic free Ca^{2+} which precedes closure (McAinsh *et al.*, 1990) might be viewed, at least in part, as inhibition of PP_i-driven K^+ translocation into the vacuole, since closure is accompanied by a massive loss of cellular K^+ (MacRobbie & Lettau, 1980).

Calcium channels at the tonoplast

Despite the importance of the vacuole as an intracellular store for Ca^{2+}, the molecular properties of the Ca^{2+} release pathways across the tonoplast remain relatively poorly characterised. We turn now to a consideration of the biochemical and biophysical aspects of these pathways.

The $InsP_3$-gated Ca^{2+} channel

Microsomal vesicles from beet exhibit Ca^{2+} transport properties which appear to be exclusively those of tonoplast (Brosnan & Sanders, 1990). Thus, ATP induces Ca^{2+} loading of the vesicles; this accumulation process is insensitive to vanadate (an inhibitor of Ca^{2+}-ATPases at the plasma membrane and endoplasmic reticulum) but is completely abolished by protonophores such as carbonylcyanide p-trifluoromethoxyphenylhydrazone (FCCP) or V-type H^+-ATPase inhibitors such as nitrate. Hence, Ca^{2+} accumulation proceeds by H^+/Ca^{2+} exchange, with the protonmotive force generated by the tonoplast H^+-ATPase.

On application of $InsP_3$, Ca^{2+} is released from preloaded vesicles (Brosnan & Sanders, 1990). Release does not exceed 20% of the total accumulated Ca^{2+}, and this observation might be taken to imply some heterogeneity in membrane origin. However, an alternative explanation appears more reasonable in the context of the uniform loading properties described above. Marked disparities in the turnover rates of channels (high turnover) and carriers (low turnover) result in a low abundance of channels in most membranes. In the particular case of beet vacuole, the density of the $InsP_3$-gated Ca^{2+} channel has been estimated from a comparison of macroscopic (whole membrane) $InsP_3$-gated Ca^{2+} currents and microscopic (single channel) currents as only 1 channel molecule per 5 μm^2 membrane (Alexandre *et al.*, 1990). Taking the surface area of a typical vesicle as 0.6 μm^2, only 12% of the vesicles could be expected to possess an $InsP_3$-gated channel, even though the whole population will possess the capacity to accumulate Ca^{2+}. Thus, considering the population as a whole, limited $InsP_3$-induced Ca^{2+} release is actually anticipated (Brosnan, 1990).

With respect to its Ca^{2+} transport properties, the $InsP_3$-gated Ca^{2+} channel of plant tonoplast appears remarkably similar to its mammalian counterpart at the endoplasmic reticulum (Table 1). Both channel types exhibit a high specificity for $InsP_3$ over other inositol phosphates, the channels operate with a $K_{0.5}$ for $InsP_3$ below 1 μM and they are potently inhibited by low molecular weight heparin and by 8-(N,N-diethylamino)-octyl 3,4,5-trimethoxybenzoate (TMB-8).

Table 1. *Comparison of InsP$_3$-gated Ca^{2+} release from plant and animal membranes*

	Plants	Animals
Phosphoinositide specificity	Specific for InsP$_3$[a, b]	Specific for InsP$_3$[c]
$K_{0.5}$ (InsP$_3$) (nM)	540[b]–600[a]	240[d]–310[e]
Inhibition by TMB-8	Yes[a]	Yes[f]
$K_{0.5}$ (heparin) (nM)	86[b]	40[e]–160[g]

Data for plants are from tonoplast enriched membranes, those for animals are from ER. Superscripts refer to references.
References: [a]Schumaker & Sze, 1987; [b]Brosnan & Sanders, 1990; [c]Prentki *et al.*, 1984; [d]Guillemette *et al.*, 1987; [e]Ghosh *et al.*, 1988; [f]Clapper & Lee, 1985; [g]Chopra *et al.*, 1989.

We have also studied the ligand-binding properties of the InsP$_3$ receptor from beet tonoplast (J. M. Brosnan and D. Sanders, unpublished data). Specific InsP$_3$ binding to tonoplast membranes cannot be detected because associated phosphatase activity results in rapid hydrolysis of the ligand. However, solubilisation of the receptor in 1% Triton X-100 enables equilibrium binding analysis in non-hydrolytic conditions. The results of these studies have furthered the parallels between the plant and animal receptors.

Receptor affinity has been quantified by displacement of [^{3}H]-InsP$_3$ from the solubilised receptor. InsP$_3$ is alone among the inositol phosphates tested in exhibiting a K_d <1 μM, with inositol (1,3,4,5)-tetrakisphosphate, inositol hexakisphosphate and inositol (1,4)-bisphosphate exhibiting progressively lower binding affinities. Heparin, however, competes effectively with InsP$_3$ for binding, and has a K_d of 295 nM.

Scatchard plots of InsP$_3$ binding are linear, indicating a single binding site. The K_d for InsP$_3$ binding (56 nM) is one order of magnitude lower than the $K_{0.5}$ for InsP$_3$-induced Ca^{2+} release from vesicles. A significant part of this discrepancy can be accounted for by the presence of ATP in the transport experiments. Thus, although ATP competes only weakly (K_d = 980 μM), the concentration of ATP used to load Ca^{2+} into the vesicles is sufficiently high (3 mM) to raise the calculated $K_{0.5}$ for Ca^{2+} release to 0.28 μM, which is within a factor of two of the observed value (Table 1).

Sensitivity to sulphydryl reagents is a further property shared between the plant and animal InsP$_3$ receptors. Thus, InsP$_3$ binding to the animal receptor is inhibited by both *N*-ethylmaleimide (NEM) (Guillemette &

Segui, 1988) and *p*-chloromercuribenzenesulphonic acid (PCMBS) (Supattapone *et al.*, 1988), and we have observed that preincubation with 50 μM PCMBS for 10 min at 4 °C results in a 70% inhibition of subsequent InsP$_3$-specific binding to the beet receptor (J. M. Brosnan and D. Sanders, unpublished data). Furthermore, binding capacity of the beet receptor is largely protected by the presence of InsP$_3$ in the preincubation medium, suggesting that a critical cysteine residue is exposed only when the receptor is not bound to the ligand.

In summary, then, the plant and animal InsP$_3$ receptors are closely allied in all their properties characterised to date. These similarities are apparent despite the differential intracellular location of the two receptors. Since the InsP$_3$-gated Ca^{2+} channels of the ER play a role in intracellular Ca^{2+} release during signal transduction in animal cells, it seems reasonable to conclude that the vacuole fulfils a similar function in plants, at least in highly vacuolated tissue such as storage root.

A novel voltage-gated Ca^{2+} channel at the tonoplast

Despite circumstantial evidence for a role for cytosolic free Ca^{2+} in mediating many types of stimulus–response coupling in plant cells, concerted efforts to detect changes in InsP$_3$ level have met with failure (Irvine, 1990). While such findings might be ascribed to the inherent difficulties of trace metabolite detection in phosphatase-rich tissue, they might also indicate that intracellular Ca^{2+} release occurs via some other, as yet undetected, pathway.

The first indications that such a pathway might exist were obtained from experiments on membrane vesicles. An inside-positive membrane potential can be imposed on vesicles with a lipophilic cation such as triphenylmethylphosphonium (TPMP$^+$), which diffuses rapidly and passively across the bilayer. Ca^{2+}-loaded vesicles from beet will typically release 25% of their free Ca^{2+} in response to addition of 10 mM TPMP$^+$. This voltage-sensitive release is apparent even in the presence of a saturating concentration (10 μM) of InsP$_3$, suggesting that the voltage sensitive release pathway is InsP$_3$-independent (Johannes *et al.*, 1992).

Confirmation of the existence of discrete Ca^{2+} release pathways can be obtained with inhibitors. Thus, in contrast to the InsP$_3$-gated Ca^{2+} channel (Table 1), the voltage-sensitive release mechanism is not sensitive to heparin. Furthermore, the voltage-sensitive pathway is effectively inhibited by Zn^{2+}, to which the InsP$_3$-gated Ca^{2+} channel is insensitive, and is potently inhibited by the lanthanide Gd^{3+}, with a $K_{0.5}$ of 20 μM.

All biological channels exhibit gating: they switch between discrete closed and open states. While bulk-phase measurements of ion release through channels, as exemplified by vesicle assays, provide a convenient experimental system for rapid screening of putative channel activity, there are many essential questions concerning channel behaviour that can be answered only with more refined techniques. In the case of a voltage-sensitive channel, for example, one especially critical point concerns the nature of voltage-sensitivity itself: is the response to transmembrane potential merely a reflection of an increased driving force which favours Ca^{2+} exit, or is the channel opening frequency additionally increased by voltage (as is the case for many plasma membrane Ca^{2+} channels from animal cells)? Supplementary questions abound: (i) is voltage-sensitive Ca^{2+} release through a specific Ca^{2+} channel, or through a relatively non-selective pathway? (ii) how does Gd^{3+} inhibit Ca^{2+} release?

These questions are most appropriately answered with the patch clamp technique, applied at the level of the single channel, rather than the whole membrane. Single channel resolution of Ca^{2+} currents enables the current flowing through the channel in its open state to be monitored directly as a function of a defined, imposed voltage, and the selectivity of the channel can be determined from the degree to which the ideal Nernstian response of the single channel reversal potential is compromised by addition of other ions. Most importantly, the propensity of the channel to open with imposed voltage can be determined in a straightforward manner by determination of the proportion of time which, in any given recording, the channel spends in its open state. This function is known as the 'open state probability'. And, with respect to the mechanism of action of inhibitors, one can surmise whether the channel is directly blocked (in which case the current flowing through the open channel will be decreased), or whether the open state probability of the channel is attenuated (with no attendant decrease in current when the channel is in its open state).

Calcium-dependent channel currents can indeed be recorded in patches of membrane from individual isolated vacuoles of sugar beet. The channel shows a high selectivity for Ca^{2+}: over the vacuolar concentration range from 5 to 20 mM Ca^{2+}, the reversal potential exhibits a near-Nernstian dependence on the Ca^{2+} gradient, such that selectivity for Ca^{2+} over K^+ can be estimated as about 20:1 (Johannes *et al.*, 1992). An example of the response of single channel currents to voltage is shown in Fig. 4 for an 'inside-out' patch in which the vacuolar side of the membrane faces the bathing medium. The recording at a holding potential of 50 mV demonstrates that the membrane patch contains several Ca^{2+} channels, each passing a uniform current when open. The current through single open channels is, as anticipated from simple considerations of driving force, increased by progressive positive-going voltages

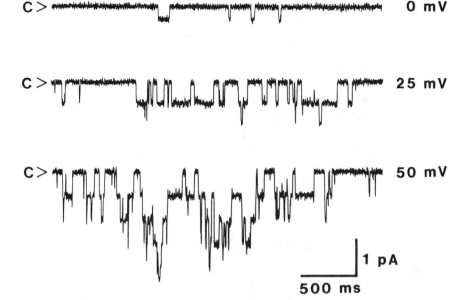

Fig. 4. Recordings of single Ca^{2+} channels in the tonoplast. Recording was from a membrane patch in the inside-out configuration (i.e. vacuolar side facing the bathing medium), and membrane potentials (shown, right) are given with the vacuolar side referenced to the cytosolic side. C> = closed state (no current). The pipette/cytoplasmic side contained 50 mM K^+, 0.5 mM Mg^{2+}, 0.5 mM Ca^{2+}; the bath comprised 10 mM Ca^{2+}. pH was adjusted to 7.3 on both sides with Hepes/dimethylglutarate. Note the increase in both single-channel current and channel opening frequency as the membrane voltage is stepped to more positive values. The patch has at least 4 equivalent channels in it, as revealed by the quantised opening events at the most positive holding potential.

over the physiological range. More intriguing, however, is the clear incidence of increased opening frequency as the holding voltage is stepped from 0 through 25 to 50 mV: opening events are rare at 0 mV, but common at the more positive potentials. The channel is therefore a bona fide voltage-gated channel.

The voltage-sensitivity of gating, supplemented by the voltage-dependence of the open-state current, results in overall Ca^{2+} efflux via this channel (integrated with respect to membrane area and time) exhibiting a steep dependence on the trans-tonoplast electrical potential. Indeed, for the conditions used in the patch clamp experiments, vacuolar

Ca^{2+} leakage via this channel would probably occur at unacceptably high rates even at rather low positive membrane potentials of the order $+20$ mV. We therefore anticipate that the channel is subject to physiological control by at least one regulator in addition to voltage.

The mechanism of Gd^{3+} inhibition of this voltage-dependent Ca^{2+} channel is, perhaps, surprising. Thus, the magnitude of the single channel current is not influenced by the presence of Gd^{3+}, whereas the open-state probability of the channel is strongly depressed. Indeed, if open-state probability is plotted as a function of $[Gd^{3+}]$, a $K_{0.5}$ for Gd^{3+} can be determined as 30 μM – a value which is in close agreement with that determined from membrane vesicles (Johannes *et al.*, 1992). It seems likely that Gd^{3+} acts mainly on the channel gate rather than obstructing the flow of Ca^{2+} through the channel.

Why two Ca^{2+} channels at the tonoplast?

Taken together, these biochemical and biophysical observations firmly point to the existence of a voltage-gated pathway for Ca^{2+} release from vacuoles which is completely independent of the $InsP_3$-gated channel. At present we can only speculate on the significance of these two channels with respect to signal transduction. It is possible that the tonoplast represents the point at which independent primary stimuli converge to elicit a common second messenger response: elevation of cytosolic free Ca^{2+}. Such independent routes to a common end might be the result of evolutionary convergence, or might have more meaning in relation to signal transduction mechanisms. By analogy with animal cells, for example, $InsP_3$-mediated responses are anticipated to be relatively short-lived as a result of hydrolysis of $InsP_3$ itself. In contrast, Ca^{2+} release through the voltage-gated channel could potentially be controlled and sustained for long periods through modulation of trans-tonoplast $\Delta\psi$ mediated by changes in the activities of electrogenic pumps or of other channels.

Acknowledgements

We gratefully acknowledge the support of the Agricultural and Food Research Council, the Science and Engineering Research Council and the Nuffield Foundation.

References

Alexandre, J., Lassalles, J.P. & Kado, R.T. (1990). Opening of Ca^{2+} channels in isolated red beet root vacuole membrane by inositol-1,4,5-trisphosphate. *Nature* **343**, 567–70.

Bates, G.W. , Goldsmith, M.H.M. & Goldsmith, T.H. (1982). Separation of tonoplast and plasma membrane potential and resistance in cells of oat coleoptiles. *Journal of Membrane Biology* **66**, 15–23.

Berridge, M.J. & Irvine, R.F. (1989). Inositol phosphates and cell signalling. *Nature* **341**, 197–205.

Bertl, A., Felle, H. & Bentrup, F.-W. (1984). Amine transport in *Riccia fluitans*. Cytoplasmic and vacuolar pH recorded by a pH-sensitive microelectrode. *Plant Physiology* **76**, 75–8.

Blackford, S., Rea, P.A. & Sanders, D. (1990). Voltage sensitivity of H^+/Ca^{2+} antiport in higher plant tonoplast suggests a role in vacuolar Ca^{2+} accumulation. *Journal of Biological Chemistry* **265**, 9617–20.

Blumwald, E. & Poole, R.J. (1985). Na^+/H^+ antiport in isolated tonoplast vesicles from storage tissues of *Beta vulgaris*. *Plant Physiology* **78**, 163–7.

Boller, T. & Wiemken, A. (1986). Dynamics of vacuolar compartmentation. *Annual Review of Plant Physiology* **37**, 137–64.

Boudet, A.M. & Ranjeva, R. (1989). The vacuole: A potential store for second messengers in plants. In *Second Messengers in Plant Growth and Development*, ed. W.F. Boss & D.J. Morré, pp. 213–25. New York: Alan R. Liss.

Briskin, D.P., Thornley, W.R. & Wyse, R.E. (1985). Membrane transport in isolated vesicles from sugarbeet taproot II. Evidence for a sucrose/H^+-antiport. *Plant Physiology* **78**, 871–5.

Britten, C.J., Turner, J.C. & Rea, P.A. (1989). Identification and purification of substrate-binding subunit of higher plant H^+-translocating inorganic pyrophosphatase. *FEBS Letters* **256**, 200–6.

Brosnan, J.M. (1990). Opening plant calcium channels. *Nature* **344**, 593.

Brosnan, J.M. & Sanders, D. (1990). Inositol trisphosphate-mediated Ca^{2+} release in beet microsomes is inhibited by heparin. *FEBS Letters* **260**, 70–2.

Chopra, L.C., Twort, C.H.C., Ward, J.P.T. & Cameron, I.R. (1989). Effects of heparin on guanosine 5-*O*-(3-thiotriphosphate) induced calcium release in cultured smooth muscle cells from rabbit trachea. *Biochemical and Biophysical Research Communications* **163**, 262–8.

Clapper, D.L. & Lee, H.C. (1985). Inositol trisphosphate induces calcium release from nonmitochondrial stores in sea urchin egg homogenates. *Journal of Biological Chemistry* **260**, 13947–54.

Dainty, J. (1968). The structure and possible function of the vacuole. In *Plant Cell Organelles* ed. J. B. Pridham, pp. 40–6. London: Academic Press.

186   D. SANDERS *ET AL.*

Davies, J.M., Rea, P.A. & Sanders, D. (1991). Vacuolar proton-pumping pyrophosphatase in *Beta vulgaris* shows vectorial activation by potassium. *FEBS Letters* **278**, 66–8.

Ghosh, T.K., Eis, P.S., Mullaney, J.M., Ebat, C.L. & Gill, D.L. (1988). Competitive, reversible and potent antagonism of inositol trisphosphate-activated calcium release by heparin. *Journal of Biological Chemistry* **263**, 11075–9.

Gilroy, S., Fricker, M., Read, N.D. & Trewavas, A.J. (1991). Role of calcium in signal transduction of *Commelina* guard cells. *The Plant Cell* **3**, 333–44.

Guillemette, G., Balla, T., Baukal, A.J. & Catt, K. (1987). Inositol 1,4,5 trisphosphate binds to a specific receptor and releases microsomal calcium in the anterior pituitary gland. *Proceedings of the National Academy of Sciences, USA* **84**, 8195–9.

Guillemette, G. & Segui, J. (1988). Effects of pH, reducing and alkylating reagents on the binding and Ca^{2+} release activities of inositol 1,4,5 trisphosphate in the bovine adrenal cortex. *Molecular Endocrinology* **2**, 1249–55.

Hamill, O.P., Marty, A., Neher, E., Sakmann, B. & Sigworth, F.J. (1981). Improved patch-clamp techniques for high-resolution current recording from cells and cell-free membrane patches. *Pflüger's Archiv* **391**, 85–100.

Hansen, U.-P., Gradmann, D., Sanders, D. & Slayman, C.L. (1981). Interpretation of current-voltage relationships for 'active' ion transport systems: I. Steady-state reaction-kinetic analysis of Class-I mechanisms. *Journal of Membrane Biology* **63**, 165–90.

Hedrich, R. & Kurkdjian, A. (1988). Characterization of an anion-permeable channel from sugar beet vacuoles: effect of inhibitors. *EMBO Journal* **7**, 3661–6.

Hedrich, R., Kurkdjian, A., Guern, J. & Flügge, U.I. (1989). Comparative studies on the electrical properties of the H^+-translocating ATPase and pyrophosphatase of the vacuolar–lysosomal compartment. *EMBO Journal* **8**, 2835–41.

Hedrich, R. & Neher, E. (1987). Cytoplasmic calcium regulates voltage-dependent ion channels in plant vacuoles. *Nature* **329**, 833–6.

Homeyer, U., Litek, K., Huchzermeyer, B. & Schultz, G. (1989). Uptake of phenylalanine into isolated barley vacuoles is driven by both tonoplast adenosine triphosphatase and pyrophosphatase. *Plant Physiology* **98**, 1388–93.

Homeyer, U. & Schultz, G. (1988). Transport of phenylalanine into vacuoles isolated from barley mesophyll protoplasts. *Planta* **176**, 378–82.

Irvine, R.F. (1990). Foreword. In *Inositol Metabolism in Plants*, ed. D. J. Morré, W.F. Boss & F.A. Loewus, pp. xiii–xviii. New York: Wiley–Liss.

Johannes, E., Brosnan, J.M. & Sanders, D. (1991). Calcium channels and signal transduction in plant cells. *BioEssays* **13**, 331–6.

Johannes, E., Brosnan, J.M. & Sanders, D. (1992). Parallel pathways for intracellular calcium release from the vacuole of higher plants. *The Plant Journal* 2, 97–102.

Johannes, E. & Felle, H. (1990). Proton gradient across the tonoplast of *Riccia fluitans* as a result of the joint action of two electroenzymes. *Plant Physiology* 93, 412–17.

Leigh, R.A. & Wyn Jones, R.G. (1984). A hypothesis relating critical potassium concentrations for growth to the distribution and function of this ion in the plant cell. *New Phytologist* 97, 1–14.

McAinsh, M.R., Brownlee, C. & Hetherington, A.M. (1990). Abscisic acid-induced elevation of guard cell cytosolic Ca^{2+} precedes stomatal closure. *Nature* 343, 186–8.

MacRobbie, E.A.C. & Lettau, J. (1980). Ion content and aperture in 'isolated' guard cells of *Commelina communis* L. *Journal of Membrane Biology* 53, 199–205.

Matile, P. (1987). The sap of plant cells. *New Phytologist* 105, 1–26.

Miller, A.J. & Sanders, D. (1987). Depletion of cytosolic free calcium induced by photosynthesis. *Nature* 326, 397–400.

Parry, R.V., Turner, J.C. & Rea, P.A. (1989). High purity preparations of higher plant vacuolar H^+-ATPase reveal additional subunits. Revised subunit composition. *Journal of Biological Chemistry* 264, 20025–32.

Poovaiah, B.W. & Reddy, A.S.N. (1987). Calcium messenger system in plants. *CRC Critical Reviews in Plant Science* 6, 47–103.

Prentki, M., Biden, T., Janjic, D., Irvine, R.F., Berridge, M.J. & Wollheim, C. (1984). Rapid mobilization of Ca^{2+} from rat insulinoma microsomes by inositol-1,4,5-trisphosphate. *Nature* 309, 562–4.

Ranjeva, R., Carrasco, A. & Boudet, A.M. (1988). Inositol-trisphosphate stimulates the release of calcium from intact vacuoles from *Acer* cells. *FEBS Letters* 230, 137–41.

Raven, J.A. (1987). The role of vacuoles. *New Phytologist* 106, 357–422.

Rea, P. A. & Poole, R.J. (1985). Proton-translocating inorganic pyrophosphatase in red beet (*Beta vulgaris* L.) tonoplast vesicles. *Plant Physiology* 77, 46–52.

Rea, P.A. & Poole, R.J. (1992). Vacuolar proton pumps. *Annual Review of Plant Physiology and Plant Molecular Biology* 43 (in press).

Rea, P.A. & Sanders, D. (1987). Tonoplast energization: two H^+ pumps, one membrane. *Physiologia Plantarum* 71, 131–41.

Rincón, M. & Boss, W.F. (1990). Second messenger role of phosphoinositides. In *Inositol Metabolism in Plants*, ed. D.J. Morré, W.F. Boss & F.A. Loewus, pp. 173–200. New York: Wiley-Liss.

Schumaker, K.S. & Sze, H. (1985). A Ca^{2+}/H^+ antiport system driven by the proton electrochemical gradient of a tonoplast H^+-ATPase from oat roots. *Plant Physiology* 79, 1111–17.

Schumaker, K.S. & Sze, H. (1987). Inositol 1,4,5-trisphosphate releases Ca^{2+} from vacuolar membrane vesicles of oat roots. *Journal of Biological Chemistry* 252, 3944–6.

Spanswick, R.M. & Williams, E.J. (1964). Electrical potentials and Na, K and Cl concentrations in the vacuole and cytoplasm of *Nitella translucens. Journal of Experimental Botany* 15, 193–200.

Streb, H., Bayerdorffer, E., Haase, W., Irvine, R.F. & Schulz, I. (1984). Effect of inositol-1,4,5-trisphosphate on isolated subcellular fractions of rat pancreas. *Journal of Membrane Biology* 81, 241–53.

Supattapone, S., Worley, P.F., Baraban, J.M. & Snyder, S.H. (1988). Solubilization, purification and characterization of an inositol trisphosphate receptor. *Journal of Biological Chemistry* 263, 1530–4.

Sze, H. (1985). H^+-translocating ATPases. Advances using membrane vesicles. *Annual Review of Plant Physiology* 36, 175–208.

Wang, Y., Leigh, R.A., Kaestner, K.H. & Sze, H. (1986). Electrogenic H^+-pumping pyrophosphatase in tonoplast vesicles of oat roots. *Plant Physiology* 81, 497–502.

Weiner, H., Stitt, M. & Heldt, H.W. (1987). Subcellular compartmentation of pyrophosphate and alkaline pyrophosphatase in leaves. *Biochimica et Biophysica Acta* 893, 13–21.

A.L. MOORE, J.N. SIEDOW, A.C. FRICAUD,
V. VOJNIKOV, A.J. WALTERS and D.G.
WHITEHOUSE

Regulation of mitochondrial respiratory activity in photosynthetic systems

The role of mitochondria in cellular energy metabolism in both photosynthetic and non-photosynthetic tissues has been the subject of much study. In non-photosynthetic tissues most of the cell's demand for ATP is met by respiration and oxidative phosphorylation whereas in photosynthetic tissues photophosphorylation can be a major contributor to cellular ATP requirements. The pathways involved are central to metabolism and interact with many other metabolic systems. An understanding of the mechanisms that control respiration is therefore vital for a fuller comprehension of cellular metabolism and efficiency. Before attempting to answer the question of what controls respiratory activity it is important to define the system in question and its limits since there are many different levels at which control can be said to occur. Such a definition allows discrimination between internal controls of the system and the effect on the system imposed by external causes.

The scope of this chapter is restricted to the system of oxidative phosphorylation and respiration in plant mitochondria. The general properties and characteristics of plant mitochondria are well documented and the reader is therefore referred to these articles (Douce, 1985; Moore & Rich, 1985; Douce & Neuburger, 1989; Moore & Siedow, 1991) for a fuller description of their structural and functional properties. The system is defined as comprising the electron transport chain, the intramitochondrial NAD^+ and phosphate pools, the adenine nucleotide translocator, the ATP synthase, the protonmotive force and the proton conductance of the inner membrane. The system interacts with its environment at four points, the matrix NAD^+ pool, the ubiquinone pool, oxygen and the extramitochondrial phosphorylation status. Thus the rate of respiration will be regulated by (i) inputs into the system either at the intramitochondrial NAD^+ or ubiquinone pool, (ii) the kinetic properties of the components, (iii) the redox state of the respiratory intermediates, (iv) the protonmotive force and (v) the proton conductance.

Society for Experimental Biology Seminar Series 50: *Plant organelles*, ed. A. K. Tobin.
© Cambridge University Press 1992, pp. 188–210.

Control of respiratory activity in isolated mitochondria

Regulation by adenylates

It has been known for many years that the oxidation of reduced substrates and the synthesis of ATP are coupled processes. Chance & Williams (1955, 1956) demonstrated that in the absence of ADP (state 4) the respiratory rate of isolated mitochondria was low and was markedly stimulated by the addition of ADP (state 3) resulting in the introduction of the term respiratory control. Similar evidence was obtained by Klingenberg (1964) who suggested that the cytosolic phosphorylation potential (ATP/ADP.P_i and termed ΔGp) controlled respiration. Although both of these observations suggest that respiratory activity is regulated by extramitochondrial ATP turnover they differ with respect to whether regulation is at the level of the extramitochondrial ATP/ADP ratio or attributable to changes in the phosphorylation potential.

ATP/ADP ratio

The adenine nucleotide translocator connects the extramitochondrial adenine nucleotide pool with the intramitochondrial pool by catalysing the electrogenic exchange of ATP^{4-} for ADP^{3-} (Vignais, 1976; Vignais & Lunardi, 1985). This led to the suggestion by Heldt & Klingenberg (1968) that the activity of this carrier could act as a rate-limiting step in the control of respiration. Since that date it has been studied in detail by several groups and the consensus of opinion is that the adenine nucleotide carrier does exert significant control over the rate of respiration in phosphorylating mitochondria, particularly at high concentrations of adenine nucleotides (Davis & Lumeng, 1975; Lemasters & Sowers, 1979; Gellerich *et al.*, 1983; Davis & Davis-Van Thienen, 1984; Bohnensack *et al.*, 1990). This conclusion, however, has been generally based upon the results of experiments in which the extramitochondrial ATP/ADP ratio has been varied by the inclusion of an ADP-regenerating system such as glucose/hexokinase, creatine/creatine kinase or soluble F_1-ATPase and inhibiting the translocator with atractyloside. Whilst these conclusions are valid when the ADP-regenerating system is in excess, anomalies occur at limiting activities of the external ATPase. For instance, the translocator can be shown to exhibit high or low control depending upon the type of system used. If the ADP-regenerating system is insensitive to changes in the extramitochondrial ATP/ADP ratio, partial inhibition of the translocator will not result in any response of the system and hence the activity of any remaining uninhibited translocators will be accelerated, thus buffering any net change in flux through the translocator. If the ADP-regenerating system is sensitive to the ATP/ADP ratio, inhibition of the translocator will result in a decrease in activity of the ADP-regenerating system, limiting the drop in the ATP/ADP ratio and hence preventing any significant stimulation of uninhibited translocators. Thus net flux through the system will be inhibited and control by the translocator high. When precautions are taken effectively to reduce limitations by the ADP-regenerating system, it can be concluded that at high ATP/ADP ratios (and with high adenylate concentrations) the activity of the translocator exerts significant control over the rate of respiration (Wanders *et al*, 1984; Bohnensack *et al*, 1990). At low levels of adenylates, however, respiration appears to be more dependent upon the absolute concentration of ADP since respiratory activity can be stimulated with increasing ADP concentration despite a rise in the ATP/ADP ratio (Dry & Wiskich, 1982).

Phosphorylation potential

Dependence of respiratory activity on the phosphorylation status stems from the observation that control by the ATP/ADP ratio varies with the amount of phosphate present, suggesting it is more dependent upon ΔGp than the ATP/ADP ratio (Wilson *et al*, 1977). A similar dependence upon phosphate, however, has not always been demonstrated to be the case in other systems (Davis & Lumeng, 1975; Kunz *et al*, 1981; Davis & Davis-Van Thienen, 1984) and it is conceivable that phosphate dependence is merely a reflection, at low phosphate concentrations, upon the interconversion of $\Delta \psi$ to ΔpH (see Brand & Murphy, 1987; Dry *et al*, 1987). Thus, because of the electrogenicity of the adenine nucleotide translocator, a fall in $\Delta \psi$ would result in reduced adenine nucleotide exchange with a consequent restriction of respiratory activity. Therefore a dependence upon phosphate concentration does not necessarily imply that the translocator does not limit oxidative phosphorylation. Overall it is more than likely that both the phosphorylation potential and the ATP/ADP ratio play major regulatory roles in the control of respiratory activity, the individual significance of each being dependent upon experimental conditions.

Regulation in chemiosmotic terms

Within the framework of the chemiosmotic hypothesis (Mitchell, 1961), respiratory control is considered to be attributable to the disequilibrium between the respiratory chain and the protonmotive force. The protonmotive force (Δp, also known as PMF) is generated by the proton pumping activity of complexes I, III and IV and dissipated by way of membrane leaks, substrate carriers or the mitochondrial ATPase (Moore & Rich, 1985). In plant mitochondria Δp is mainly composed of the membrane potential component since the ΔpH gradient across the inner mitochondrial membrane is continuously dissipated owing to the presence of a K^+/H^+ antiporter (Moore & Rich, 1985). Stimulation of respiration is considered to be caused by a decrease in Δp as a result of an increased flow of protons either through the ATPase (in the presence of ADP) or by way of leaks across the inner membrane (such as induced by uncouplers: Whitehouse *et al*, 1989). The overall respiratory rate is therefore considered to be a function of the total thermodynamic driving force on the electron transport chain, $2\Delta Eh - n\Delta p$ where ΔEh is the redox potential difference across the chain, Δp the protonmotive force and n the H^+/O stoichiometry. Considerable controversy surrounds the relationship between the respiratory rate, Δp and ΔEh. In its simplest form the relationship would be expected to be linear; however, it has been shown in a number of cases that the magnitude of Δp is unrelated to the rate of electron flux since the rate of respiration can be substantially inhibited without major depression of Δp (Nicholls, 1977; Sorgato & Ferguson, 1979; Mandolino *et al*, 1983; Cotton *et al*, 1984; Diolez & Moreau, 1987). This has been variously attributed to the non-ohmic behaviour of the inner membrane (Nicholls, 1977; Krishnamoorthy & Hinkle, 1984; Brown & Brand, 1987; Murphy, 1989; Brand, 1990*a*, *b*), to heterogeneity of the coupling of the mitochondrial population (Duszynski & Wojtczak, 1985; Wojtczak *et al*, 1990), to localised protonic coupling (Westerhof *et al*, 1984) and to variations in the stoichiometry of the redox-driven proton pumps (redox slippage: Pietrobon *et al*, 1983). Further discussion on the possible role of the most important of these parameters (i.e. the proton conductance) in the regulation of electron transport will be considered within the context of the metabolic control theory.

In summary, the chemiosmotic hypothesis states that electron flux is controlled by Δp and ΔEh which themselves may be controlled by the cytosolic phosphorylation potential (through the activity of the adenine nucleotide translocator and the ATPase) and substrate supply (including activity of the various dehydrogenases and substrate carriers).

Regulation in terms of the control theory

A most useful advance in recent years has been the application of the metabolic control theory (Kacser & Burns, 1973, 1979; Heinrich & Rapoport, 1974) to respiration and oxidative phosphorylation in isolated mitochondria. The metabolic control theory was developed for identifying controlling steps within a metabolic pathway and determining the quantitative distribution of this control. A major aspect of this theory is that it concentrates attention on the system and individual steps within the system have importance only if changes in their activity result in an alteration of the properties of the system as a whole. It expounds the idea that if a step in the pathway has control over flux through the pathway then decreasing the activity of that step will decrease the overall flux. An important advantage of the theory over other approaches to metabolic control is that it is quantitative. It clearly defines the system under consideration, allows for control by all elements within the system and measures the relative importance of these controls. For instance, if lowering the activity of an enzyme has no effect over a flux, then that enzyme has no control over that flux. If the enzyme has a large effect upon flux, then that enzyme has a high control. The amount of control is described by the flux control coefficient and the sum of all the flux control coefficients of a system is unity. It can be determined experimentally using specific inhibitors of individual enzyme steps.

The control theory has been used successfully to investigate the regulation of mitochondrial respiratory activity in animal systems (Groen *et al*, 1982; Tager *et al*, 1983; Brand *et al*, 1988). Groen *et al*. (1982) used it to quantify the amount of control exerted by different steps on mitochondrial oxidative phosphorylation in rat liver mitochondria. This was achieved by titrating with inhibitors or uncouplers in a range of states from state 3 to state 4 created by adding different amounts of hexokinase to regulate the level of ATP. In state 4 it was found that virtually all the control of respiration was at the proton leak through the inner mitochondrial membrane. Between states 3 and 4 control was distributed between the leak, the dicarboxylate carrier, the adenine nucleotide carrier and hexokinase. Under state 3 conditions, respiration was controlled by the adenine nucleotide carrier, the dicarboxylate carrier and cytochrome oxidase. Thus respiration is controlled in phosphorylating mitochondria by both ATP turnover and the electron supply. In the absence of ADP, the distribution of control alters drastically residing almost entirely in the proton leak. However, in contrast to this result, Brand *et al* (1988) found that even in the absence of ADP control was not solely limited to the proton leak since some control was still exerted by the respiratory chain.

To date there has been only one analysis of the control of respiratory activity in plant mitochondria: Padovan *et al.* (1989) found that under state 3 conditions, in isolated turnip mitochondria, control was distributed between cytochrome oxidase, the bc_1 complex and the H^+-ATPase in decreasing order of importance. Interestingly, the adenine nucleotide translocator exerted no control on flux under these conditions in contrast to that reported for rat liver mitochondria. A similar lack of control by the translocator has, however, been reported for rat heart (Doussiere *et al*, 1982) and yeast mitochondria (Mazat *et al*, 1986) and may be attributed to the much higher content of this carrier within these preparations. Under ADP-limited conditions both cytochrome oxidase and the bc_1 complex exerted little control, suggesting that under these conditions control in plant systems is located in other steps of the respiratory process such as the dehydrogenases, substrate carriers and the proton leak (Padovan *et al*, 1989).

In view of the importance of the proton leak as a possible major control point and the likelihood of the respiratory state of plant mitochondria *in vivo* being between state 3 and state 4 (Douce & Neuburger, 1989; Douce *et al*, 1991), a knowledge of the proton leak and its dependence upon the protonmotive force is of paramount importance. Whitehouse *et al.* (1989) used a TPMP$^+$-sensitive electrode to measure the maximal rate of dissipation (J_{diss}) of $\Delta\psi$ at steady state upon addition of a respiratory inhibitor as an indicator of the membrane ionic (H$^+$) conductance in isolated mitochondria. The technique was based upon the assumption that at static head the rate of H$^+$ efflux (via the respiratory chain) exactly matches the rate of H$^+$ influx (via H$^+$ leaks and the H$^+$-ATPase), and consequently the initial rate of H$^+$ influx upon addition of a respiratory inhibitor should give an indication of the proton conductance across the inner membrane. It is also assumed that membrane capacitance remains constant during the initial period of measurement, and consequently the technique gives a measure of the current flowing across the membrane and allows H$^+$ conductance to be measured without complications owing to heterogeneity of coupling and/or redox slippage. It can be seen from Fig. 1 that there is a non-linear relationship between $\Delta\psi$ and the respiratory rate and between the leak of protons across the inner membrane. This implies that the H$^+$ conductance increases at high values of the membrane potential. Since transition from state 3 to state 4 is accompanied by an increase in membrane potential towards these high values (Moore *et al*, 1978; Moore & Bonner, 1982; Moore & Rich, 1985) this result has important physiological implications.

Although changes in the H$^+$ conductance of the inner membrane can quantitatively account for the non-ohmic relationship between the respiratory rate and the protonmotive force, it does not necessarily imply that all control resides in this particular step (see Brand *et al*, 1988). For instance, it can be seen from Fig. 2 that under non-phosphorylating conditions control of respiration is shared between the respiratory chain and the proton leak. These results were obtained by titrating with malonate to inhibit the electron transport chain or with uncoupler to

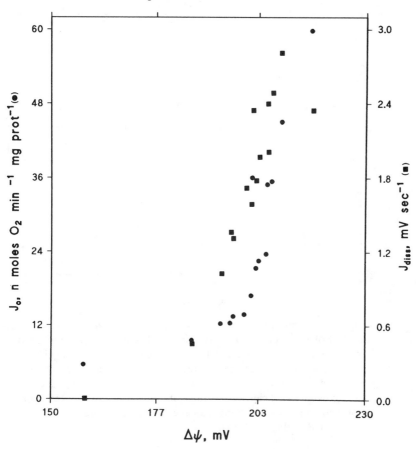

Fig. 1. The dependence of the initial rate of dissipation and respiratory rate on the steady-state membrane potential in potato mitochondria. Membrane potentials and respiratory activity were measured in the presence of 5 mM succinate and 0–5 mM malonate and data points taken from Whitehouse *et al.* (1989). The membrane potential was dissipated by addition of 0.5 mM KCN.

increase the H^+ conductance. Using the flux control summation and connectivity theorems of the metabolic control theory to calculate the control over non-phosphorylating respiration, it is apparent that although the proton leak does have a large flux control coefficient (0.66) there is also a significant contribution to control by the respiratory chain (0.34) even under ADP-limited conditions. Although the analysis does not indicate how control is distributed between the individual steps in the chain it

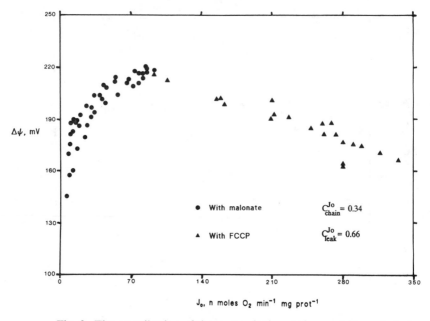

Fig. 2. The contribution of the proton leak and the respiratory chain to control of respiration in potato mitochondria under non-phosphorylating conditions. The flux control coefficients of the respiratory chain and proton leak were determined by titration with malonate (●) (to inhibit respiratory activity) and FCCP (▲) (to increase proton conductance), respectively. Coefficients were calculated using the flux control summation and connectivity theorems of the metabolic control theory as described by Brand *et al.* (1988).

is apparent from the results of Padovan *et al.* (1989) that significant control must lie with either the dehydrogenases, the ubiquinone pool or cytochrome *c* diffusion.

Regulation by the ubiquinone pool

The possibility that the redox poise of the ubiquinone pool could be a significant control point in the regulation of respiratory activity is of particular interest for not only is it a major component of the respiratory chain (Moore & Rich, 1985; Douce & Neuburger, 1989), but it also plays a pivotal role in the partitioning of electron flux between the main respiratory chain and the alternative oxidase (Moore & Siedow, 1991). Although it is well established that the branch point of the alternative oxidase from the main respiratory chain is at the level of the ubiquinone pool (Rich & Moore, 1976), the mechanism by which this occurs is unresolved. Bahr & Bonner (1973) postulated that the alternative pathway only becomes engaged when the ubiquinone pool becomes sufficiently reduced to allow the oxidation of ubiquinol by the alternative oxidase to become thermodynamically favourable; furthermore, the latter occurred only when electron flux through the main chain became saturated. In isolated mitochondria such a situation is envisaged to occur under ADP-limited conditions whereas in plant cells this could be achieved when cellular ADP levels are low and mitochondrial electron transfer becomes limiting. When the ubiquinone pool is oxidised, for instance under state 3 conditions, ubiquinol oxidation via the alternative oxidase becomes thermodynamically unfavourable and, although present, the pathway is no longer engaged. An alternative model (de Troostemberg & Nyns, 1978), based on the 'Q-pool' model of Kroger & Klingenberg (1973) suggests that the distribution of reducing equivalents between the two pathways is in direct proportion to their relative kinetic capacities. This model makes the assumption that when two oxidases operate simultaneously they compete for electrons from a single homogeneous ubiquinone pool. As long as neither pathway becomes saturated electron flow would proceed through both in a constant ratio. Unfortunately, both models run counter to experimental observations. The Bahr & Bonner model cannot account for the switching of electron flux from the alternative pathway to the main respiratory chain (as reported by Wilson, 1988) since it assumes that the main pathway is already saturated. The second model cannot accommodate the well-documented situation where the alternative pathway is present but not operative (i.e. under state 3 conditions: see Lance *et al*, 1985).

Two recent technical advances could help to resolve the question of how electron flux is regulated between the two pathways. Guy *et al.* (1989) used a differential oxygen discrimination technique as a non-invasive assay for assessing the contribution of the alternative pathway to plant respiration (in both intact tissues and isolated mitochondria) in the absence of any added inhibitors. When oxygen was reduced exclusively via complex IV a discrimination against ^{18}O of 17–19‰ was found whereas when oxygen was reduced by the alternative pathway a greater degree of discrimination (24–26‰) was observed. In the absence of inhibitors, intermediate values were observed, suggesting that uninhibited respiration included contributions from both pathways. This technique is of particular importance since it provides the first opportunity to measure the actual contributions of the alternative pathway to plant respiration in intact tissues in the absence of any added inhibitors.

The second technical advance is the application of a voltametric technique to measure continuously the steady-state redox poise of the ubiquinone pool during electron turnover in isolated mitochondria (Moore *et al*, 1988). It was found that in cyanide-sensitive tissues the respiratory rate, under either state 3 or state 4 conditions, was directly proportional to the redox poise of the ubiquinone pool, confirming the original suggestions of Kroger & Klingenberg (1973). When partially cyanide-resistant mitochondria were used, however, the relationship between the respiratory rate and the redox state of the ubiquinone pool, under state 4 conditions, became distinctly non-linear. The non-linearity was thought to be caused by the engagement of the alternative pathway (Moore *et al*, 1988). This was confirmed by the finding that in highly cyanide-resistant mitochondria significant engagement of the alternative pathway was not apparent until the ubiquinone pool was significantly reduced (35–40%) and increased disproportionately on further reduction (Dry *et al*, 1989). An example of this type of relationship is shown in Fig. 3. It is apparent in soybean cotyledons, at least, that a significant proportion of the uninhibited state 4 respiratory rate results from electron flow via the alternative pathway, consistent with the suggestions of Bahr & Bonner (1973) and the recent observations of Guy *et al.* (1989). Under state 3 conditions the ubiquinone pool remains relatively oxidised, even at the maximal respiratory rate, thus preventing engagement of the alternative pathway to any significant extent. Obviously the degree to which the alternative pathway is engaged, even under state 3 conditions, will be dependent upon the relative rates of electron input and output from the Q-pool.

Such results at first approximation support the Bahr & Bonner model, in which alternative pathway activity is regulated by the existence of a thermodynamically unfavourable step with a redox potential at least 35 mV more negative than ubiquinone. However, when the experimentally observed alternative pathway activity, at different ubiquinone redox poises, was compared with predicted activity based on the Bahr & Bonner model it was found that the experimental data did not fit the model (Dry *et al*, 1989). Engagement could therefore not operate along the lines of an 'on/off switch' which was dependent upon the cytochrome chain reaching near or complete saturation. Although there is some similarity between the observed and predicted curves it is clear that a model based solely on a single, limiting equilibrium cannot account for the relationship observed experimentally.

In an attempt to obtain a closer fit to the experimentally observed curves, a kinetic model has been developed (Moore & Siedow, 1991) based on a theoretical equation in which the reduction of the alternative oxidase by ubiquinol is considered to occur by the following steps:

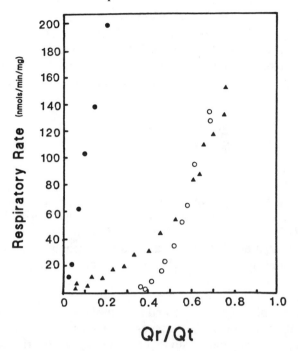

Fig. 3. Dependence of the respiratory rate on the redox poise of the ubiquinone pool in soybean cotyledon mitochondria. Respiratory activity and steady-state reduction of ubiquinone was measured voltametrically as described by Moore *et al.* (1988) and Dry *et al.* (1989). Succinate-dependent respiratory activity was progressively inhibited by the addition of malonate either in the presence of ADP (●), or under state 4 conditions (▲), or in the presence of myxothiazol (○). Qr/Qt represents the proportion of quinone in the reduced state.

$$Q_r + E_o \underset{k_{-1}}{\overset{k_{+1}}{\rightleftharpoons}} Q_o + E_r$$

$$Q_r + E_r \underset{k_{-2}}{\overset{k_{+2}}{\rightleftharpoons}} Q_o + E_r'$$

$$E_r' \overset{k_{sub}}{\longrightarrow} E_o \tag{1}$$

where E_o, E_r and E_r', are the oxidised, semi-reduced and fully reduced forms of the oxidase, Q_o and Q_r the oxidised and reduced forms of ubiquinone, k_{+1} and k_{-1} are rate constants governing the first reversible electron transfer from Q_r to the oxidised oxidase (E_o), k_{+2} and k_{-2} are rate constants governing the second reversible step from Q_r to the semi-reduced oxidase (E_r) and k_{sub} is a pseudo-first-order rate constant covering the reduction of oxygen by the fully reduced enzyme. The steady-state rate (v) through the alternative oxidase complex is then given by an equation of the form:

$$v = \frac{Q_r^2 V_o[k_{+1}(k_{+2}Q_t + k_{sub}) + k_{sub}k_{+2}]}{Q_t[Q_r(k_{+1}k_{+2}Q_r + k_{sub}k_{+2} + k_{+1}k_{sub} + k_{-2}k_{+1}(Q_t-Q_r)) + k_{-1}(Q_t-Q_r)(k_{sub} + k_{-2}(Q_t-Q_r)]} \quad (2)$$

In this equation, v is the velocity and V_o is the maximum rate of electron flux when $Q_r = Q_t$. (Q_r/Q_t represents the proportion of Q reduced under steady state, relative to total reducible Q). The behaviour of this equation is illustrated in Fig. 4 and compared with that observed experimentally using a variety of mitochondria that vary in cyanide resistance. It is apparent from Fig. 4 that there is a good fit between the observed and predicted curves even though the mitochondria vary considerably in their degree of cyanide resistance (from 40 to 100% cyanide-resistant). The predicted curves were generated merely by varying k_{-1} and k_{-2} whilst maintaining k_{+1}, k_{+2} and k_{sub} constant. It is apparent that the dependence of v/V_o on Q_r/Q_t becomes more linear as k_{-1} and k_{-2} are decreased which is comparable with that obtained in the kinetic model of Reed & Ragan (1987). In other words, the overall rate of electron transport via the alternative oxidase is not only dependent upon the redox poise of the ubiquinone pool but also upon the magnitude of the equilibrium constants k_{+1}/k_{-1} and k_{+2}/k_{-2}. When these values are low the rate of electron transport becomes largely insensitive to changes in the redox state of the ubiquinone pool at low values of Q_r and as they increase velocity becomes much more dependent upon the degree of reduction of the pool. Interestingly, the shape of the curve is largely

---

Fig. 4. Comparison of observed alternative pathway activity with predicted activity (A) based on a kinetic model. The dependence of alternative pathway activity on the redox poise of the ubiquinone pool was determined experimentally in B, *Arum maculatum*, C, soybean cotyledon and D, sweet potato mitochondria under state 4 conditions in the presence of myxothiazol. Solid lines represent that predicted by the kinetic model using Equation 2 and the values for the kinetic constants as indicated. In B–D, k_{-1} and k_{-2} were varied as indicated whilst maintaining k_{+1}, k_{+2} and k_{sub} as in A. v is the steady-state respiratory rate in the presence of malonate, Vo is the uninhibited rate and Qr/Qt is the proportion of ubiquinone reduced under steady-state conditions relative to total reducible ubiquinone.

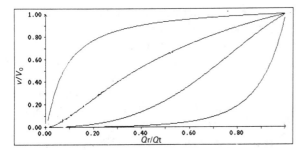

$k+1 = 0.35$
$k+2 = 0.7$
$k_{sub} = 0.14$

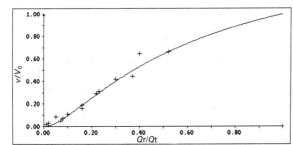

$k-1 = 0.1$
$k-2 = 0.1$

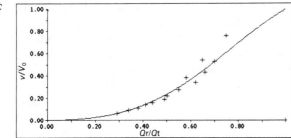

$k-1 = 1.2$
$k-2 = 0.35$

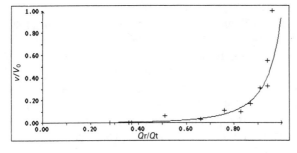

$k-1 = 1.6$
$k-2 = 20$

insensitive to variations in k_{sub} in agreement with the observation that, *in vitro*, alternative oxidase activity is not rate-limited by variations in oxygen concentration (Bendall & Bonner, 1971). The model can also account for the situation where the pathway is present but not engaged. For instance, in sweet potato the alternative pathway does not appear to be engaged until approximately 80% of the ubiquinone pool has been reduced, owing to the low value of the equilibrium constants. Such a case may also explain why selectively pruned barley plants appear unable to engage the alternative pathway even when the carbohydrate status is high (see Farrar & Williams, 1991).

In summary, the kinetic model outlined quantitatively accounts for deviations from ubiquinone pool behaviour predicted by the Kroger & Klingenberg model and, furthermore, predicts with greater precision than other models the observed dependence of velocity on the ubiquinone reduction state. When coupled with information provided by the differential oxygen discrimination technique, these two techniques could well clarify much that is currently unclear about the partitioning of electron flux between the main and alternative pathways in plant mitochondria.

Control of respiratory activity under *in vivo* conditions

From the foregoing discussion it is obvious that a considerable number of factors can regulate respiratory activity. However, perhaps one of the most important regulators, *in vivo*, will be the level of the adenylates in the cytosol. Whether or not control of mitochondrial respiratory activity is mediated by the ATP/ADP ratio, the cytosolic phosphorylation potential or merely the absolute concentration of ADP, it is apparent that increases in these parameters (or decrease in ADP concentration) will result in severe restraints upon the efficient operation of the mitochondrial respiratory chain. Although previous reports on the level of cytosolic adenylates during illuminating and non-illuminating conditions suggested there was either only brief transient change (Hampp *et al*, 1982) or a net decrease (Stitt *et al*, 1982) in the ATP/ADP ratio upon transition to the light, recent measurements by Gardeström (1987) and Gardeström & Malmberg (1990) suggest that under limiting CO_2 conditions, a situation more likely to exist *in vivo*, the cytosolic ATP/ADP ratio is considerably higher in the light compared with darkness. If this proves to be the case, it will result in a transition of the mitochondrial respiratory state towards a state 4 (ADP-limited) situation. There is some evidence in favour of this idea since Valles *et al*. (1984) have previously demonstrated that illumination of wheat protoplasts results in an increased mitochondrial Δp. Such increases in Δp would result in increased intramitochondrial NADH/NAD ratios and influence the redox status of the electron transport carriers which in turn would dictate the engagement of non-phosphorylating pathways.

It should be stressed, however, that although increased cytosolic ATP/ADP ratios may significantly decrease the amount of respiratory activity that is coupled to ATP synthesis, it may not necessarily dramatically inhibit overall respiration because of the presence of non-phosphorylating pathways. For instance, it is well established that considerable increases in respiratory activity can be achieved by the addition of a second substrate even under ADP-limited situations (Dry *et al*, 1983; Douce *et al*, 1984; Day *et al*, 1985, 1987; Dry & Wiskich, 1985; Bryce *et al*, 1990; Moore *et al*, 1991). Such respiratory stimulations are the result of an increased redox driving force which in turn increases both Δp and the redox poise of the ubiquinone pool (Moore *et al*, 1990, 1991). A rise in Δp will have a dual effect for not only will it induce a considerable rise in membrane conductance (because of non-ohmicity) and therefore allow a faster rate of respiration (Whitehouse *et al*, 1989), but it will oppose some of the increase in driving force caused by the additional substrate since the rate of electron transport is a function of $2\Delta E_h - n\Delta p$ (see above). The dramatic increase in the level of steady-state redox level of the ubiquinone pool observed under these conditions will, however, tend to reduce this opposition and allow an increase in the flux through the bc_1 complex. It may also allow engagement of the alternative oxidase, but this will be dependent upon the overall capacity of the oxidase (which in most photosynthetic tissues is low) (Lance *et al*, 1985). Increases in respiratory flux can also be achieved if other non-electrogenic pathways, such as the rotenone-insensitive bypass, are engaged (Day *et al*, 1987; Bryce *et al*, 1990; Moore *et al*, 1990, 1991). As a result of bypassing complex I the total number of protons pumped will be reduced thereby increasing the overall driving force (see above). This appears to be the case in pea leaf mitochondria, at least, since rotenone addition only causes a partial re-oxidation of the ubiquinone pool, suggesting that the bypass is operative and possibly a major pathway of electron flow to the ubiquinone pool under ADP-limited situations (Moore *et al*, 1991). *In vivo*, engagement of the rotenone-insensitive pathway may occur as a result of an increased substrate supply to the mitochondria, possibly because of changes in the carbohydrate content of the plant (Farrar & Williams, 1991).

A knowledge of the contribution of non-phosphorylating pathways and modulation of ionic conductance to the overall respiratory rate of the cell is of some importance particularly under photorespiratory conditions.

During photorespiration, glycine is converted to serine by the dual action of glycine decarboxylase and serine hydroxymethyl transferase (see Dry *et al*, 1987; Douce & Neuburger, 1989). Continued operation of this cycle is dependent upon the re-oxidation of NADH which has been suggested to occur either via substrate shuttles or via the respiratory chain (see Dry *et al*, 1987 and Wiskich & Meidan, this volume, for discussion). Although the exact extent to which both or either of these processes are involved in the re-oxidation of NADH is still debatable, recent results using inhibitors of either the ATP synthase (Kromer *et al*, 1988; Kromer & Heldt, 1991) or glycine decarboxylase (Gardeström & Wigge, 1988) suggest that mitochondrial oxidative phosphorylation contributes ATP to the cytosol even under illuminating conditions. At first sight these results appear contradictory with the idea of adenylate control upon mitochondrial respiratory activity in the light. However, as suggested earlier, increases in the ATP/ADP radio (which appear to occur upon illumination) will result in an increased mitochondrial Δp and, as a consequence of non-ohmicity, a lowered yield of ATP in the light. Thus although mitochondria may contribute to the cytosolic ATP pool in the light it may be at a reduced efficiency of ATP production compared with that observed in the dark.

Acknowledgements

This work was supported in part by grants from the SERC (ALM), The Royal Society (ALM) and The National Science Foundation (JNS). We thank Drs I.B. Dry, A.M. Wagner and J.T. Wiskich for discussions of some of the ideas presented in the kinetic model.

References

Bahr, J.T. & Bonner, W.D. (1973). Cyanide-insensitive respiration. II Control of the alternative pathway. *Journal of Biological Chemistry* **248**, 3446–50.

Bendall, D.S. & Bonner, W.D. (1971). Cyanide-insensitive respiration in plant mitochondria. *Plant Physiology* **47**, 236–45.

Bohnensack, R., Gellerich, F.N., Schild, L. & Kunz, W. (1990). The function of the adenine nucleotide translocator in the control of oxidative phosphorylation. *Biochimica et Biophysica Acta* **1018**, 182–4.

Brand, M.D. (1990*a*). The contribution of the leak of protons across the mitochondrial inner membrane to standard metabolic rate. *Journal of Theoretical Biology* **145**, 267–86.

Brand, M.D. (1990*b*). The proton leak across the mitochondrial inner membrane. *Biochimica et Biophysica Acta* **1018**, 128–33.

Brand, M.D., Hafner, R.P. & Brown, G.C. (1988). Control of respiration in non-phosphorylating mitochondria is shared between the proton leak and the respiratory chain. *Biochemical Journal* 255, 535–9.

Brand, M.D. & Murphy, M.P. (1987). Control of electron flux through the respiratory chain in mitochondria and cells. *Biological Reviews* 62, 141–93.

Brown, G.C. & Brand, M.D. (1987). Changes in permeability to protons and other cations at high protonmotive force in rat liver mitochondria. *Biochemical Journal* 234, 75–81.

Bryce, J.H., Azcón-Bieto, J., Wiskich, J.T. & Day, D.A. (1990). Adenylate control of respiration in plants: the contribution of rotenone-insensitive electron transport to ADP-limited oxygen consumption by soybean mitochondria. *Physiologia Plantarum* 78, 105–11.

Chance, B. & Williams, G.R. (1955). Respiratory enzymes in oxidative phosphorylation. I. Kinetics of oxygen utilisation. *Journal of Biological Chemistry* 217, 383–93.

Chance, B. & Williams, G.R. (1956). The respiratory chain and oxidative phosphorylation. *Advances in Enzymology* 17, 65–134.

Cotton, N.P.J., Clarke, A.J. & Jackson, B.J. (1984). Changes in membrane ionic conductance, but not changes in slip, can account for the non-linear dependence of the electrochemical proton gradient upon the electron transport rate in chromatophores. *European Journal of Biochemistry* 142, 193–8.

Davis, E.J. & Davis-Van Thienen, W.I.A. (1984). Rate control of phosphorylation-coupled respiration by rat liver mitochondria. *Archives of Biochemistry and Biophysics* 233, 573–81.

Davis, E.J. & Lumeng, L. (1975). Relationships between the phosphorylation potentials generated by liver mitochondria and respiratory state under conditions of adenosine diphosphate control. *Journal of Biological Chemistry* 250, 2275–82.

Day, D.A., Neuburger, M. & Douce, R. (1985). Interactions between glycine decarboxylase, the tricarboxylic acid cycle and the respiratory chain in pea leaf mitochondria. *Australian Journal of Plant Physiology* 12, 119–30.

Day, D.A., Wiskich, J.T., Bryce, J.H. & Dry, I.B. (1987). Regulation of ADP-limited respiration in isolated plant mitochondria. In *Plant Mitochondria: Structural, Functional and Physiological Aspects*, ed. A.L. Moore & R.B. Beechey, pp. 59–66. New York: Plenum Press.

de Troostemberg, J.-C. & Nyns, E.J. (1978). Kinetics of the respiration of cyanide-insensitive mitochondria from the yeast *Saccharomycopsis lipolytica*. *European Journal of Biochemistry* 85, 423–32.

Diolez, P. & Moreau, F. (1987). Relationships between membrane potential and oxidation rate in potato mitochondria. In *Plant Mitochondria: Structural, Functional and Physiological Aspects*, ed. A.L. Moore & R.B. Beechey, pp. 17–26. New York: Plenum Press.

Douce, R. (1985). *Mitochondria in Higher Plants. Structure, Function and Biogenesis.* New York: Academic Press.

Douce, R., Bligny, R., Brown, D., Dorne, A.-J., Genix, P. & Roby, C. (1991). In *Compartmentation of Plant Metabolism in Non-photosynthetic Tissues*, Society for Experimental Biology Seminar Series, No. 42, ed. M.J. Emes, pp. 127–45. Cambridge: Cambridge University Press.

Douce, R., Day, D.A. & Neuburger, M. (1984). Influx and efflux of NAD in plant mitochondria. *Current Topics in Plant Biochemistry and Physiology* 3, 329–39.

Douce, R. & Neuburger, M. (1989). The uniqueness of plant mitochondria. *Annual Review of Plant Physiology and Plant Molecular Biology* 40, 371–414.

Doussiere, J., Ligeti, E., Brandolin, G. & Vignais, P.V. (1982). Control of oxidative phosphorylation in rat heart mitochondria. The role of the adenine nucleotide carrier. *Biochimica et Biophysica Acta* 766, 492–500.

Dry, I.B., Bryce, J.H. & Wiskich, J.T. (1987). Regulation of mitochondrial respiration. In *The Biochemistry of Plants*, Vol. 11, ed. D.D. Davis, pp. 213–52. New York: Academic Press.

Dry, I.B., Day, D.A. & Wiskich, J.T. (1983). Preferential oxidation of glycine by the respiratory chain of pea leaf mitochondria. *FEBS Letters* 158, 154–8.

Dry, I.B., Moore, A.L., Day, D.A. & Wiskich, J.T. (1989). Regulation of alternative pathway activity in plant mitochondria: Nonlinear relationship between electron flux and the redox poise of the quinone pool. *Archives of Biochemistry and Biophysics* 273, 148–57.

Dry, I.B. & Wiskich, J.T. (1982). Role of the external adenosine triphosphate/adenosine diphosphate ratio in the control of plant mitochondrial respiration. *Archives of Biochemistry and Biophysics* 217, 72–9.

Dry, I.B. & Wiskich, J.T. (1985). Characteristics of glycine and malate oxidation by pea leaf mitochondria: Evidence of differential access to NAD and respiratory chains. *Australian Journal of Plant Physiology* 12, 329–39.

Duszynski, J. & Wojtczak, L. (1985). The apparent non-linearity of the relationship between the rate of respiration and the protonmotive force of mitochondria can be explained by heterogeneity of mitochondrial preparations. *FEBS Letters* 182, 243–8.

Farrar, J.F. & Williams, J.H.H. (1991). Control of the rate of respiration in roots: compartmentation, demand and the supply of substrate. In *Compartmentation of Plant Metabolism in Non-photosynthetic Tissues*. Society for Experimental Biology Seminar Series No. 42, ed. M. J. Emes, pp. 167–88. Cambridge: Cambridge University Press.

Gardeström, P. (1987). The effect of light on mitochondrial respiration studied by rapid fractionation of protoplasts. In *Plant Mitochondria:*

Structural, Functional and Physiological Aspects, ed. A. L. Moore & R.B. Beechey, pp. 161–9. New York: Plenum Press.

Gardeström, P. & Malmberg, G. (1990). Subcellular ATP/ADP ratios in barley protoplasts during photosynthesis induction. *Physiologia Plantarum* **79**, A60.

Gardeström, P. & Wigge, B. (1988). Influence of photorespiration on ATP/ADP ratios in the chloroplasts, mitochondria, and cytosol, studied by rapid fractionation of barley (*Hordeum vulgare*) protoplasts. *Plant Physiology* **88**, 69–76.

Gellerich, F.N., Bohnensack, R. & Kunz, W. (1983). Control of mitochondrial respiration; the contribution of the adenine nucleotide translocator depends on the ATP- and ADP-consuming enzymes. *Biochimica et Biophysica Acta* **722**, 381–91.

Groen, A.K., Wanders, R.J.A., Westerhof, H. V., van de Meer, R. & Tager, J. M. (1982). Quantification of the contribution of various steps to the control of mitochondrial respiration. *Journal of Biological Chemistry* **257**, 2754–7.

Guy, R.D., Berry, J.A., Fogel, M.L. & Hoerig, T.C. (1989). Differential fractionation of oxygen isotopes by cyanide-resistant and cyanide-sensitive respiration in plants. *Planta* **177**, 483–91.

Hampp, R., Goller, M. & Zeigler, H. (1982). Adenylate levels, energy charge and phosphorylation potential during dark–light and light–dark transition in chloroplasts, mitochondria and cytosol of mesophyll protoplasts from *Avena sativa* L. *Plant Physiology* **69**, 448–55.

Heinrich, R. & Rapoport, T.A. (1974). A linear steady-state treatment of enzymatic chains. General properties, control and effector strength. *European Journal of Biochemistry* **42**, 89–95.

Heldt, H.W. & Klingenberg, M. (1968). Differences between the reactivity of endogenous and exogenous adenine nucleotides in mitochondria as studied at low temperature. *European Journal of Biochemistry* **4**, 1–8.

Kacser, H. & Burns, J.A. (1973). The control of flux. *Symposia of the Society for Experimental Biology* **32**, 65–104.

Kacser, H. & Burns, J.A. (1979). Molecular democracy: who shares controls? *Biochemical Society Transactions* **7**, 1149–60.

Klingenberg, M. (1964). Reversibility of energy transformations in the respiratory chain. *Angewandte Chemie Int. Ed.* **3**, 54–61.

Krishnamoorthy, G. & Hinkle, P.C. (1984). Nonohmic proton conductance of mitochondria and liposomes. *Biochemistry* **23**, 1640–5.

Kroger, A. & Klingenberg, M. (1973). The kinetics of the redox reactions of ubiquinone related to the electron-transport activity in the respiratory chain. *European Journal of Biochemistry* **24**, 358–68.

Kromer, S. & Heldt, H.W. (1991). On the role of mitochondrial oxidative phosphorylation in photosynthesis metabolism as studied by the effect of oligomycin on photosynthesis in protoplasts and leaves of barley (*Hordeum vulgare*). *Plant Physiology* **95**, 1270–6.

Kromer, S., Stitt, M. & Heldt, H.W. (1988). Mitochondrial oxidative phosphorylation participating in photosynthesis metabolism of a leaf cell. *FEBS Letters* **226**, 352–6.

Kunz, W., Bohnensack, R., Bohme, G., Kuster, U., Letko, G. & Schoenfeld, P. (1981). Relations between extramitochondrial and intramitochondrial adenine nucleotide systems. *Archives of Biochemistry and Biophysics* **209**, 219–29.

Lance, C., Chaveau, M. & Dizengremel, P. (1985). In *Encyclopedia of Plant Physiology*, Vol. 18, ed. R. Douce & D.A. Day, pp. 202–47. Berlin: Springer-Verlag.

Lemasters, J.J. & Sowers, A.E. (1979). Phosphate dependence and atractyloside inhibition of mitochondrial oxidative phosphorylation. *Journal of Biological Chemistry* **254**, 1248–51.

Mandolino, G., Desantis, A. & Melandri, B.A. (1983). Localised coupling in oxidative phosphorylation by mitochondria from Jerusalem artichoke (*Helianthus tuberosus*). *Biochimica et Biophysica Acta* **890**, 279–85.

Mazat, J.M., Jean-Bart, E., Rigoulet, M. & Guerin, B. (1986). Control of oxidative phosphorylation in yeast mitochondria. Role of the phosphate carrier. *Biochimica et Biophysica Acta* **849**, 7–15.

Mitchell, P. (1961). Coupling of phosphorylation to electron and hydrogen transfer by a chemi-osmotic type of mechanism. *Nature* **191**, 144–8.

Moore, A.L. & Bonner, W.D. (1982). Measurements of membrane potentials in plant mitochondria with the safranine method. *Plant Physiology* **70**, 1271–6.

Moore, A.L., Day, D.A., Dry, I.B. & Wiskich, J.T. (1990). Regulation of electron transport activity in plant mitochondria by the redox poise of the quinone pool. In *Highlights in Ubiquinone Research*, ed. G. Lenaz, O. Barnebei, A. Rabbi & M. Battino, pp. 170–3. London: Taylor & Francis.

Moore, A.L., Dry, I.B. & Wiskich, J.T. (1988). Measurement of the redox state of the ubiquinone pool in plant mitochondria. *FEBS Letters* **235**, 76–80.

Moore, A.L., Dry, I.B. & Wiskich, J.T. (1991). Regulation of electron transport in plant mitochondria under state 4 conditions. *Plant Physiology* **95**, 34–40.

Moore, A.L. & Rich, P.R. (1985). The organisation of the respiratory chain and oxidative phosphorylation. In *Encyclopedia of Plant Physiology*, Vol. 18, ed. R. Douce & D.A. Day, pp. 134–72. New York: Springer-Verlag.

Moore, A.L., Rich, P.R. & Bonner, W.D. (1978). Factors influencing the components of the total protonmotive force in mung bean mitochondria. *Journal of Experimental Botany* **108**, 1–12.

Moore, A.L. & Siedow, J.N. (1991). The regulation and nature of the cyanide-resistant alternative oxidase of plant mitochondria. *Biochimica et Biophysica Acta* **1059**, 121–40.

Murphy, M.P. (1989). Slip and leak in mitochondrial oxidative phosphorylation. *Biochimica et Biophysica Acta* **977**, 123–41.

Nicholls, D.G. (1977). The effective proton conductance of the inner membrane of mitochondria from brown adipose tissue. Dependency on proton electrochemical gradient. *European Journal of Biochemistry* **77**, 349–56.

Padovan, A.C., Dry, I.B. & Wiskich, J.T. (1989). An analysis of the control of respiration in isolated plant mitochondria. *Plant Physiology* **90**, 928–33.

Pietrobon, D., Zoratti, M. & Azzone, G.F. (1983). Molecular slipping in redox and ATPase proton pumps. *Biochimica et Biophysica Acta* **723**, 317–21.

Reed, J.S. & Ragan, C.I. (1987). The effect of rate limitation by cytochrome c on the redox state of the ubiquinone pool in reconstituted NADH:cytochrome c reductase. *Biochemical Journal* **247**, 657–62.

Rich, P.R. & Moore, A.L. (1976). The involvement of the protonmotive ubiquinone cycle in the respiratory chain of higher plants and its relation to the branch point of the alternate pathway. *FEBS Letters* **65**, 339–44.

Sorgato, M.C. & Ferguson, S.J. (1979). Variable proton conductance of submitochondrial particles. *Biochemistry* **18**, 5737–42.

Stitt, M., Lilley, R. McC. & Heidt, H.W. (1982). Adenine nucleotide levels in the cytosol, chloroplasts and mitochondria of wheat leaf protoplasts. *Plant Physiology* **70**, 971–7.

Tager, J.M., Wanders, R.J.A., Groen, A.K., Kunz, W., Bohnensack, R., Kuster, U., Letko, G., Bohme, G., Duszynski, J. & Wojtczak, L. (1983). Control of mitochondrial respiration. *FEBS Lett.* **151**, 1–9.

Valles, K.L.M., Proudlove, M.O., Beechey, R.B. & Moore, A.L. (1984). Membrane potential measurements in C_3 protoplasts. In *Advances in Photosynthesis Research*, Vol. II., ed. C. Sybesma, pp. 337–40. The Hague: Martinus Nijhoff/Dr W. Junk Publishers.

Vignais, P.V. (1976). Molecular and physiological aspects of adenine nucleotide transport in mitochondria. *Biochimica et Biophysica Acta* **456**, 1–38.

Vignais, P.V. & Lunardi, J. (1985). Chemical probes of the mitochondrial ATP synthesis and translocation. *Annual Review of Biochemistry* **54**, 977–1014.

Wanders, R.J.A., Groen, A.K., Van Roermund, C.W.T. & Tager, J.M. (1984). Factors determining the relative contribution of the adenine-nucleotide translocator and the ADP-regenerating system to the control of oxidative phosphorylation in isolated rat-liver mitochondria. *European Journal of Biochemistry* **142**, 417–24.

Westerhof, H.V., Melandri, B.A., Venturoli, G., Azzone, G.F. & Kell,

D.B. (1984). Mosaic protonic coupling hypothesis for free energy transduction. *Biochimica et Biophysica Acta* **768**, 257–92.

Whitehouse, D.G., Fricaud, A.C. & Moore, A.L. (1989). Role of nonohmicity in the regulation of electron transport in plant mitochondria. *Plant Physiology* **91**, 487–92.

Wilson, D.F., Owen, C.S. & Holian, A. (1977). Control of mitochondrial respiration: a quantitative evaluation of the roles of cytochrome c and oxygen. *Archives of Biochemistry and Biophysics* **182**, 749–62.

Wilson, S.B. (1988). The switching of electron flux from the cyanide-insensitive oxidase to the cytochrome pathway in mung bean (*Phaseolus aureus* L.) mitochondria. *Biochemical Journal* **249**, 301–3.

Wojtczak, L., Bogucka, K., Duszynski, J., Zablocha, B. & Zolkiewska, A. (1990). Regulation of mitochondrial resting state respiration: slip, leak, heterogeneity? *Biochimica et Biophysica Acta* **1018**, 177–81.

A.R SLABAS, R. SWINHOE, D. SLABAS, W.
SIMON and T. FAWCETT

Biosynthesis and assembly of the enzymes involved in lipid metabolism in plants

Fatty acids have an important central role in plant metabolism: this is shown schematically in Fig. 1. The major functions of fatty acids may be divided into three categories: structural, biosynthetic and storage. They also have an important role in signalling.

Structural role of fatty acids in plant cells

Fatty acids provide, amongst other things, the necessary membrane systems in which enzymes function, and the waxy coats (Mikkelsen & Von Wettstein-Knowles, 1984) of plants. Their importance in providing the correct environment for enzymes to function is exemplified by the reconstitution of the product of a cloned purified glycerol 3-phosphate acyltransferase gene, which required insertion into a biological membrane before biological activity was regained (Green *et al.*, 1981). Membranes occur extensively within cells and form the boundaries for the various subcellular organelles such as mitochondria, chloroplasts and the nucleus. Specific arrays of proteins associated with them have various

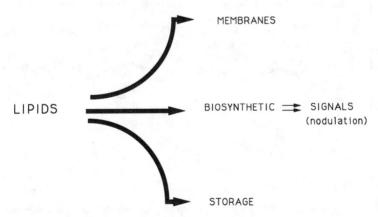

Fig. 1. Schematic representation of the role of lipids in metabolism.

Society for Experimental Biology Seminar Series 50: *Plant organelles*, ed. A. K. Tobin.
© Cambridge University Press 1992, pp. 211–28.

functions: catalytic, receptors, ion channels, permeases and electron transport components. The metabolic contents of these compartments are separated by the membrane system to a greater or lesser degree. The lipid composition, when assayed in detail, differs for each of these membrane systems. We are, however, far from having a complete analysis of all of the membrane lipids as there are few laboratories with the facilities (i) to perform this type of lipid analysis and (ii) rigorously to purify these membrane systems. It is well documented that the fatty acid composition of the membrane lipids may change during environmental stress and that the degree of unsaturation can vary according to the temperature (Murata & Kurisu, 1984). Murata has elegantly demonstrated this in Cyanobacteria and has used the phenomenon to clone a delta-12 desaturase by complementing a mutant strain of *Synechocystis* PCC6803 using selection at low temperatures (Wada *et al.*, 1990).

Biosynthetic function of fatty acids

Fatty acids, in the form of acyl CoAs and acyl carrier proteins (ACPs) act as intermediates in complex lipid biosynthesis and as substrates for fatty acid desaturation (McKeon & Stumpf, 1982; Frentzen *et al.*, 1983). New compounds have been isolated which are acyl derivatives and are important components in nodulation (Lerouge *et al.*, 1990); Fig. 2 shows the structure of one of these compounds, which is involved in the symbiotic host specificity of *Rhizobium melioti*. The biosynthesis of acylated compounds such as these must be mediated by components which share the fundamental reactions of central *de novo* fatty acid biosynthesis. Recent

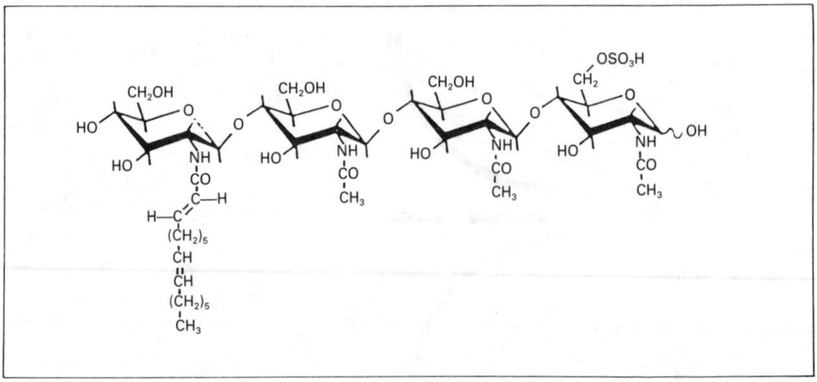

Fig. 2. Symbiotic host specificity is determined by this compound, which contains a long chain fatty acid moiety.

experiments in this laboratory support this view as there is amino acid sequence homology between NodG and β-ketoacylACP reductase (Sheldon *et al.*, 1990).

Fatty acids as storage compounds in plant cells

There is a wide variety of fatty acids which may be used as storage compounds. These range from small molecules, such as acetate (C2) in *Euonymus* to tetracus-15-enol (C24:1) which is found in jojoba oil. Such storage materials have important commercial applications primarily owing to their variety. They constitute a major storage reserve of seeds, mainly in the form of triglycerides (Gurr, 1980). In time there is little doubt that their importance will increase because of concerns about renewable raw materials and social pressures that potential agricultural land should be fully utilised.

Biosynthesis of fatty acids

Plant lipids are thus diverse in structure and function; they are, however, synthesised from basic building blocks. There are a number of excellent reviews on the biosynthesis of fatty acids (e.g. Stumpf, 1987; Harwood, 1988). We will attempt in this chapter to distil some of the essential elements from them and (i) add our interpretation, (ii) present pertinent new data from recent publications and (iii) select questions which we believe will require answers in the future.

Site of *de novo* biosynthesis

The site of *de novo* fatty acid biosynthesis is the plastid (see Stumpf, 1987). Isolated plastids when incubated with radioactive acetate synthesise complex lipids (see Wood *et al.*, this volume, for a discussion). Much of the early pioneering work in this area was performed by Stumpf and his colleagues at Davis. It is not surprising therefore that enzymes of lipid biosynthesis are located in plastids. Figure 3 shows an electron micrograph of an immunogold localisation of one of these components, enoyl ACP reductase (ER), using rabbit anti-ER antibodies.

It is important to point out that not all plastids are equivalent and that the metabolic activities of those from different plant organs, such as leaf and seed, could well be substantially different, despite similarities in morphology.

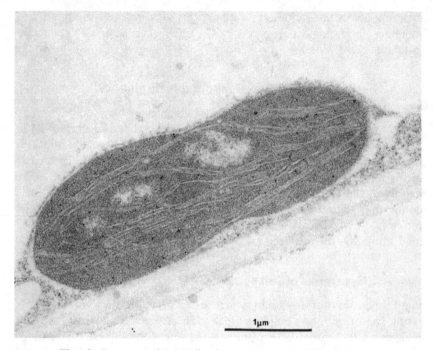

Fig. 3. Immunogold localisation of enoyl (ACP) reductase to the chloroplast of *Brassica napus*.

Enzymes involved in lipid biosynthesis

Major advances in our understanding of these reactions have come from studies of systems which synthesise large quantities of lipid and therefore provide a rich source of enzymes for purification purposes.

Two complex enzyme systems are involved in *de novo* fatty acid biosynthesis. These are acetyl CoA carboxylase (ACC) and fatty acid synthetase (FAS). A single report of a barley mutant has suggested that the FAS components are nuclear coded (Dorne *et al.*, 1982) and hence poly(A)$^+$ RNA has been used as starting material for gene cloning (Safford *et al.*, 1988).

Acetyl CoA carboxylase

This enzyme catalyses the two-step, ATP-driven carboxylation of acetyl CoA to form malonyl CoA, the essential substrate for FAS. Early reports in the literature have stated that the enzyme has variable subunit size dependent on the source from which it was isolated (Hellyer *et al.*, 1986). Suggestions have been made that a lower subunit size occurs in leaf material than in seed material (Wurtele & Nikolau, 1990). Two lines of evidence argue against this. (i) The protein has been purified to electrophoretic homogeneity from leaves and seed of oilseed rape and the subunit size when analysed by SDS/PAGE was 220000 in both cases (Slabas & Hellyer, 1985; Slabas *et al.*, 1986). Thus for rape the seed and the leaf form have a similar high molecular weight when due care is taken and rapid isolation procedures employed. (ii) In wheat an antibody has been raised against the homogeneous 220 kDa purified ACC from wheat germ (Slabas *et al.*, 1988) and on Western blotting a crude extract of wheat leaf and wheatgerm a dominant 220 kDa band is recognised in both (see Fig. 4).

ACC is a biotin-containing enzyme and evidence is emerging that in plants, as in animals, there is a family of biotin-containing enzymes (Chandler & Ballard, 1986; Turnham & Northcote, 1983). Great care will have to be exercised when purifying these enzymes to overcome problems of proteolysis. ACC activity and lipid biosynthesis rates have been measured in developing rape seeds (Turnham & Northcote, 1983). ACC activity and fatty acid synthesis *in vivo* are both subject to a high degree of regulation (Slabas *et al.*, 1987b). Although very little is known about the molecular control of ACC expression in higher plants, the gene has been cloned from animals (Dai *et al.*, 1986) where there are constitutive and inducible forms of the enzyme showing different regulation (Luo & Kim, 1990). We think that it is highly probable that there will be a similar situation in plants.

Fatty acid synthetase

Fatty acid synthetases in plants, animals and yeast differ considerably in organisation; this is shown schematically in Fig. 5. Both the animal and yeast cDNA genes have been cloned (Schweizer *et al.*, 1984, 1989). For a functional catalytic component in animals a dimer is required whilst in yeast a hexamer is required for catalytic activity ($\alpha_6\beta_6$]. Thus at least two ACP domains are necessary for the enzyme to function.

Early work in three independent laboratories established that the system in plants, as in *E. coli*, exists as a large multi-enzyme complex (Shimakata & Stumpf, 1982a; Caughey & Kekwick, 1982; Hoj & Mikkelsen, 1982). At least seven separate polypeptides are present as shown in the simplified model in Fig. 6. This is only a *basal model* and requires expansion in detail. We have as yet no true appreciation of the exact nature or number of components, or the stoichiometry of each, in plant FAS.

The plant-type dissociable fatty acid synthetase is termed a type II FAS. The synthesis of fatty acids *in vitro* is highly stimulated by the

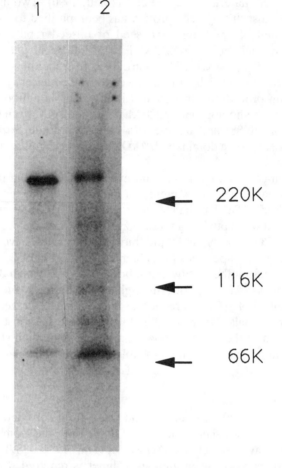

Fig. 4. Western blot of (1) wheatgerm (2) wheat leaf total protein extracts using an antibody raised against wheatgerm acetyl CoA carboxylase.

addition of ACP, presumably due to substrate dilution effects on homogenisation, as has been observed for the ferredoxin requirement for the reconstituted chloroplast system of Walker (Walker & Slabas, 1976). Yeast and animal FAS are high molecular weight, non-dissociable and have been termed type I. There is no clear prokaryotic/eukaryotic difference in structure as the bacteria *Mycobacterium smegmatis* (Wood *et al.*, 1978) and *Brevibacterium ammoniagenes* (Morishima *et al.*, 1982) have all the necessary catalytic domains in one polypeptide like the

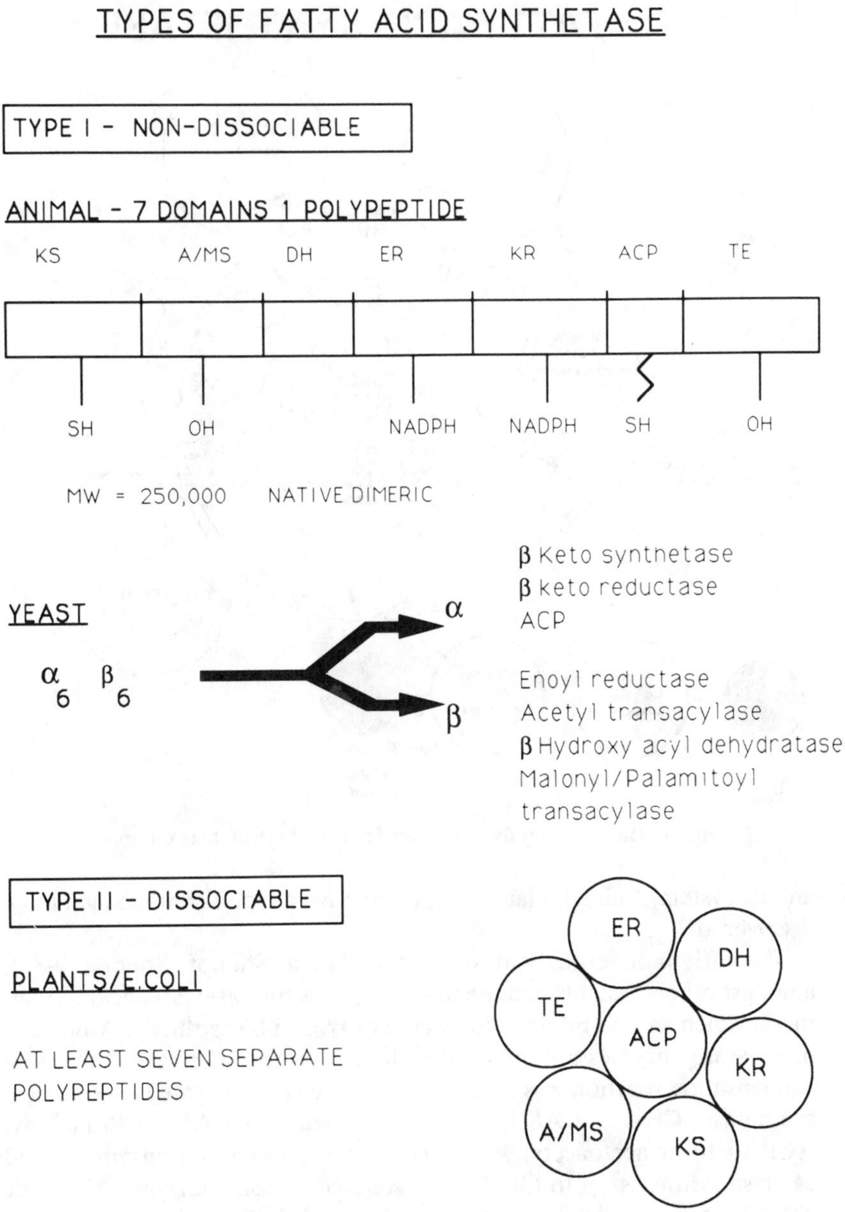

TYPES OF FATTY ACID SYNTHETASE

TYPE I - NON-DISSOCIABLE

ANIMAL - 7 DOMAINS 1 POLYPEPTIDE

| KS | A/MS | DH | ER | KR | ACP | TE |

SH   OH   NADPH   NADPH   SH   OH

MW = 250,000   NATIVE DIMERIC

YEAST

$\alpha_6\ \beta_6$ → α / β

β Keto synthetase
β keto reductase
ACP

Enoyl reductase
Acetyl transacylase
β Hydroxy acyl dehydratase
Malonyl/Palamitoyl
transacylase

TYPE II - DISSOCIABLE

PLANTS/E.COLI

AT LEAST SEVEN SEPARATE
POLYPEPTIDES

ER DH TE ACP KR A/MS KS

Fig. 5. Representation of the types of fatty acid synthetase: KS, β-keto synthetase; A/MS, acetyl/malonyl transferase; DH, β-hydroxy acyl dehydrase; ER, enoyl reductase; KR, β-ketoacyl reductase; ACP, acyl carrier protein; TE, thioesterase.

FATTY ACID SYNTHETASE

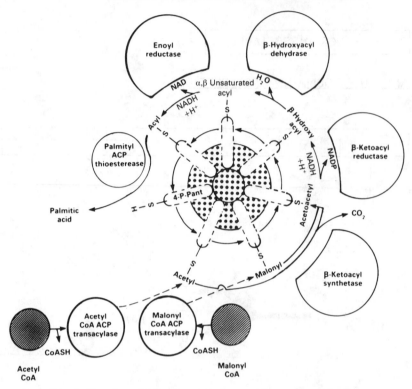

Fig. 6. Basal model for the plant fatty acid synthetase complex.

animal system. Indeed, plants might also have a type I complex, yet to be discovered!

The early pioneering work of Shimakata and Stumpf (Stumpf, 1984), amongst others, established the following facts for basic fatty acid biosynthesis which will be referred to as **core fatty acid biosynthesis**. A number of separate enzymes are involved with ACP as the substrate. The initial condensation reaction was catalysed by condensing enzyme I (β-keto-acyl synthetase, CE I or KAS I) which condensed acetyl ACP with malonyl ACP to form acetoacetyl ACP. This same condensing enzyme would catalyse synthesis up to C16:0. A second condensing enzyme (CE II or KAS II) was required to catalyse C16 to C18 (Shimakata & Stumpf, 1982*b*). Hence schematically we have biosynthesis proceeding as below:

$$C2:0 \xrightarrow[\text{KAS I}]{\text{CE I}} C16:0 \xrightarrow[\text{KAS II}]{\text{CE II}} C18:0$$

Complications with the basal model for fatty acid synthetase

Several complications require expansion of the simple model. The following need to be considered:

1. Nature of the initial reaction.
2. Reasons for isoforms of ACP.
3. Types of substrate specificity.
4. Termination mechanisms.
5. Export and activation.
6. Membrane-bound enzymes.
7. Stoichiometry.
8. Evidence for potential complex and how to take it further.
9. Future with microspore cultures/*Arabidopsis*.
10. Tissue and temporal specificity.

(1) Nature of initial reaction – condensing enzymes and acetyl CoA ACP transacylase limitation

The initial reaction in FAS was until very recently believed to be catalysed by acetyl CoA ACP transacylase which adds an acetyl group onto ACP. This activity was measured in many tissues and believed to be rate-limiting, although *in vitro* experiments may not necessarily represent the situation *in vivo*. However, a new thiolactomycin-sensitive aceto-acetyl ACP synthetase activity (KAS III) was discovered in *E. coli*, which would proceed when KAS I had been inactivated by cerulenin (Jackowski & Rock, 1987). This activity was looked for in plants and duly found (Jaworski *et al.*, 1989; Walsh *et al.*, 1990). It now appears that the initial reaction of plant FAS is catalysed by KAS III, a condensation between acetyl CoA and malonyl ACP. This enzyme is believed also to have acetyl CoA:ACP transacylase (ACT) activity but the evidence is not conclusive (Jaworski *et al.*, 1990). Proof requires the purification of KAS III to homogeneity and assessment of ACT activity in the final preparation. There are precedents for an enzyme having two activities (Cooper *et al.*, 1989). The situation could turn out to be much more complex, as in *E. coli* where mutation of KAS III still allows the synthesis of fatty acids to proceed (C.O. Rock, personal communication). Under these circumstances KAS II together with ACT mediates the initial reactions in fatty acid synthesis. The whole question of ACT/KAS III and rate limitation requires re-evaluation by purification of the component enzymes and also by gene deletion studies in transgenic plants.

There are at least two condensation enzymes, KAS I, KAS II. Recently

KAS I has been purified to homogeneity from oilseed rape (Mackintosh *et al.*, 1989) and the cDNA gene from barley has been cloned (Siggaard-Andersen *et al.*, 1990). Data from this laboratory, on the active site of KAS I, show that there is strong sequence conservation with retroval synthase and chalcone synthase, both of which catalyse condensation-type reactions (Schroder *et al.*, 1988).

(2) Isoforms of ACP

Of all the components of FAS, the most extensively studied is ACP. The protein from *seeds* was first isolated from rape (Slabas *et al.*, 1987*a*) but it had substantial instability problems. However, attention to a specific acylation assay, using *E. coli* acyl ACP synthetase, allowed the protein to be purified to homogeneity and sequenced despite total loss of biological activity. Over 10 cDNA genes were isolated from a rape embryo library (Safford *et al.*, 1988) and de Silva *et al.* have isolated seed-specific genomic clones (de Silva *et al.*, 1990).

There are a number of different types of ACP. Two forms of ACP have been detected immunologically in leaf material of spinach and castor bean. These are termed ACP I and ACP II (Ohlrogge & Kuo, 1985). A protein which migrates on SDS-PAGE to the same position as ACP II is also present in seed material but one corresponding to a mobility of ACP I is absent. The interpretation is limited by the analytical tools in that antibodies are not absolute probes for amino acid sequence but only for epitopes which could be shared by proteins which have different primary sequences. It would seem that the leaf has 2 ACPs and the seed has 1 ACP which is somewhat incongruous as seeds have a major storage function and a structural function whereas the leaf has mainly a structural function. One would have predicted it to be the other way around. Substrate specificity studies on acyltransferase activity (required for MGDG synthesis), acyl ACP thioesterase activity (export from plastid) and TAG synthesis have been made using the two ACPs (Guerra *et al.*, 1986). The results are the opposite to what one would have anticipated considering their major roles. We believe the solution to this problem resides in the differences not detected immunologically and an over-simplification of ACP types by immunological typing. From Northern hybridisation studies leaf ACP cDNA probes do not cross-hybridise with seed mRNA (Scherer & Knauf, 1987) and seed ACP cDNA probes do not hybridise with leaf mRNA (Safford *et al.*, 1988). At this level the two molecules are not equivalent. Because ACP has not been purified and sequenced from seeds and leaves of the same plant it is dangerous to assume that the proteins have the same amino acid sequence. This situation also remains to be resolved. Further complications arise from a report that an ACP is present in mitochondria (Chuman & Brody, 1989) and that an ACP III has been found in barley (Hansen, 1987). It should, however, be noted that in the case of mitrochondrial ACP the metabolic state of the tissue seems to be of importance. There are many forms of ACP which differ at the primary structure level and we have yet to ascribe specific functions to all of them. Again the use of antisense/gene deletion strategies will probably be required to clarify the situation.

(3) Isoforms of catalytic proteins and the complexity of substrate specificity

The existence of at least three forms of KAS for substrate specificity purposes has been mentioned. With β-ketoreductase, enoyl reductase and malonyl CoA ACP transacylase at least two isoforms, with different specificities have been reported (Harwood, 1988) at the level of **core fatty acid biosynthesis**. Their precise function needs to be determined. FAS has 7 catalytic domains and 1 ACP in the basal model; there could easily be as many as 12 catalytic domains and 3 ACPs. However, the situation may appear to be more complicated than it actually is: upon homogenisation there will be a mixing of protein pools which are normally separated *in vivo*.

(4) Termination mechanisms

In plants the termination mechanism is believed to be via acyl-ACP thioesterase (McKeon & Stumpf, 1982); this enzyme has been recently purified to homogeneity from oilseed rape (Hellyer & Slabas, 1990). It has a specificity for 18:1, and a balance between thioesterase activity and acyl-ACP transferase activity is believed to control the ratio of prokaryotic vs eukaryotic lipids (Roughan & Slack, 1984).

An alternative thioesterase (Pollard *et al.*, 1990) with a specificity for medium-chain fatty acids has been partially purified from California bay. This specificity resembles that of acyl ACP thioesterase from rat mammary gland (Slabas *et al.*, 1983; Safford *et al.*, 1987). A comparison of the properties of the two plant thioesterases would give valuable insight into the nature of the molecular basis of substrate specificity. Detailed progress in this area will, however, require the use of plant ACPs, which are now starting to become available.

(5) Export and activation/movement

There is no concrete evidence to the molecular identity of the exported species from plastids or the mechanism of export. Free fatty acids could traverse the membrane but this is still not proven – application of silicone oil techniques and inhibitors of active transport could resolve this problem (see Wood *et al.*, this volume, for a detailed discussion). Once exported and activated to acyl CoAs it would be expected that perhaps an acyl CoA binding protein would be involved in moving the acyl CoA about the cytoplasm, as in animals, but no such specific protein has been identified in seeds during lipid deposition. Non-specific lipid carrier proteins have been identified during germination (Watanabe & Yamada, 1986). Recent studies, in yeast, point to phospholipid transfer proteins having an important role in the compartment-specific stimulation of protein secretion (Bankaitis *et al.*, 1990).

(6) Membrane-bound enzymes
Elongation enzymes external to the plastid will undergo the same basic reactions as **core fatty acid biosynthesis**. Hence it is to be expected that probes designed from protein sequence considerations would enable eventual identification of membrane-bound elongases. These enzymes apparently use acyl CoAs as substrates; however, they may represent a type I complex. Major advances in this area have been slow due to the extreme difficulty in working with membrane-bound systems.

(7) Stoichiometry
The stoichiometry of FAS components is unknown, but with successful procedures for enzyme purification, genetic cloning and analysis with antibodies, it should be possible to answer this question in the near future. The situation in rape looks promising as a number of enzymes have been purified to homogeneity from this source.

(8) Evidence for a potential complex and how to take it further
It seems likely that there will be a close association of FAS components *in vivo*. A number of the components can be purified using ACP affinity chromatography. However, since this represents one-half of the reacting species it is not perhaps surprising. To date the strongest evidence for a complex in plants was obtained fortuitously: enoyl reductase activity co-purified throughout the preparation of acyl ACP thioesterase. These enzymes represent the terminal two reactions of FAS. It will be interesting to see if antibodies raised against enoyl reductase will also precipitate other components of the FAS complex under mild conditions. The complex, if it exists, could be visualised by immunogold techniques using different sized gold particles attached to different antibodies.

(9) The future with microspore cultures and *Arabidopsis*

Microspores from rape in culture synthesise lipids in response to abscisic acid (Taylor *et al.*, 1990). Using seed-specific promoters coupled to appropriate reporter genes it should be possible to elucidate intermediates in such a signal transduction pathway. Under normal circumstances the endogenous structural gene product, and its response to abscisic acid, would be very difficult to monitor. However, by coupling the promotor to a suitably sensitive reporter gene such as GUS it should be easy to monitor at the single cell level. Mutations could be constructed in the plant signal transduction pathway and these would be identified as cells which no longer gave a GUS response. Such experiments are currently being performed in the temperature-sensing pathway in Cyanobacteria (N. Murata, personal communication).

It would be unfair to look to the future of plant lipid research and not mention the large contribution made using *Arabidopsis*, primarily from Somerville's group (Kunst *et al.*, 1989; Browse *et al.*, 1990). They have elegantly isolated a number of mutants of fatty acid synthesis and this will eventually allow them to clone these genes by gene walking which is currently being pursued in that laboratory. It should however be borne in mind that mutations in core FAS are conditionally lethal, and if these are of interest then specialised selection procedures will have to be developed. Since oilseed rape is closely related to *Arabidopsis* it should be possible to isolate core FAS enzymes from *Arabidopsis* using rape cDNA as probes. We have recently used a β-ketoreductase cDNA gene from rape to isolate such a core enzyme from *Arabidopsis* and in fact these two contain over 90% absolute nucleotide sequence homology.

(10) Tissue and temporal specificity

A number of the components of plant FAS are regulated in both a temporal and a tissue-specific manner (Safford *et al.*, 1988; Slabas *et al.*, 1987*b*; de Silva *et al.*, 1990). Such information is obviously encoded in both *cis*-elements in the promoter and *trans*-acting factors which are involved in their regulation (Mitchell & Tijian, 1989). Studies on the promoters of a number of FAS components will help us eventually to understand those elements which they have in common. These elements will include both embryo-specific and temporal-specific factors. As the biological activities of the FAS components involved in storage lipid biosynthesis appear to have the same temporal regulation it is likely that the study of the genes will advance our understanding of temporal expression. If eventually one wishes to modify lipid composition by the introduction of a foreign gene into plants it is important not only to have tissue-specific expression but also correct temporal expression. Exciting times lie ahead in unravelling both the structure and regulation of these genes.

References

Bankaitis, V.A., Aitken, J.R., Cleves, A.E. & Dowhan, W. (1990). An essential role for a phospholipid transfer protein in yeast Golgi function. *Nature* **347**, 561–2.

Browse, J., Miquel, M. & Somerville, C. (1990). Genetic approaches to understanding plant lipid metabolism. In *Plant Lipid Biochemistry, Structure and Utilization*, ed. P.J. Quinn & J.L. Harwood, pp. 431–8. London: Portland Press.

Caughey, I. & Kekwick, R.G.O. (1982). The characteristics of some components of the fatty acid synthetase system in the plastids from the mesocarp of Avocado (*Persea americana*) Fruit. *European Journal of Biochemistry* **123**, 553–61.

Chandler, C.S. & Ballard, F.J. (1986). Multiple biotin-containing proteins in 3T3-LI cells. *Biochemical Journal* **237**, 123–30.

Chuman, L. & Brody, S. (1989). Acyl carrier protein is present in mitrochondria of plants and in eukaryotic micro-organisms. *European Journal of Biochemistry* **184**, 643–9.

Cooper, C.L., Hsu, L., Jackowski, S. & Rock, C.O. (1989). 2-acylglycerol phosphoethanolamine acyltransferase/acyl-acyl carrier protein synthetase is a membrane associated acyl carrier protein binding protein. *Journal of Biological Chemistry* **264**, 7384–9.

Dai, D.H., Pape, M.E., Lopez-Casillas, F., Luo, X.C., Dixon, J.E. & Kim, K.H. (1986). Molecular cloning of cDNA for Acetyl-coenzyme A carboxylase. *Journal of Biological Chemistry* **261**, 12395–9.

de Silva, J., Loader, N.M., Jarman, C., Windust, J.H.C., Hughes, S.G. & Safford, R. (1990). The isolation and sequence analysis of two seed-expressed acyl carrier protein genes from *Brassica napus*. *Plant Molecular Biology* **14**, 537–48.

Dorne, A.J., Carde, J.P., Joyard, J., Borner, T. & Douce, R. (1982). Polar lipid composition of a plastid ribosome-deficient barley mutant. *Plant Physiology* **69**, 1467–70.

Frentzen, M., Heinz, E., McKeon, T. & Stumpf, P.K. (1983). Specificities and selectivities of glycerol-3-phosphate acyltransferase and monoglycerol-3-phosphate acyltransferase from pea and spinach chloroplasts. *European Journal of Biochemistry* **129**, 629–36.

Green, P.R., Merrill, A.H. & Bell, R.M. (1981). Membrane phospholipid synthesis in *Escherichia coli* purification, reconstitution, and characterization of sn-glycerol-3-phosphate acyltransferase. *Journal of Biological Chemistry* **256**, 11151–9.

Guerra, D.J., Ohlrogge, J.B. & Frentzen, M. (1986). Activity of acyl carrier protein isoforms in reactions of plant fatty acid metabolism. *Plant Physiology* **82**, 448–53.

Lipid metabolism in plants 225

Gurr, M.I. (1980). The biosynthesis of triacylglycerols. In *The Biochemistry of Plants*, Vol. 4, ed. P.K. Stumpf & E.E. Conn, pp. 205–48. London: Academic Press.

Hansen, L. (1987). Three cDNA clones for barley leaf acyl carrier proteins I and III. *Carlsberg Research Communications* 52, 381–92.

Harwood, J.L. (1988). Fatty acid metabolism. *Annual Review of Plant Physiology and Molecular Biology* 38, 101–38.

Hellyer, A., Bambridge, H.E. & Slabas, A.R. (1986). Plant acetyl-CoA carboxylase. *Biochemical Society Transactions* 14, 565–8.

Hellyer, A. & Slabas, A.R. (1990). Acyl-[acyl-carrier protein] thioesterase from oil seed rape – purification and characterization. In *Plant Lipid Biochemistry, Structure and Utilization*, ed. B.J. Quinn & J.L. Harwood, pp. 157–9. London: Portland Press.

Hoj, P.B. & Mikkelsen, J.D. (1982). Partial separation of individual enzyme activities of an ACP-dependent fatty acid synthetase from barley chloroplasts. *Carlsberg Research Communications* 47, 119–41.

Jackowski, S. & Rock, C.O. (1987). Acetoacetyl-acyl carrier protein synthease, a potential regulator of fatty acid biosynthesis in bacteria. *Journal of Biological Chemistry* 262, 7927–31.

Jaworski, J.G., Clough, R.C. & Barnum, S.R. (1989). A cerulenin insensitive short chain 3-ketoacyl-acyl carrier protein synthease in *Spinacia oleracea* leaves. *Plant Physiology* 90, 41–4.

Jaworski, J.G., Clough, R.G., Barnum, S.R., Post-Beittenmiller, D. & Ohlrugge, J.B. (1990). Initial reactions of fatty acid synthesis and their regulation. In *Plant Lipid Biochemistry, Structure and Utilization*, ed. B.J. Quinn & J.L. Harwood, pp. 97–104. London: Portland Press.

Kunst, L., Browse, J. & Somerville, C. (1989). A mutant of *Arabidopsis* deficient in desaturation of palmitic acid in leaf lipids. *Plant Physiology* 90, 943–7.

Lerouge, P., Roche, P., Faucher, C., Maillet, F., Truchet, G., Prome, J.C. & Denarie, J. (1990). Symbiotic host-specificity of *Rhizobium melioti* is determined by a sulphated and acylated glucosanine oligosaccharide signal. *Nature* 344, 781–4.

Luo, X. & Kim, K.H. (1990). An enhancer element in the housekeeping promoter for acetyl CoA carboxylase. *Nucleic Acids Research* 18, 3249–54.

McKeon, T.A. & Stumpf, P.K. (1982). Purification and characterization of the stearoyl-acyl carrier protein desaturase and the acyl-acyl carrier protein thioesterase from maturing seeds of safflower. *Journal of Biological Chemistry* 257, 12141–7.

Mackintosh, R.W., Hardie, D.G., Slabas, A.R. (1989). A new assay procedure to study the induction of β-ketoacyl-ACP synthease I and II, and the complete purification of β-ketoacyl-ACP synthease I from developing seeds of oil seed rape (*Brassica napus*). *Biochimica et Biophysica Acta* 1002, 114–24.

Mikkelsen, J.D. & von Wettstein-Knowles, P. (1984). Biosynthesis of esterified alkan-2-ols in barley spike epicuticular wax. In *Structure, Function and Metabolism of Plant Lipids*, ed. P. Siegenthaler & W. Eichenberger, pp. 517–24. Amsterdam: Elsevier Science Publishers.

Mitchell, P.J. & Tjian, R. (1989). Transcriptional regulation in mammalian cells by sequence-specific DNA binding proteins. *Science* **245**, 371–8.

Morishima, N., Ikai, A., Noda, H. & Kawaguchi, A. (1982). Structure of bacterial fatty acid synthetase from *Brevibacterum ammoniagenes*. *Biochimica et Biophysica Acta* **708**, 305–12.

Murata, N. & Kurisu, K. (1984). Fatty acid composition of phosphatidylglycerols from plastids in chilling-sensitive and chilling-resistant plants. In *Structure, Function and Metabolism of Plant Lipids*, ed. P. Siegenthaler & W. Eichenberger, pp. 551–4. Amsterdam: Elsevier Science Publishers.

Ohlrogge, J.B. & Kuo, T.M. (1985). Plants have isoforms of acyl carrier proteins that are expressed differently in different tissues. *Journal of Biological Chemistry* **260**, 8032–7.

Pollard, M.P., Anderson, L., Fan, C., Hawkins, D.J. & Davies, H.M. (1990). A specific acyl-ACP thioesterase implicated in medium-chain fatty acid production in immature cotyledons of *Umbellularia californica*. *Archives of Biochemistry and Biophysics* **284**, 306–12.

Roughan, G. & Slack, R. (1985). Glycerolipid synthesis in leaves. *Trends in Biological Sciences* **9**, 383–6.

Safford, R., de Silva, J., Lucas, C., Windust, J.H.C., Shedden, J., James, C.M., Sidebottom, C.M., Slabas, A.R., Tombs, M.P. & Hughes, S.G. (1987). Molecular cloning and sequence analysis of complementary DNA encoding rat mammary gland medium-chain S-acyl fatty acid synthetase thioester hydrolase. *Biochemistry* **26**, 1358–64.

Safford, R., Windust, J.H.C., Lucas, C., de Silva, J., James, C.M., Hellyer, A., Smith, C.G., Slabas, A.R. & Hughes, S.G. (1988). Plastid-localized seed acyl-carrier protein of *Brassica napus* is encoded by a distinct, nuclear multigene family. *European Journal of Biochemistry* **174**, 287–95.

Scherer, D.E. & Knauf, V.C. (1987). Isolation of complementary DNA clone for the acyl carrier protein-1 of spinach. *Plant Molecular Biology* **9**, 127–34.

Schroder, S., Brown, J.W.S. & Schroder, J. (1988). Molecular analysis of retroval synthase cDNA, genomic clones and relationship with chalcone synthase. *European Journal of Biochemistry* **172**, 161–9.

Schweizer, M., Lebert, C., Holtke, J., Roberts, L.M. & Schweizer, E. (1984). Molecular cloning of the yeast fatty acid synthetase genes, FAS 1 and FAS 2: Illustrating the structure of the FAS 1 cluster gene by transcript mapping and transformation studies. *Molecular and General Genetics* **194**, 457–65.

Schweizer, M., Takabayashi, K., Laux, T., Beck, K.F. & Schreglmann, R. (1989). Rat mammary gland fatty acid synthetase: localization of the constituent domains and two functional polyadenylation/termination signals in the cDNA. *Nucleic Acids Research* **17**, 567–86.

Sheldon, P.S., Kekwick, R.G.O., Sidebottom, C., Smith, C.G. & Slabas, A. R. (1990). 3-oxoacyl-(acyl-carrier protein) reductase from avocado (*Persea americana*) fruit mesocarp. *Biochemical Journal* **271**, 713–20.

Shimakata, T. & Stumpf, P.K. (1982*a*). The prokaryotic nature of the fatty acid synthetase of developing *Carthornus tinctorius* L. (safflower) seeds. *Archives of Biochemistry and Biophysics* **2A**, 144–54.

Shimakata, T. & Stumpf, P.K. (1982*b*). Isolation and function of spinach leaf -ketoacyl-[acyl-carrier-protein] synthetases. *Proceedings of the National Academy of Sciences, USA* **79**, 5808–12.

Siggaard-Andersen, M., Kauppinen, S. & von Wettstein-Knowles, P. (1990). Condensing enzymes in fatty acid and polyketide biosynthesis. *24th NIBB Conference*, Okazaki, Japan, Jan. 15–17, 1990.

Slabas, A.R., Davies, C., Hellyer, A., Mackintosh, R.W., Shelden, P., Hardie, D.G., Kekwick, R.G. O. & Safford, R. (1988). Molecular structure of fatty acid synthesizing enzymes from developing seeds of oil seed rape. In *Plant Lipids: Targets for Manipulation*, Monograph A, ed. N.G. Pinfield & A.K. Stobart, pp. 1–9. Bristol: British Plant Growth Regulator Group.

Slabas, A.R., Harding, J.J., Hellyer, A., Roberts, P. & Bambridge, H.E. (1987*a*). Induction, purification and characterization of acyl carrier protein from developing seeds of oil seed rape (*Brassica napus*). *Biochimica et Biophysica Acta* **921**, 50–9.

Slabas, A.R. & Hellyer, A. (1985). Rapid purification of a high molecular weight subunit polypeptide form of rape seed acetyl CoA carboxylase. *Plant Science Letters* **39**, 177–82.

Slabas, A.R., Hellyer, A. & Bambridge, H.E. (1986). The basic polypeptide subunit of rape leaf acetyl-CoA carboxylase is a 220 kDa protein. *Biochemical Society Transactions* **14**, 714–15.

Slabas, A.R., Hellyer, A., Sidebottom, C., Bambridge, H., Cottingham, I.R., Kessell, R., Smith, C.G., Sheldon, P., Kekwick, R.G. O., de Silva, J., Windust, J., James, C.M., Hughes, S.G. & Safford, R. (1987*b*). Molecular structure of plant fatty acid synthesis enzymes. In *Plant Molecular Biology*, ed. D. von Wettstein & N.H. Chua, pp. 265–77. NATO ASI Series A, Vol. 140.

Slabas, A.R., Ormesher, J., Roberts, P.A., Sidebottom, C.M., Tombs, M.P., Jeffcoat, R.J. & James, A.T. (1983). The interaction of mammalian medium-chain hydrolase with yeast fatty acid synthetase. *European Journal of Biochemistry* **134**, 27–32.

Stumpf, P.K. (1984). Fatty acid synthesis in higher plants. In *Fatty Acid Metabolism and its Regulation*, ed. S. Numa, pp. 155–79. Amsterdam: Elsevier.

Stumpf, P.K. (1987). The biosynthesis of saturated fatty acids. In *The Biochemistry of Plants*, Vol. 9, ed. P.K. Stumpf & E.E. Conn, pp. 121–36. London: Academic Press.

Taylor, D.C., Weber, N., Underhill, E.W., Pomeroy, M.V., Keller, W.A., Scowcruft, W.R., Wilben, R.W., Moloney, M.M. & Holbrook, L.A. (1990). Storage-protein regulation and light accumulation in microspore embryos of *Brassica napus* L. *Planta* **181**, 18–26.

Turnham, E. & Northcote, D.H. (1983). Changes in the activity of acetyl CoA carboxylase during rape-seed formation. *Biochemical Journal* **212**, 223–9.

Wada, H., Gombos, Z. & Murata, N. (1990). Enhancement of chilling tolerance of a cyanobacterium by genetic manipulation of fatty acid desaturation. *Nature* **347**, 200–3.

Walker, D.A. & Slabas, A.R. (1976). Stepwise generation of the natural oxidant in a reconstituted chloroplast system. *Plant Physiology* **57**, 203–8.

Walsh, M.C., Klopfenstein, W.E. & Harwood, J.L. (1990). The short chain condensing enzyme has a wide spread occurrence in the fatty acid synthetases from higher plants. *Phytochemistry* **29**, 3797–9.

Watanabe, S. & Yamada, M. (1986). Purification and characterization of a non-specific lipid transfer protein from germinating castor bean endosperm which transfers phospholipids and galactolipids. *Biochimica et Biophysica Acta* **876**, 116–23.

Wood, W.I., Peterson, D.O. & Bloch, K. (1978). Subunit structure of *Mycobacterium smegmatis* fatty acid synthetase. *Journal of Biological Chemistry* **253**, 2650–6.

Wurtele, E.S. & Nikolau, B.J. (1990). Plants contain multiple biotin enzymes: Discovery of 3-methylcrotonyl-CoA carboxylase, proprionyl-CoA carboxylase and pyruvate carboxylase in the plant kingdom. *Archives of Biochemistry and Biophysics* **278**, 179–86.

CLIFFORD WOOD, CHRISTINE MASTERSON
and DAVID R. THOMAS

The role of carnitine in plant cell metabolism

Carnitine is widely distributed in the tissues of animals, plants and microorganisms. However, whilst there is an abundance of literature on the role of carnitine in animal metabolism, its role in plant metabolism has been less well studied. This chapter reviews work carried out on carnitine in plants and discusses its possible roles in plant metabolism, in particular the shuttling of activated acyl groups between membrane-bound organelles.

General background

Carnitine (3-hydroxy-4-N-trimethyl ammonium butyrate) is a highly polar compound that is widely distributed in nature. It was discovered over 85 years ago independently by two groups, Kutscher (1905) and Gulewitsch & Krimberg (1905). The latter group found the empirical formula to be $C_7H_{15}NO_3$. Crawford & Kenyon (1927) identified its structure as being $(CH_3)_3N^+CH_2CHOHCH_2COO^-$. Carnitine has two low-energy conformers; an extended conformer (Fig. 1A) and a folded conformer (Fig. 1B). Theoretical data show that the preferred conformation for carnitine is the extended form (Murray $et~al.$, 1980).

The carbon atom at the 3 position is an asymmetric carbon atom; thus, carnitine exists as D- and L-optical isomers, the L-form being the naturally occurring, biologically active isomer.

After being considered for many years as merely a constituent of vertebrate muscle, a role for carnitine was established when Fraenkel & Blewett (1947) demonstrated that it is an essential growth factor for larvae of the mealworm, $Tenebrio~molitor$. Parallel investigations by Bremer's group (Bremer, 1983) and Fritz's group (Fritz, 1963) established a role for carnitine in the β-oxidation of long-chain fatty acids. In mammals carnitine is not metabolised, except by bacteria in the gut (Bieber, 1988). Instead, several roles for carnitine in mammalian metabolism have been proposed that involve conjugation of acyl residues to the

Society for Experimental Biology Seminar Series 50: $Plant~organelles$, ed. A. K. Tobin.
© Cambridge University Press 1992, pp. 229–63.

A

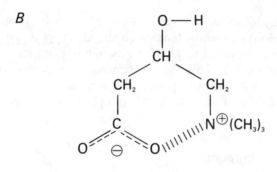

B

C

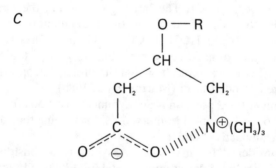

Fig. 1. Conformations of carnitine/acylcarnitines. *A*, The extended conformer of carnitine, where the charged onium head and carboxylate anion are separated – the preferred conformer of free carnitine. *B*, The folded conformer of carnitine. Net electrostatic attraction between the onium head and carboxylate anion forms an internal ionic bond, giving a folded conformer represented by a 'quasi' six-membered ring. *C*, The folded conformer of acylcarnitine (where R = any acyl group) – the preferred conformer of acylcarnitines.

3-hydroxyl group as illustrated in Fig. 1C, with subsequent translocation of these acylcarnitines from one cellular compartment to another. The preferred conformation of these acylcarnitines is the folded conformer (Fig. 1C; Murray *et al.*, 1980). The enzymes responsible for the formation of acylcarnitines are the carnitine acyltransferases. These enzymes catalyse the general reaction

$$\text{acyl CoA} + \text{L-carnitine} \rightleftharpoons \text{L-acylcarnitine} + \text{CoASH} \qquad (1)$$

These reactions are freely reversible, reflecting the high-energy nature of L-acylcarnitines, capable of donating acyl groups to CoASH.

There are two main types of carnitine acyltransferase: carnitine short-chain acyltransferase, also known as carnitine acetyltransferase (CAT; EC 2.3.1.7), and carnitine long-chain acyltransferase, also known as carnitine palmitoyltransferase (CPT; EC 2.3.1.21). Mammalian CATs show a rather broad acyl specificity, with optimum activities for propionyl CoA, with acetyl and butyryl CoA being slightly lower, followed by progressively lower activities up to about C_8 or C_{10} acyl chain lengths (Bieber, 1988). CPT also has a broad acyl specificity from approximately 6 to 20 or more carbons in chain length (Bieber, 1988). There is some dispute as to whether a mid-chain acyltransferase (carnitine octanoyltransferase, COT) also exists. In brief, carnitine acyltransferase activity has been found in the mitochondria and peroxisomes of liver which utilise mid-chain length acyl CoAs. However, attempts to isolate and demonstrate the existence of both COT and CPT activity in mitochondria have led to the conclusion that mitochondrial CPT is responsible for both activities, whilst the broad acyl specificity of the peroxisomal enzyme has caused dispute as to whether it should be designated a COT or a CPT enzyme (for a detailed discussion and references, see review by Bieber, 1988).

The essential role that has been established for carnitine in mammalian metabolism is to act as a carrier of activated acyl groups across membrane barriers. Acyl CoAs permeate membranes at rates too slow to supply metabolic demands at any source or sink. Acylcarnitines, however, can freely permeate membranes via integral membrane translocases, which act as antiport exchange systems for the import and export of carnitine/acylcarnitines. This translocase system has been extensively investigated by Pande and co-workers (Pande, 1975; Pande & Parvin, 1980) and by Tubbs and co-workers (Ramsay & Tubbs, 1975, 1976). In addition the carnitine acyltransferase reactions serve to buffer CoA/acyl CoA pools, affecting both the availability of activated acyl residues and availability of CoASH (Bieber *et al.*, 1982).

Distribution of carnitine in plant tissues

Carnitine is known to be present in several plant tissues. Using the *Tenebrio* bioassay, it was found in corn, wheat germ, wheat seeds and alfalfa seedlings (Fraenkel, 1951, 1953). Using a spectrophotometric DTNB (5,5'-dithio-bis-(2-nitrobenzoic acid)) assay (Marquis & Fritz, 1964) the presence of carnitine was confirmed in wheat seeds and it was also found in oat seedlings, cauliflower inflorescences, avocado mesocarp and peanut seed (Panter & Mudd, 1969). Employing the radioactive assay of Cederblad & Linstedt (1972), carnitine was also found in germinating pea cotyledons (McNeil & Thomas, 1975) and barley leaves (Ariffin *et al.*, 1982). The known distribution and levels of carnitine in plant tissues are summarised in Table 1.

Table 1. *Carnitine content of some higher plants*

	L–carnitine (μg g^{-1})	
Plant material	Fresh weight	Dry weight
Wheat[a]		7–14
Corn[a]		<0.5
Alfalfa[a]		20
Wheat seed[b]		4
Wheat germ[b]		12
Oat seed[b]		1
Oat seedling[b]		14
Peanut seed embryo[b]		8
Avocado mesocarp[b]		48
Cauliflower inflorescence[b]		14
Pea cotyledons[c]		
0 h germinated	<0.2	<0.5
24 h germinated	1	2.5
48 h germinated	0·5–1	2
72 h germinated	0·5–1	2
Barley seed[d]	<0.2	
7 day-old barley leaves[d]		
0 h light	<0.25	
24 h light	0.5–1	
Green barley leaves[d]		
7-day-old	0.5–1	
14-day-old	0.5–1	

[a] from Fraenkel, 1953. [b] from Panter & Mudd, 1969. [c] from McNeil & Thomas, 1975. [d] from Ariffin *et al.*, 1982.

Distribution of carnitine acyltransferases in plant tissues

Panter (1971) and Panter & Mudd (1973) assayed CPT activity in the mitochondria of avocado mesocarp. This enzyme showed substrate stereospecificity: the D-isomers of carnitine and palmitoylcarnitine failed to act as substrates, and the enzyme favoured L-palmitoylcarnitine formation.

Thomas & McNeil (1976) demonstrated carnitine short-chain (CAT) and long-chain (CPT) acyltransferase activity in washed preparations of pea cotyledon mitochondria. A mid-chain (COT) activity was also suggested. Their results are shown in Fig. 2. The three peaks could indicate the presence of three transferase enzymes, namely short-, mid- and long-chain carnitine acyltransferases. If the short- and long-chain enzymes, however, have a broad spectrum of chain length specificity, then the mid-length fatty acids would be utilised by both short- and long-chain carnitine acyltransferases simultaneously, giving a high rate of oxidation for these mid-chain length fatty acids, thus producing the middle peak. The evidence for a mid-chain carnitine acyltransferase is therefore inconclusive.

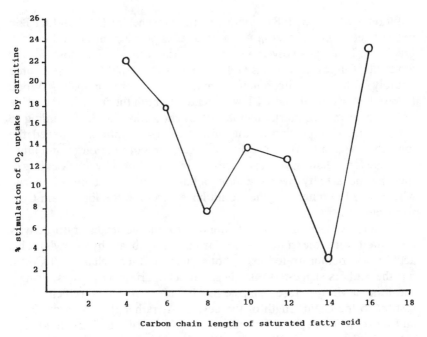

Fig. 2. Stimulation by carnitine of O_2 uptake during oxidation of fatty acids of varying chain lengths by pea cotyledon mitochondria (adapted from Thomas & McNeil, 1976).

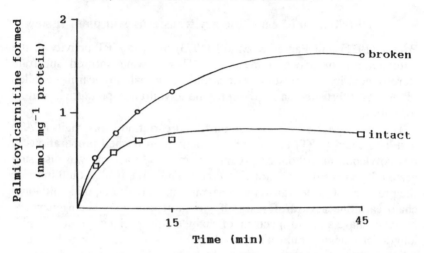

Fig. 3. Carnitine palmitoyltransferase activity in pea cotyledon mitochondria. □—□, intact mitochondria; ○—○, ruptured mitochondria. (From Thomas & Wood, 1986; reproduced by kind permission of Springer-Verlag, Berlin.)

Wood et al. (1983, 1984) confirmed the presence of CAT and CPT in pea cotyledon mitochondria using organelles purified on Percoll density gradients. Using pea cotyledon mitochondria purified on sucrose density gradients, Burgess & Thomas (1986) demonstrated that CAT was located entirely within the mitochondrial matrix. However, Thomas & Wood (1986) demonstrated that CPT was located to both the inside and outside of the inner mitochondrial membrane as shown in Fig. 3. According to their data, the rupturing of the mitochondrial membrane allowed the long-chain (palmitoyl) CoA substrate access to both the outer and inner carnitine long-chain acyltransferase enzymes, as shown by the doubling of the enzyme activity over the rate obtained with intact mitochondria, where only the outer enzyme would have access to the long-chain CoA substrate.

Gerbling & Gerhardt (1988) found carnitine acyltransferase activity associated with the mitochondria from mung bean hypocotyls. This activity was demonstrated using both short- and long-chain acyl CoAs, but the authors suggested that these mitochondria may possess a single carnitine acyltransferase enzyme of broad substrate specificity with respect to the chain length of the acyl CoA, rather than separate long- and short-chain enzymes. Further attempts to purify the enzyme(s) indicated that carnitine acyltransferase activity was confined to the inner mitochondrial membrane fraction, and attempts to solubilise this membrane-bound activity failed (Gerbling *et al.*, 1991). Inhibitor studies suggested resemblance of the protein(s) to a short-chain rather than a long-chain carnitine acyltransferase (Gerbling *et al.*, 1991).

Carnitine long-chain and short-chain acyltransferases have been found in barley etioplasts (Thomas *et al.*, 1982, 1983) and in pea chloroplasts (McLaren *et al.*, 1985).

In contrast to plant mitochondria and chloroplasts, plant microbodies appear to lack any carnitine acyltransferases. Cooper & Beevers (1969) could not detect any carnitine acyltransferase activity in the glyoxysomes of castor bean endosperm. Glyoxysomes from cotton cotyledons (Miernyk & Trelease, 1981), leaf peroxisomes from spinach (Bieber *et al.*, 1981), peroxisomes from mung bean hypocotyls and glyoxysomes from sunflower cotyledons (Gerbling & Gerhardt, 1988) all showed no activity when assayed for carnitine acyltransferase.

Carnitine effects on whole tissue systems

Stobart & Thomas (1968), working on greening tissue cultures of *Kalanchoe crenata*, showed that the chloroplasts of green cultures were juvenile in character, with neither the dimensions nor the structured complexity of mature leaf chloroplasts. They also did not fix CO_2 in the light at rates high enough to support an autotrophic mode of growth (McLaren & Thomas, 1967). Detailed analysis of the components of the plastids demonstrated a deficiency in those constituents of the chloroplast derived from acetyl CoA, such as fatty acids, carotenoids, chlorophylls, plastoquinone and sterols (Stobart *et al.*, 1967; Thomas & Stobart, 1970). Compounds likely to increase metabolic traffic through acetyl CoA were added to the callus growth medium to try and boost production of these essential components required for chloroplast maturation. Of all the compounds tested, only carnitine had an effect. Some calluses doubled or trebled their chlorophyll and fatty acid content when 5 mM D,L-carnitine was added to the culture medium, although the effect was very variable (D. R. Thomas, unpublished data).

The effect of carnitine on the greening of etiolated barley leaves has also been investigated. The level of carnitine itself increased with greening, from very low levels in ungerminated grain to a maximum level in 7-day-old green leaves (Ariffin *et al.*, 1982). 10 mM D,L-carnitine supplied to etiolated barley leaves greening in the light increased the rate of chlorophyll production by 40% (Thomas *et al.*, 1981), as shown in Fig. 4.

Carnitine has also been found to affect the rate of acidification and deacidification in Crassulacean acid metabolism (CAM) plants. When carnitine was added to the water supply of detached leaves of *Kalanchoe*,

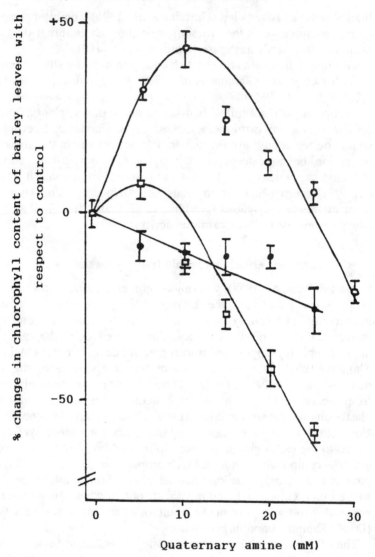

Fig. 4. The effect of quaternary amines on chlorophyll production in greening barley leaves. Plants were grown for 7 days in the dark. Detached leaves were then placed with their cut ends in a solution of carnitine (O—O), choline (□—□) or betaine (●—●) and illuminated for 24 h prior to chlorophyll determination (adapted from Thomas *et al.*, 1981).

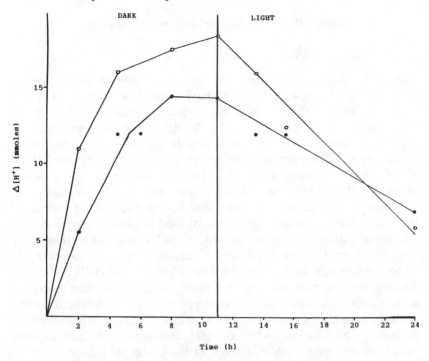

Fig. 5. The effect of carnitine on the dark acidification and light deacidi-fication of leaves of *Kalanchoe crenata*. Detached leaves were placed with their petioles in water (●—●, control) or 20 mM D,L-carnitine (○—○) and sampled and assayed for titratable acidity at intervals over a 24 h period.

the rate of dark acidification was increased over that of the controls. When the leaves were subsequently returned to the light, the deacidifica-tion rate was also accelerated by carnitine treatment, the net values of acidification being higher in the carnitine treatment. The results of C. Wood and D. R. Thomas (unpublished data) are shown in Fig. 5. These results were confirmed by Brown and Griffiths, who repeated the experi-ments a few years later (Griffiths, 1988). No physiological significance can be attached to these data as yet, as both Wood & Thomas and Brown & Griffiths used D,L-carnitine at 20 mM, but neither 10 mM nor 40 mM produced any significant difference between treated and control leaves.

Carnitine effects on isolated organelle systems

Mitochondria

Panter & Mudd (1973) observed that the addition of L-carnitine stimulated oxidation of a range of fatty acids (from butyric to stearic acid) by avocado slices, whilst D-carnitine inhibited oxidation of all but octanoic and palmitic acid. O_2 uptake in the presence of L-palmitoylcarnitine was more than double the endogenous rate of O_2 uptake by avocado mitochondria, whereas D-palmitoylcarnitine had no effect on this endogenous rate. Palmitic acid was not oxidised unless ATP, CoASH and $MgCl_2$ were added. The oxidation of L-palmitoylcarnitine, by contrast, was inhibited by the presence of these cofactors. Palmitoyl CoA was oxidised at about half the rate of L-palmitoylcarnitine. Panter & Mudd (1973) concluded that since L-palmitoylcarnitine was oxidised, an enzyme in the mitochondria was either converting palmitoylcarnitine into palmitoyl CoA (CPT activity) or forming free palmitic acid plus carnitine (palmitoylcarnitine hydrolase activity; EC 3.1.1.28). They considered the second alternative very unlikely since L-palmitoylcarnitine was oxidised faster than palmitic acid. Moreover, addition of ATP, CoASH and $MgCl_2$ did not increase palmitoylcarnitine oxidation, though having a large stimulatory effect on palmitate oxidation. Since D-palmitoylcarnitine was not oxidised by the mitochondria and D-carnitine failed to stimulate fatty acid oxidation by slices, the substrate stereospecificity of the carnitine palmitoyltransferase in avocado was presumably similar to that of the animal enzyme. Panter (1971) found that added carnitine was not essential for optimum oxidation of fatty acids by avocado mitochondria but that acylcarnitines were oxidised. CPT activity was assayed in these mitochondria and the enzyme found to favour L-palmitoylcarnitine formation.

McNeil & Thomas (1976) investigated the effect of carnitine on palmitate oxidation by pea cotyledon mitochondria. The requirement, to provide maximum O_2 uptake by pea cotyledon mitochondria for the cofactors Mg^{2+}, CoASH, ATP and the 'sparker' acid malate during the oxidation of palmitate, suggested that β-oxidation of the fatty acid was occurring with subsequent entry of acetyl groups into the TCA cycle. 1 mM L-carnitine stimulated O_2 uptake following ADP addition (i.e. state 3) by 28%. When ATP was omitted from the incubation medium, the addition of carnitine hardly affected the rate of O_2 uptake. This suggested that carnitine was exerting its effect subsequent to the formation of palmitoyl CoA. When malate was omitted from the incubation medium, the absolute rate of O_2 uptake was reduced, but the magnitude of the carnitine-induced stimulation of O_2 uptake in the presence of malate (28% stimulation) was hardly affected when malate was withheld (24.9% stimulation). Thus, although the absence of the 'sparker' acid malate limited acetyl CoA entry into the TCA cycle, thereby reducing the absolute rate of O_2 uptake, carnitine could still exert its stimulatory effect on O_2 uptake, probably by enhancing the availability of acyl CoA derivatives to β-oxidation sites. The carnitine enhancement of palmitate oxidation required intact mitochondrial membranes. Disruption of the mitochondrial membranes, by sonic and osmotic shock, or by detergent treatment, resulted in disappearance of the carnitine stimulation. Thus carnitine was not acting as a detergent. It was suggested that carnitine exerted its effect through the formation of palmitoylcarnitine, which penetrates a membrane barrier more easily than does palmitoyl CoA, and obtains access to β-oxidation sites within the pea cotyledon mitochondria.

Wood *et al.* (1984) demonstrated that palmitoylcarnitine was rapidly oxidised by pea mitochondria. L-carnitine was an essential requirement for the oxidation of palmitate or palmitoyl CoA. When palmitate was sole substrate, ATP and Mg^{2+} were also necessary in order to achieve maximal rates of oxidation. CoASH inhibited the oxidation of palmitate and it was shown that CoASH was acting as a competitive inhibitor of the carnitine-stimulated O_2 uptake as shown by the Lineweaver–Burk plots in Fig. 6. Wood *et al.* (1984) suggested that reactions 2 and 3 occurred outside the mitochondrial membrane barrier, i.e. on the cytosol side, to form palmitoylcarnitine:

$$\text{palmitate} + \text{CoASH} + \text{ATP} \xrightarrow{Mg^{2+}} \text{palmitoyl CoA} + \text{AMP} + \text{PPi} \qquad (2)$$
$$\text{palmitoyl CoA synthetase}$$

$$\text{palmitoyl CoA} + \text{L-carnitine} \ \rightleftharpoons \ \text{L-palmitoylcarnitine} + \text{CoASH} \qquad (3)$$
$$\text{CPT}$$

Palmitoylcarnitine then entered the mitochondria via an intrinsically located translocase. Once inside the mitochondria, the palmitoyl group could then undergo β-oxidation in the mitochondrial matrix.

In further support of this hypothesis, Thomas & Wood (1986) demonstrated that CPT was located on both the inside and the outside of the inner mitochondrial membrane of pea cotyledon mitochondria (see Fig. 3). Thomas *et al.* (1988) also found that the long-chain acyl CoA synthetase of these mitochondria was located on the outer mitochondrial membrane, and that palmitoyl CoA could not permeate the inner mitochondrial membrane, but first had to be converted to palmitoylcarnitine.

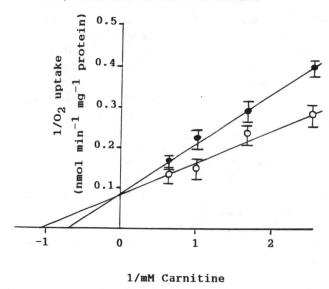

1/mM Carnitine

Fig. 6. Lineweaver–Burk plots demonstrating competitive inhibition by CoASH of the L-carnitine effect on palmitate oxidation by pea cotyledon mitochondria. O−O, 0.04 mM CoASH; ●−●, 0.06 mM CoASH. (From Wood *et al.*, 1984; reproduced by kind permission of Springer-Verlag, Berlin.)

Thus, as in animal cells, carnitine has a role to play in the β-oxidation of long-chain fatty acids by plants.

L-carnitine also affects the short-chain acyl CoA metabolism of plant mitochondria. Thomas & Wood (1982), using washed mitochondrial preparations, showed that acetylcarnitine was rapidly oxidised by pea cotyledon mitochondria and that L-carnitine was essential for the oxidation of acetate, or acetyl CoA. When acetate was the sole substrate, ATP and Mg^{2+} were also required in order to obtain maximum rates of oxidation, presumably being needed for the activation of acetate to acetyl CoA catalysed by acetyl CoA synthetase:

$$\text{acetate + ATP + CoASH +} \xrightarrow[\text{acetyl CoA synthetase}]{Mg^{2+}} \text{acetyl CoA + AMP + PPi} \quad (4)$$

The requirement for L-carnitine was suggested to be for the reaction catalysed by carnitine acetyltransferase:

$$\text{acetyl CoA + L-carnitine} \underset{\text{CAT}}{\rightleftharpoons} \text{L-acetylcarnitine + CoASH} \quad (5)$$

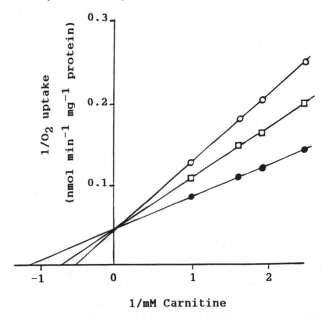

Fig. 7. Lineweaver–Burk plots demonstrating competitive inhibition by CoASH of the L-carnitine effect on acetate oxidation by pea cotyledon mitochondria. O–O, 0.12 mM CoASH; □–□, 0.08 mM CoASH; ●–●, 0.04 mM CoASH. (From Thomas & Wood, 1982; reproduced by kind permission of Springer-Verlag, Berlin.)

CoASH additions inhibited the oxidation of acetate, acetyl CoA and acetylcarnitine and it was shown that CoASH was acting as a competitive inhibitor of carnitine-stimulated O_2 uptake (see Fig. 7). Thomas & Wood (1982) suggested that enough endogenous CoASH remained outside the mitochondrial membrane barrier in their preparations to satisfy the requirements for reaction 4. On addition of CoASH, the equilibrium of reaction 5 was displaced to the left, leading to the formation of acetyl CoA outside the mitochondrial membrane barrier. Since acetyl CoA does not permeate the membrane barrier, O_2 uptake was inhibited. Thomas & Wood (1982) postulated that reactions 4 and 5 occurred outside the mitochondrial membrane barrier, i.e. on the cytosol side, to form acetylcarnitine. The acetylcarnitine then readily entered the matrix of the pea mitochondria via an integral membrane translocase, where an internal CAT transferred the acetyl groups to CoASH. The acetyl groups were then able to enter the TCA cycle.

In later work, Burgess & Thomas (1986) purified mitochondria from pea cotyledons on sucrose density gradients so that they were free from contamination by soluble enzymes from other parts of the cell. Acetate and acetyl CoA were not oxidised by these preparations, even in the presence of all cofactors. L-acetylcarnitine, however, was oxidised. Assays carried out on intact, purified mitochondria did not detect any carnitine acetyltransferase activity, but when the mitochondria were ruptured, CAT activity was detected using several procedures. The authors concluded that carnitine acetyltransferase was located entirely within the mitochondrial matrix and did not occur on the outside (cytosol side).

Wood *et al.* (1983), however, detected CAT activity in intact mitochondria purified on Percoll gradients. Interestingly, though, this activity of the carnitine short-chain acyltransferase enzyme was *c.* 1000-fold less than that reported for carnitine long-chain acyltransferase in intact pea mitochondria (Wood *et al.*, 1984). Thus, any CAT activity associated with the outside of the mitochondrial membrane barrier, is very low when compared with the activity of the long-chain transferase.

Thomas & Wood (1982, 1986), Wood *et al.* (1983), Thomas *et al.* (1983), McLaren *et al.* (1985) and Burgess & Thomas (1986) proposed that the acetylcarnitine formed in the mitochondrial matrix by an internal mitochondrial CAT may well be exported to extramitochondrial sites for anabolism, e.g. the chloroplast stroma where it may be used for fatty acid synthesis. In support of this hypothesis, Burgess & Thomas (1986) and Budde *et al.* (1991) have demonstrated that 1 mM carnitine diminished both the state 3 and state 4 rate of O_2 uptake of mitochondria oxidising pyruvate. This they attributed to acetyl CoA, formed from the pyruvate by the action of the pyruvate dehydrogenase complex (PDC), being acted on by the matrix CAT to form acetylcarnitine which was exported, instead of the acetyl CoA entering the TCA cycle. This interpretation was supported by data showing the effect of carnitine on malonate inhibition of state 3 O_2 uptake. Malonate slows down the supply of oxaloacetate, which combines with acetyl CoA to form citrate, and thus the TCA cycle is slowed down or stopped. In the presence of carnitine, however, this malonate inhibition of O_2 uptake was halved, presumably because carnitine removed acetyl groups from acetyl CoA via matrix CAT to form acetylcarnitine. The $NADH_2$ produced by the action of PDC could thus continue to be oxidised, maintaining O_2 uptake rates.

Chloroplasts

Wood (1984) investigated the effect of carnitine on the incorporation of acetyl moieties into chloroplast lipids. L-carnitine stimulated the incorporation of acetyl CoA into the saponifiable fraction of barley etioplasts. In the light and in the presence of 1 mM L-carnitine, incorporation of ^{14}C from (1-^{14}C) acetyl CoA into the saponifiable fraction trebled that incorporation obtained in the dark or in the absence of L-carnitine in the light. Wood (1984) concluded that this was attributable to an external CAT converting acetyl CoA, which cannot permeate the chloroplast envelope (Brooks & Stumpf, 1965), to acetylcarnitine, which can (see reaction 5), thus rendering the acetyl groups available to the stromal lipid synthesising sites. In the presence of Mg^{2+}, CoASH and ATP, the addition of 1 mM L-carnitine also doubled the incorporation of (1-^{14}C) acetate into the saponifiable fraction of barley etioplasts. This was accounted for by an external acetyl CoA synthetase, in the etioplast preparations converting some of the supplied acetate to acetyl CoA (see reaction 4). As before, L-carnitine would then allow this acetyl CoA access to the chloroplast stroma via conversion to L-acetylcarnitine by the external CAT enzyme. In the presence of 0.1 mM CoASH, ^{14}C incorporation from (1–^{14}C) acetate was reduced in comparison to that measured when 0.04 mM CoASH was used, suggesting that CoASH was acting as an inhibitor of the carnitine effect. To test this inhibitory effect, incorporation of ^{14}C from (1-^{14}C) acetate was measured at two concentrations of CoASH with varying L-carnitine concentrations. The resultant Lineweaver–Burk plots (see Fig. 8) showed that CoASH was behaving as a competitive inhibitor of L-carnitine action in exactly the same way as had been observed for pea mitochondria oxidising acetate (see Fig. 7; Thomas & Wood, 1982) or palmitate (see Fig. 6; Wood *et al.*, 1984). Wood (1984) concluded that an increase in CoASH concentration would promote the formation of acetyl CoA (see reaction 5) which does not penetrate the chloroplast envelope and would therefore be unavailable for lipid synthesis. L-carnitine stimulation and subsequent inhibition by CoASH of acetyl group incorporation into saponifiable lipids was also shown for pea chloroplasts supplied with acetyl CoA as substrate (Wood, 1984).

The effect of carnitine specifically on fatty acid synthesis was then investigated using isolated, purified pea chloroplasts. Masterson *et al.* (1990*a*) showed the effect of L-carnitine on acetyl CoA incorporation into fatty acids. There was a negligible incorporation of acetyl CoA into fatty acids, just 25 pmol mg^{-1} chlorophyll min^{-1}. This would be expected if acetyl CoA cannot penetrate the chloroplast envelope. In the presence of 1 mM L-carnitine, however, acetyl CoA incorporation into fatty acids was increased 9-fold to 222 pmol mg^{-1} chlorophyll min^{-1}. This was concluded to be due to an external CAT converting acetyl CoA to acetylcarnitine which can penetrate the chloroplast membrane barrier.

L-acetylcarnitine itself was an excellent substrate for fatty acid synthesis, being incorporated into fatty acids at a rate of 172 nmol mg^{-1} chlorophyll min^{-1}, at a saturating substrate concentration of 0.15 mM

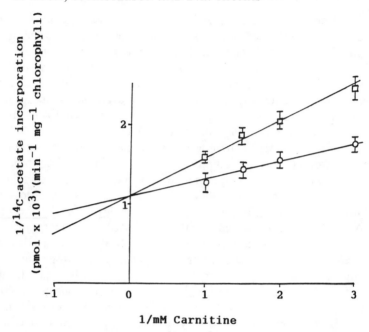

Fig. 8. Lineweaver–Burk plots demonstrating competitive inhibition by CoASH of the carnitine effect on acetate incorporation into saponifiable lipids by barley etioplasts. □—□, 0.1 mM CoASH; O—O, 0.04 mM CoASH. (From Wood, 1984.)

(Masterson *et al.*, 1990*a*); of four substrates tested, L-acetylcarnitine gave higher incorporation rates into fatty acids than acetate, pyruvate or citrate at all substrate concentrations, and so was the preferred substrate for fatty acid synthesis in pea chloroplasts, as is clearly shown in Fig. 9. As well as the quantity of incorporation being higher, the labelling pattern of the fatty acids also differed. When L-acetylcarnitine was supplied as substrate, 65% of the incorporated label appeared in monoenoic fatty acids and 35% in saturated fatty acids, whereas when acetate, pyruvate and citrate were supplied as substrate, the amounts of label recovered in both the saturated and monoenoic fatty acids were 50% each (see Table 2).

D-acetylcarnitine did not act as a substrate for fatty acid synthesis, indicating that the L-acetylcarnitine was not being hydrolysed to free acetate before incorporation into fatty acids (Masterson *et al.*, 1990*b*). Furthermore, equimolar quantities of deoxycarnitine and D-carnitine inhibited the incorporation of L-acetylcarnitine into fatty acids by 66%

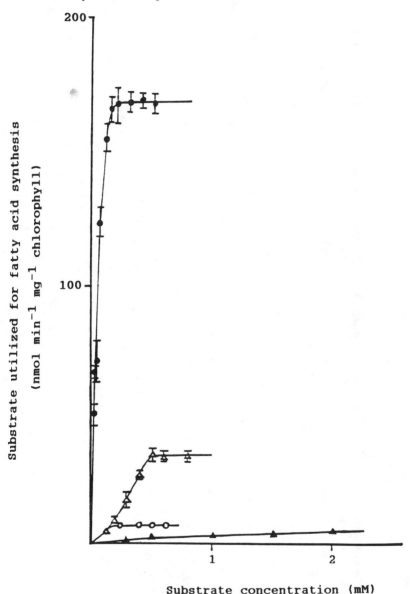

Fig. 9. A comparison of the rates of incorporation of varying concentrations of (^{14}C)-labelled substrates into fatty acids by isolated pea chloroplasts. ●—●, L-acetylcarnitine; △—△, acetate; ○—○, pyruvate; ▲—▲, citrate. (From Masterson *et al.*, 1990*a*; reproduced by kind permission of Blackwell Scientific Publications Limited.)

Table 2. *A comparison of the fatty acid labelling patterns after 15 min incubations of pea chloroplasts with* ^{14}*C-labelled substrates*

Substrate	Incorporation rate into total fatty acids ($nmol\ min^{-1}\ mg^{-1}$ chlorophyll)	Percentage distribution of incorporated substrate into different fatty acid types[a] %			
		S	M	D	T
(1,5–^{14}C)-citrate	4.5±0.25	52	48	0	0
(2–^{14}C)-pyruvate	6.8±0.3	48	52	0	0
(1–^{14}C)-acetate	35.4±3.1	56	44	0	0
L–(1–^{14}C acetyl) carnitine	162.0±8.0	35	65	0	0

[a] S, saturated; M, monoenoic; D, dienoic; T, trienoic fatty acids.
(From Masterson *et al.*, 1990a; reproduced by kind permission of Blackwell Scientific Publications Limited.)

and 76% respectively, whilst having no effect on the rate of acetate incorporation into fatty acids, as shown in Tables 3 and 4. These results not only confirmed that L-acetylcarnitine and acetate were incorporated into fatty acids by different routes, i.e. that L-acetylcarnitine was not hydrolysed to free acetate prior to incorporation, but also showed that L-acetylcarnitine incorporation into fatty acids was being competitively inhibited at two sites. The two sites were interpreted by Masterson *et al.* (1990*b*) to be the integral membrane carnitine:acylcarnitine translocator and the internal chloroplastic carnitine acetyltransferase enzyme. Masterson *et al.* (1990*a*) proposed a scheme whereby chloroplasts could be supplied with L-acetylcarnitine by the mitochondria. They postulated that acetyl CoA in the mitochondria could be converted to L-acetylcarnitine by CAT in the mitochondrial matrix. The L-acetylcarnitine could then be exported from the mitochondria via a carnitine:acylcarnitine translocator in the mitochondrial inner membrane. L-acetylcarnitine could then enter the chloroplast via a chloroplastic carnitine:acylcarnitine translocator to be converted to acetyl CoA by the chloroplastic carnitine acetyltransferase.

The nature of carnitine action

When L-(^{14}C-methyl)carnitine was fed to greening barley leaves for 1 h in the light, 93.6% of the total ^{14}C recovered was present as unchanged carnitine and 6.1% was recovered as palmitoylcarnitine (Thomas *et al.*, 1981), demonstrating that carnitine acted neither as a carbon nor as a nitrogen source (see Table 5). Carnitine was deemed unlikely to act as a methyl group donor because little ^{14}C was incorporated into other components. Other 'onium' compounds, choline and betaine, did not substitute for carnitine in promoting chlorophyll synthesis (see Fig. 4).

McNeil & Thomas (1976) demonstrated that carnitine enhancement of palmitate oxidation by pea cotyledon mitochondria required intact mitochondrial membranes. Disruption of the mitochondrial membranes by sonic and osmotic shock or by detergent treatment resulted in a disappearance of the carnitine stimulation. Thus carnitine was not acting as a detergent and was not causing rupture of membranes. The cofactor requirements also indicated that carnitine was exerting its effect subsequent to the formation of palmitoyl CoA.

Masterson *et al.* (1990*a*) demonstrated that L-acetylcarnitine was the best substrate for fatty acid synthesis by isolated pea chloroplasts (see Fig. 9). Inhibitor studies suggested that the L-acetylcarnitine was not being hydrolysed to free acetate prior to incorporation into fatty acids, but was being converted to acetyl CoA via carnitine acetyltransferase

Table 3. *The effect of deoxycarnitine additions on the incorporation of* L–(1–^{14}C acetyl) carnitine and (1–^{14}C)-acetate into the fatty acids of pea leaf chloroplasts

Substrate	Deoxycarnitine concentration (mM)	Substrate incorporation into fatty acids (nmol min^{-1} mg^{-1} chlorophyll)	Percentage inhibition
L–(1–^{14}C acetyl) carnitine	0.0 (control)	165.2±8.1	–
	0.3	72.7±5.4	56
	0.5	56.2±1.6	66
	1.5	46.3±2.9	72
(1–^{14}C)-acetate	0.0 (control)	34.9±2.4	–
	500	33.4±1.9	4

Adapted from Masterson *et al.*, 1990*b*.

Table 4. *The effect of D-carnitine additions on the incorporation of L–(1–^{14}C acetyl) carnitine and (1–^{14}C)-acetate into the fatty acids of pea leaf chloroplasts*

Substrate	D–carnitine concentration (mM)	Substrate incorporation into fatty acids (nmol min^{-1} mg^{-1} chlorophyll)	Percentage inhibition
L–(1–^{14}C acetyl) carnitine	0.0 (control)	161.0±5.6	–
	0.5	38.0±2.1	76
	5.0	0	100
(1–^{14}C)-acetate	0.0 (control)	33.9±2.9	–
	0.5	33.5±1.5	1.1
	5.0	34.3±1.6	0

Adapted from Masterson *et al.*, 1990*b*.

Table 5. *Carbon recovery following feeding of ^{14}C-labelled carnitine, CO_2 and glucose to green barley leaves*

Fraction	Substrate		
	CO_2 (pmol g^{-1} fresh wt)	Glucose (pmol g^{-1} fresh wt)	Carnitine (pmol g^{-1} fresh wt)
CO_2 evolved	–	5430	40
Ethanol extract	13 600	761	233
Acidic fraction	8010	167	8
Basic fraction	3270	41	189
Neutral fraction	2260	536	21
Solid residue	299	8	4

Each solution contained 40 μatoms carbon. Green barley leaves were supplied isotope for 1 h in the light prior to ethanol extraction and subsequent fractionation. From Thomas *et al.*, 1981; reproduced by kind permission of Pergamon Press, UK.

(Masterson *et al.*, 1990*b*). Rupturing of the chloroplast membrane reduced the efficiency of L-acetylcarnitine as a substrate (Masterson *et al.*, 1990*a*).

All these results suggest that carnitine exerts its effect by facilitating the transfer of activated acyl groups across intact membranes via the carnitine acyltransferase enzymes and integral membrane carnitine:acylcarnitine translocators. Figure 10 demonstrates the situation for any membrane-bound compartments in a plant cell and shows how carnitine alleviates the rigid compartmentation of acyl CoA:CoASH. CoASH:acyl CoA couples can permeate membranes only at rates insufficient to meet metabolic demands at any source or sink. In each membrane-bound compartment, a finite pool of CoASH:acyl CoA exists, and rapid changes in this total pool size cannot be visualised logically. However, the proportion of CoASH to acyl CoA may vary in each of the membrane-bound pools. Also the type of acyl CoA within each pool may vary.

Acylcarnitine:carnitine couples are freely translocated between the membrane-bound compartments of the plant cell. An event or perturbation occurring in any one single compartment would momentarily affect cell metabolism until equalisation is re-established via the movement by mass action of acylcarnitine:carnitine couples between all the compartments of the cell, thereby restoring the original ratios of acyl CoA: CoASH couples in each compartment pertaining before the perturbation. Carnitine thus acts homeostatically in the plant cell. Since the carnitine

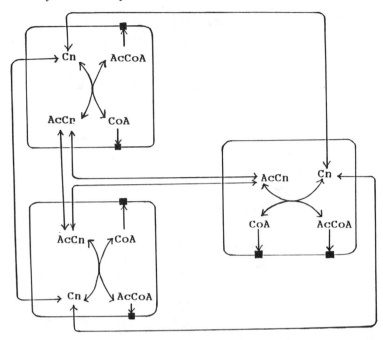

Fig. 10. Alleviation of the rigid compartmentation of acyl CoA/CoASH by carnitine. Cn, L-carnitine; AcCn, L-acylcarnitine; CoA, coenzyme A; AcCoA, acyl coenzyme A; ▮, no or very little transport. (Adapted from Masterson *et al.*, 1990*b*.)

acyltransferase enzyme reactions are freely reversible, by simple mass action they will serve to buffer acyl CoA:CoASH pools within a cell compartment, affecting both the availability of activated acyl residues and availability of CoASH. Likewise, by mass action acylcarnitines/carnitine will freely move between the various cell compartments in exchange for acylcarnitines/carnitine, thus maintaining equilibrium throughout the cell in a strictly controlled manner. Some specific examples of the carnitine-mediated shuttling of metabolites between organelles for which there is evidence are presented below.

Carnitine is essential for the entry of long-chain fatty acids to mitochondrial β-oxidation sites (McNeil & Thomas, 1976; Wood *et al.*, 1984; Thomas & Wood, 1986; Thomas *et al.*, 1988). The scheme is shown diagrammatically in Fig. 11. Activation of free fatty acids occurs at the outer mitochondrial membrane; the inner mitochondrial membrane is a barrier to acyl CoAs (Thomas *et al.*, 1988). A carnitine long-chain acyltransferase located on the outside of the mitochondrial inner mem-

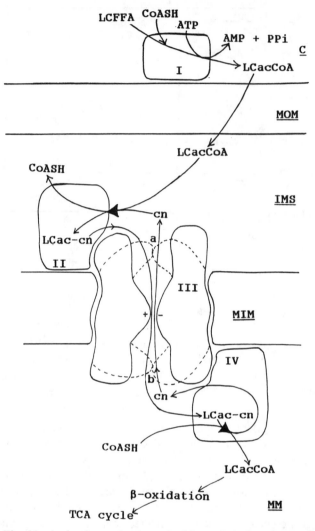

Fig. 11. Activation and transport of long-chain free fatty acids into the mitochondrial matrix for β-oxidation. When the IMS side of the translocator (a) is open and the MM side of the translocator (b) is closed, the positive quaternary nitrogen and the negative carboxyl anion of the long-chain acylcarnitine form ionic bonds with the (−) and (+) sites in the channel of the translocator (III). It is envisaged that (a) then closes and (b) opens. By mass action, matrix carnitine exchanges with the long-chain acylcarnitine which then is available to the active site of enzyme IV which reforms long-chain acyl CoA and releases carnitine. The released carnitine may exchange when (b) is open and (a) is closed for another

brane (Thomas & Wood, 1986; Thomas *et al.*, 1988) allows the formation of the permeant species, long-chain acylcarnitine. This passes through the mitochondrial inner membrane via an intrinsic translocator (Thomas & Wood, 1986) and is reconverted to long-chain acyl CoA by a carnitine long-chain acyltransferase located on the inside of the inner mitochondrial membrane. This internal CPT has been demonstrated by Thomas & Wood (1986). The long-chain acyl CoA then enters β-oxidation in the mitochondrial matrix. β-oxidation in mitochondria may not proceed to completion, but may break fatty acids down to smaller units which are then exported and used in biosynthesis. Export of such fatty acids could occur by a reversal of the above scheme (Fig. 11).

Thomas & Wood (1982) and Burgess & Thomas (1986) proposed a similar scheme for the import/export of short-chain acyl groups by mitochondria (see Fig. 12). CAT is located on the inside of the inner membrane in pea mitochondria. In the presence of carnitine there would be competition between carnitine acetyltransferase and citrate synthase for the acetyl CoA pool of the mitochondrial matrix. Any acetylcarnitine formed would be freely available for export via the integral membrane translocase. Acetylcarnitine is known to be a supplier of acetyl groups for fatty acid synthesis in pea chloroplasts (Masterson *et al.*, 1990*a*). The scheme for entry is identical to the Thomas–Wood scheme (Thomas & Wood, 1982) proposed for mitochondria (see Fig. 12).

A possible source of this acetylcarnitine is therefore the mitochondria, as proposed in Fig. 13. Acetylcarnitine formed from acetyl CoA within the mitochondria would be exported by the acylcarnitine translocase with a concurrent import of carnitine from the cytosol. This now cytosolic acetylcarnitine would be able to penetrate the chloroplast membrane barrier, again via a translocase, and reconvert to acetyl CoA. The uptake of acetylcarnitine into the chloroplasts would be accompanied by an equivalent export of carnitine into the cytosol, where it could penetrate the mitochondria. Burgess & Thomas (1986) showed that carnitine acetyltransferase was located only in the mitochondrial matrix. Thus

Fig. 11–*cont.*
molecule of long-chain acylcarnitine ionically bonded inside the translocator (III) and the whole process is repeated. I, long-chain acyl CoA synthetase; II, external carnitine long-chain acyltransferase; III, carnitine/acylcarnitine translocase; IV, internal carnitine long-chain acyltransferase. C, cytosol; MOM, mitochondrial outer membrane; IMS, inter-membrane space; MIM, mitochondrial inner membrane; MM, mitochondrial matrix. CFFA, long-chain free fatty acid; cn, carnitine LCacCoA, long-chain acyl CoA; LCac-cn, long-chain acyl carnitine. (Adapted from Thomas & Wood, 1986.)

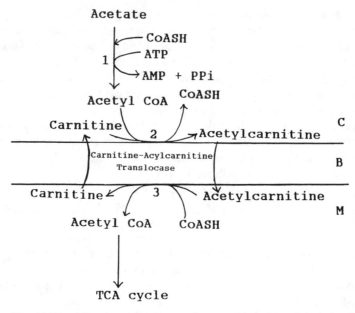

Fig. 12. Postulated reactions occurring on either side of the mitochondrial membrane barrier during the carnitine-mediated oxidation of acetate by pea mitochondria. 1, acetyl CoA synthetase; 2, cytosol side carnitine acetyltransferase; 3, matrix side carnitine acetyltransferase. C, cytosol; B, mitochondrial membrane barrier; M, mitochondrial matrix. (Adapted from Thomas & Wood, 1982.)

export of acetylcarnitine may be an important role of the mitochondrial CAT. Under conditions of high acetyl CoA concentrations, e.g. from pyruvate (not only from glycolysis: ap Rees, 1974, 1980, but also from degradation of proteins: Neal & Beevers, 1960), the formation of acetylcarnitine would proceed as the low concentration of CoASH and high concentration of acetyl CoA would displace the equilibrium of the reaction 5 in favour of acetylcarnitine formation. Excess acetylcarnitine formed may well be exported through the membrane barrier to cytosol sites and used there for biosynthetic purposes, e.g. fatty acid synthesis.

Some experimental evidence for carnitine-mediated shuttling of metabolites from mitochondria to chloroplasts has been provided. Wood (1984) produced uniformly labelled mitochondria by pretreating germinated pea cotyledon slices with ^{14}C-acetate prior to organelle isolation. The radiolabelled mitochondria were incubated with and without carnitine and the supernatant was collected after mitochondrial sedimen-

Table 6. *Recovery of ^{14}C-label in chloroplasts, CO_2, mitochondria and supernatant from incubations supplied with ^{14}C-radioactive mitochondria*

Light/dark	L-carnitine (1 mM)	^{14}C-label recovered(Bq)			
		CO_2	Mito-chondria	Super-natant	Chloro-plasts
Light	+	3.4	118	159	11.8
Light	−	4.2	153	77	2.9
Dark	+	3.5	142	57	3.0
Dark	−	4.1	151	43	1.6

From Wood, 1984.

tation. Unlabelled chloroplasts were then incubated in this supernatant medium after which they were collected by centrifugation. The results in Table 6 show that in the light, L-carnitine produced a four-fold increase in the amount of label recovered in the chloroplasts compared with the control in the absence of L-carnitine. The fact that the supernatant surrounding the chloroplasts contained twice the amount of label as the control counterpart (due to increased export of label from the mitochondria in the presence of external carnitine) would not solely account for a four-fold increase. L-carnitine must have stimulated both export of label from the mitochondria and incorporation of label from the surrounding medium into the chloroplast.

In the dark, L-carnitine doubled the amount of ^{14}C-label incorporation compared with the control. This incorporation in the dark however, was lower than the light-treated equivalent. Thus, whilst the ^{14}C-labelled transported species is not known at present, the data suggest that carnitine can and does influence movement of metabolite(s) from the mitochondria to the chloroplasts.

Carnitine may also play a role in the export of long-chain fatty acids from the chloroplasts for elongation, desaturation and modification by enzymes associated with the endoplasmic reticulum. The current view is that, whilst all chloroplasts can synthesise fatty acids up to oleate (18:1), the synthesis of polyunsaturated fatty acids can occur by one of two pathways, namely the so-called 'prokaryotic' pathway of 16:3 plants such as spinach in which desaturation from 18:1 to 18:3 occurs in complete association with the chloroplast and the 'eukaryotic' pathway of 18:3

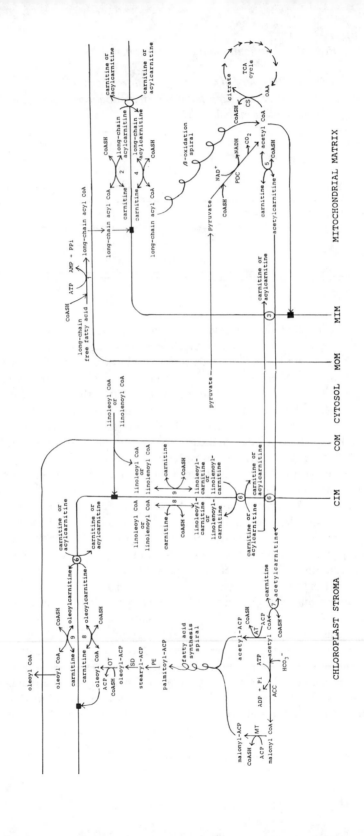

plants such as pea, in which desaturation to polyunsaturated fatty acids occurs on the endoplasmic reticulum (see review by Jaworski, 1987). In the latter case, synthesis of fatty acids by isolated chloroplasts will cease with the production of palmitate (16:0), stearate (18:0) and oleate (18:1). The results in Table 2 demonstrate that this is true for isolated pea chloroplasts.

These fatty acids are believed to be released from the fatty acyl-ACP and the free fatty acid crosses the chloroplast membrane into the cytosol where it is activated to the acyl CoA. However, as the presence of carnitine long-chain acyltransferase is confirmed in pea chloroplasts (McLaren *et al.*, 1985) and in barley etioplasts (Thomas *et al.*, 1982, 1983), it is possible to postulate an alternative mechanism for the export of palmitate and oleate (Fig. 13).

Fatty acylcarnitine formed inside the chloroplast from acyl groups released from the ACP complex may cross the chloroplast membrane barrier where it reforms acyl CoA by exchange of carnitine with cytosolic CoASH. The exchange of acylcarnitines (e.g. 18:1) would be balanced by an equimolar import of cytosolic carnitine or (18:2)/(18:3) acylcarnitines returning to the chloroplast. This role of carnitine, however, is at present

---

Fig. 13. Generalised scheme showing postulated roles of carnitine/ acylcarnitines in the movement of metabolites between the various cell compartments. 1, Long-chain acyl CoA synthetase (associated with outer mitochondrial membrane); 2, carnitine long-chain acyltransferase (associated with the cytosolic face of the mitochondrial inner membrane); 3, carnitine/acylcarnitine translocase (an integral membrane protein associated with the mitochondrial inner membrane); 4, carnitine long-chain acyltransferase (associated with the matrix side of the mitochondrial inner membrane); 5, carnitine short-chain acyltransferase (associated with the mitochondrial matrix); 6, carnitine/acylcarnitine translocase (an integral membrane protein associated with the chloroplast inner membrane); 7, carnitine short-chain acyltransferase (associated with the chloroplast stroma); 8, carnitine long-chain acyltransferase (associated with the stroma side of the chloroplast inner membrane); 9, carnitine long-chain acyltransferase (associated with the cytosolic face of the chloroplast inner membrane). PDC, pyruvate dehydrogenase complex; CS, citrate synthase; AT, acetyl transacylase; MT, malonyl transacylase; OT, oleoyl transacylase; ACC, acetyl CoA carboxylase; PE, palmitoyl-ACP elongase; SD, stearyl-ACP desaturase. CIM, chloroplast inner membrane; COM, chloroplast outer membrane; MIM, mitochondrial inner membrane; MOM, mitochondrial outer membrane; ▮, acyl CoA barrier.

a postulate, made by extrapolation of proven roles of carnitine and, as yet, few experimental data are available to support it.

All of these models for the shuttling of acyl groups between organelles as their acylcarnitine derivatives have two main advantages. First, they are energetically favourable. The acylcarnitine–ester linkage is a 'high energy bond' and the acyl group is maintained in an activated form when acyl CoA is converted to acylcarnitine. Secondly, acylcarnitines are freely mobile across membranes, but this movement is also controlled by the antiport exchange mechanism of the integral membrane translocators. Alternative models have been proposed whereby the shuttling of acyl groups between organelles is achieved by movement of the free fatty acid; for example, Murphy & Stumpf (1981), Liedvogel & Stumpf (1982) and Murphy & Walker (1982) proposed that acetyl groups are shuttled between mitochondria and chloroplasts as free acetate. Galliard & Stumpf (1966), Simcox et al. (1977) and Stumpf (1980) suggested that long-chain acyl groups are exported from the chloroplast to the endoplasmic reticulum as free fatty acids. Such schemes are energetically expensive as the acyl CoAs have first to be hydrolysed and then resynthesised. In addition, the free fatty acids enter cell compartments by simple diffusion and, as such, their transportation is not under metabolic control.

In animal mitochondria, acylcarnitines are known to traverse the inner membrane via an intrinsic translocator; so far, work on the role of carnitine in plant metabolism has suggested that this is also the case in plant systems. Figure 11 shows a diagrammatic representation of how such a translocator may operate in plant cell organelle membranes. The transporter binding site has only to recognise the quaternary nitrogen cation and the carboxyl anion of carnitine to provide a two-point attachment. Attachment would be independent of the R group (see Fig. 1C). It is proposed that the carnitine transporter system in plant mitochondria and chloroplast membranes may catalyse four modes of transport as follows:

1.   carnitine in $\rightleftharpoons$ carnitine out
2.   carnitine in $\rightleftharpoons$ acylcarnitine out
3.   acylcarnitine in $\rightleftharpoons$ carnitine out
4.   acylcarnitine in $\rightleftharpoons$ acylcarnitine out

The acyl group in modes 2, 3 and 4 may be long-chain (e.g. palmitate), mid-chain (e.g. octanoate) or short-chain (e.g. acetate).

Control of carnitine and acylcarnitine movement is not considered to be governed by the translocases in whatever membrane: the translocase proceeds to an equilibrium of the various acylcarnitines and carnitine available on either side of that membrane. It will be (i) the rate of formation and (ii) the rate of utilisation of any acylcarnitine on either or each side of the membrane which determines the direction of flow.

Conclusions

Carnitine is clearly a naturally occurring compound in plants and carnitine acyltransferases have been found in a number of tissues. L-carnitine is essential for the uptake of acyl CoAs into the mitochondria for β-oxidation, and L-acetylcarnitine is the most efficient substrate for fatty acid synthesis by chloroplasts. These processes which are affected by carnitine all involve CoASH:acyl CoA couples which are rigidly compartmented in the plant cell by membrane barriers. Carnitine affects all aspects of acyl CoA/CoASH metabolism and the various roles of carnitine and carnitine acyltransferases in metabolism can be divided into two main categories: (i) an obligatory role in which carnitine and a specific carnitine acyltransferase are required for translocating activated acyl residues across an acyl CoA barrier; (ii) a facilitative or CoASH regeneration role in which carnitine modulates the availability of CoASH and the ratio of acyl CoA/CoASH via specific carnitine acyltransferases.

References

ap Rees, T. (1974). Pathways of carbohydrate breakdown in higher plants. In *MTP International Review of Science, Plant Biochemistry*, Vol. 11, ed. D.H. Northcote, pp. 89–127. Maryland: University Park Press.

ap Rees, T. (1980). Assessment of the contributions of metabolic pathways to plant respiration. In *The Biochemistry of Plants*, Vol. 2, ed. P. K. Stumpf & E.E. Conn, pp. 1–29. New York: Academic Press.

Ariffin, A., McNeil, P.H., Cooke, R.J., Wood, C. & Thomas, D.R. (1982). Carnitine content of greening barley leaves. *Phytochemistry* **21**, 1431–2.

Bieber, L.L. (1988). Carnitine. *Annual Review of Biochemistry* **57**, 261–83.

Bieber, L.L., Emaus, R., Valkner, K.J. & Farrell, S. (1982). Possible functions of short-chain and medium-chain carnitine acyltransferases. *Federation Proceedings* **41**, 2858–61.

Bieber, L.L., Krahling, J.B., Clarke, P.R.H., Valkner, K.J. & Tolbert, N.E. (1981). Carnitine acyltransferases in rat liver peroxisomes. *Archives of Biochemistry and Biophysics* **211**, 599–604.

Bremer, J. (1983). Carnitine – metabolism and functions. *Physiological Reviews* **63**, 1420–68.

Brooks, J.L. & Stumpf, P.K. (1965). A soluble fatty acid synthesizing system from lettuce chloroplasts. *Biochimica et Biophysica Acta* **98**, 213–16.

Budde, R.J. L., Fang, T.K., Randall, D.D. & Miernyk, J.A. (1991). Acetyl coenzyme A can regulate activity of the mitochondrial pyruvate dehydrogenase complex *in situ*. *Plant Physiology* **95**, 131–6.

Burgess, N. & Thomas, D.R. (1986). Carnitine acetyltransferase in pea cotyledon mitochondria. *Planta* **167**, 58–65.

Cederblad, G. & Linstedt, S. (1972). Method for the determination of carnitine in the picomole range. *Clinica Chimica Acta* **37**, 235.

Cooper, T.G. & Beevers, H. (1969). β-oxidation in glyoxysomes from castor bean endosperm. *Journal of Biological Chemistry* **244**, 3514–20.

Crawford, J.W.C. & Kenyon, J. (1927). The constitution of carnitine. Part 1. The synthesis of α-hydroxy-γ-butyrotrimethylbetaine. *Journal of the Chemical Society* **79**, 396–402.

Fraenkel, G. (1951). Effect and distribution of Vitamin B_T. *Archives of Biochemistry and Biophysics* **34**, 457–68.

Fraenkel, G. (1953). Studies on the distribution of Vitamin B_T (carnitine). *Biological Bulletin* **104**, 359–71.

Fraenkel, G. & Blewett, M. (1947). The importance of folic acid and unidentified members of the vitamin B complex in the nutrition of certain insects. *Biochemical Journal* **41**, 469–75.

Fritz, I.B. (1963). Carnitine and its role in fatty acid metabolism. *Advances in Lipid Research* **1**, 285–334.

Galliard, T. & Stumpf, P.K. (1966). Fat metabolism in higher plants. XXX Enzymatic synthesis of ricinoleic acid by a microsomal preparation from developing *Ricinus communis* seeds. *Journal of Biological Chemistry* **241**, 5806–12.

Gerbling, H., Gandour, R.D., Moore, T.S. & Gerhardt, B. (1991). Carnitine acyltransferase activity of plant mitochondria. In *Proceedings of the 9th International Symposium on Plant Lipid Biochemistry*. London: Portland Press (in press).

Gerbling, H. & Gerhardt, B. (1988). Carnitine acyltransferase activity of mitochondria from mung-bean hypocotyls. *Planta* **174**, 90–3.

Griffiths, H. (1988). Crassulacean acid metabolism: A re-appraisal of physiological plasticity in form and function. *Advances in Botanical Research* **15**, 43–92.

Gulewitsch, W. & Krimberg, R. (1905). Zur Kenntnis der Extraktivstoffe der Muskeln. II. Mitteilung. Über das Carnitin. *Zeitschrift für Physiologische Chemie* **45**, 326–30.

Jaworski, J.G. (1987). Biosynthesis of monoenoic and polyenoic fatty acids. In *The Biochemistry of Plants*, Vol. 9, ed. P.K. Stumpf & E.E. Conn, pp. 159–74. New York: Academic Press.

Kutscher, F. (1905). Über Leibig's Fleischextrakt. *Zeitschrift im Untersuchung dieses Nahr- und Genussmittel* **10**, 528–37.

Liedvogel, B. & Stumpf, P.K. (1982). Origin of acetate in spinach leaf cell. *Plant Physiology* **69**, 897–903.

McLaren, I. & Thomas, D.R. (1967). CO_2 fixation, organic acids and some enzymes in green and colourless tissue cultures of *Kalanchoe crenata*. *New Phytologist* **66**, 683–95.

McLaren, I., Wood, C., Jalil, M.N.H., Yong, B.C.S. & Thomas, D.R. (1985). Carnitine acyltransferases in pea chloroplasts. *Planta* **163**, 197–200.

McNeil, P.H. & Thomas, D.R. (1975). Carnitine content of pea seedling cotyledons. *Phytochemistry* **14**, 2335–6.

McNeil, P.H. & Thomas, D.R. (1976). The effect of carnitine on palmitate oxidation by pea cotyledon mitochondria. *Journal of Experimental Botany* **27**, 1163–80.

Marquis, N.R. & Fritz, I.B. (1964). Enzymological determination of free carnitine concentrations in rat tissues. *Journal of Lipid Research* **5**, 184–7.

Masterson, C., Wood, C. & Thomas, D.R. (1990*a*). L-acetylcarnitine, a substrate for chloroplast fatty acid synthesis. *Plant, Cell and Environment* **13**, 755–65.

Masterson, C., Wood, C. & Thomas, D.R. (1990*b*). Inhibition studies on acetyl group incorporation into chloroplast fatty acids. *Plant, Cell and Environment* **13**, 767–71.

Miernyk, J.A. & Trelease, R.N. (1981). Control of enzyme activities in cotton cotyledons during maturation and germination. IV. β-oxidation. *Plant Physiology* **67**, 341–6.

Murphy, D.J. & Stumpf, P.K. (1981). The origin of chloroplastic acetyl coenzyme A. *Archives of Biochemistry and Biophysics* **212**, 730–9.

Murphy, D.J. & Walker, D.A. (1982). Acetyl coenzyme A biosynthesis in the chloroplast. *Planta* **156**, 84–8.

Murray, W.J., Reed, K.W. & Roche, E.B. (1980). Conformations of carnitine and acetylcarnitine and the relationship to mitochondrial transport of fatty acids. *Journal of Theoretical Biology* **82**, 559–72.

Neal, G.E. & Beevers, H. (1960). Pyruvate utilization in castor bean endosperm and other tissues. *Biochemical Journal* **74**, 409–16.

Pande, S.V. (1975). A mitochondrial carnitine-acylcarnitine translocase system. *Proceedings of the National Academy of Science, USA* **72**, 883–7.

Pande, S.V. & Parvin, R. (1980). Carnitine–acylcarnitine translocase catalyses an equilibrating unidirectional transport as well. *Journal of Biological Chemistry* **255**, 2994–3001.

Panter, R.A. (1971). The role of carnitine in β-oxidation in higher plants. PhD thesis, University of California, Riverside.

Panter, R.A. & Mudd, J. B. (1969). Carnitine levels in some higher plants. *FEBS Letters* **5**, 169–70.

Panter, R.A. & Mudd, J.B. (1973). Some aspects of carnitine metabolism in avocado (*Persea americana*). *Biochemical Journal* **134**, 655–8.

Ramsay, R.R. & Tubbs, P.K. (1975). The mechanism of fatty acid uptake by heart mitochondria: an acylcarnitine–carnitine exchange. *FEBS Letters* **54**, 21–5.

Ramsay, R.R. & Tubbs, P.K. (1976). The effects of temperature and some inhibitors on the carnitine exchange system of heart mitochondria. *European Journal of Biochemistry* **69**, 299–303.

Simcox, P.D., Reid, E.E., Canvin, D.T. & Dennis, D.T. (1977). Enzymes of the glycolytic and pentose phosphate pathways in proplastids from the developing endosperm of *Ricinus communis* L. *Plant Physiology* **59**, 1128–32.

Stobart, A.K., McLaren, I. & Thomas, D.R. (1967). Chlorophylls and carotenoids of colourless callus, green callus and leaves of *Kalanchoe crenata*. *Phytochemistry* **6**, 1467–74.

Stobart, A.K. & Thomas, D.R. (1968). Chlorophyllase in tissue culture of *Kalanchoe crenata*. *Phytochemistry* **7**, 1963–72.

Stumpf, P.K. (1980). Biosynthesis of saturated and unsaturated fatty acids. In *The Biochemistry of Plants*, Vol. 4, ed. P.K. Stumpf & E.E. Conn, pp. 177–204. New York: Academic Press.

Thomas, D.R., Ariffin, A., Noh Hj Jalil, M., Yong, B.C.S., Cooke, R. J. & Wood, C. (1981). Effect of carnitine on greening barley leaves. *Phytochemistry* **20**, 1241–4.

Thomas, D.R. & McNeil, P.H. (1976). The effect of carnitine on the oxidation of saturated fatty acids by pea cotyledon mitochondria. *Planta* **132**, 61–3.

Thomas, D.R., Noh Hj Jalil, M., Ariffin, A., Cooke, R.J., McLaren, I., Yong, B.C.S. & Wood, C. (1983). The synthesis of short- and long-chain acylcarnitine by etio-chloroplasts of greening barley leaves. *Planta* **158**, 259–63.

Thomas, D.R., Noh Hj Jalil, M., Cooke, R.J., Yong, B.C.S., Ariffin, A., McNeil, P.H. & Wood, C. (1982). The synthesis of palmitoylcarnitine by etio-chloroplasts of greening barley leaves. *Planta* **154**, 60–5.

Thomas, D.R. & Stobart, A.K. (1970). Lipids of tissue cultures of *Kalanchoe crenata*. *Journal of Experimental Botany* **21**, 274–85.

Thomas, D.R. & Wood, C. (1982). Oxidation of acetate, acetyl CoA and acetylcarnitine by pea mitochondria. *Planta* **154**, 145–9.

Thomas, D.R. & Wood, C. (1986). The two β-oxidation sites in pea cotyledons. Carnitine palmitoyltransferase: location and function in pea mitochondria. *Planta* **168**, 261–6.

Thomas, D.R., Wood, C. & Masterson, C. (1988). Long-chain acyl CoA synthetase, carnitine and β-oxidation in the pea seed mitochondrion. *Planta* **173**, 263–6.

Wood, C. (1984). The role of carnitine acyltransferases in plant acyl CoA metabolism. PhD thesis, University of Newcastle upon Tyne.

Wood, C., Noh Hj Jalil, M., Ariffin, A., Yong, B.C.S. & Thomas, D.R. (1983). Carnitine short-chain acyltransferase in pea mitochondria. *Planta* **158**, 175–8.

Wood, C., Noh Hj Jalil, M., McLaren, I., Yong, B.C.S., Ariffin, A., McNeil, P.H., Burgess, N. & Thomas, D.R. (1984). Carnitine long-chain acyltransferase and oxidation of palmitate, palmitoyl coenzyme A and palmitoylcarnitine by pea mitochondria preparations. *Planta* **161**, 255–60.

HEIDE SCHNABL

Metabolic interactions of organelles in guard cells

The opening of stomata occurs as a result of an increase in turgor of the guard cells which is coupled to an uptake of potassium ions in relatively high concentrations (up to 400 mM: Raschke & Fellows, 1971; Outlaw, 1983). The active K^+ transport has to be initiated by a process which requires a supply of energy in the form of ATP. There are two sources of ATP which are known to function in stomatal movement: in the light, photophosphorylation seems to act as a potential ATP source (Ogawa et al., 1979; Assmann et al., 1985; Gotow et al., 1985); in the dark, oxidative phosphorylation has been proposed as the energy source for the swelling of guard cells (Raghavendra, 1981; Shimazaki et al., 1983; Schwartz & Zeiger, 1984).

The role of ATP hydrolysis in ion uptake is still unresolved, although a membrane-bound Mg^{2+}-activated ATPase and a K^+-stimulated ATPase have been detected in the epidermis (Fujino (1967), Raghavendra et al. (1976), Kasamo (1979) and Lurie & Hendrix (1979)). The enzyme associated with stomatal opening was active in the light period, and was stimulated by fusicoccin and inhibited by abscisic acid (ABA). The other enzyme, coupled with stomatal closing, showed the opposite behaviour. The key enzyme for H^+ transport across the plasmalemma guard cells appears to be a vanadate-sensitive, electrogenic ATPase which is K^+-stimulated and Mg^{2+}-dependent (Serrano, 1985; Nejidat et al., 1986; Fricker & Willmer, 1987; Shimazaki & Kondo, 1987). The vanadate-sensitive ATPase isolated from whole leaf and epidermal tissue was slightly stimulated by fusicoccin at a concentration of 0.13 μM, but not by ABA (Blum et al., 1988). In contrast, Nejidat et al. (1989) found an inhibitory effect of ABA and discussed an indirect modulation by light and ABA. Further investigations of the guard cell pump are required to clarify the data.

A specific transport mechanism for K^+ across the plasmalemma of guard cells was found by Schröder et al. (1984, 1987) and Schröder (1988), who succeeded in measuring specialized K^+ channels. They

Society for Experimental Biology Seminar Series 50: *Plant organelles*, ed. A. K. Tobin.
© Cambridge University Press 1992, pp. 265–79.

identified two types of K^+ currents in guard cells, which may allow efflux of K^+ during stomatal closing and uptake of K^+ during stomatal opening. Direct evidence was gained for the hypothesis that the inward-compensating K^+ conductance channels are a pathway for K^+ accumulation by the finding that they were blocked by Al^{3+}, which are known to inhibit stomatal opening but not closing.

The positive ion charge of K^+ imported into the guard cells has to be balanced in order to maintain electroneutrality. Two possibilities have been proposed: the exchange of H^+ for K^+ and the simultaneous import of Cl^-. The proton export associated with stomatal opening was measured by Raschke & Humble (1973), who determined a 1:1 exchange of K^+ and H^+ in epidermal strips. The cytosolic proton level must, however, be kept constant in order to maintain a physiological pH. The release of protons which occurs during malic acid synthesis within the guard cells is assumed to serve as a mechanism for regulating the pH of the cytosol. In parallel to malate synthesis, chloroplast-associated starch is broken down, yielding phosphoenolpyruvate (PEP). The formation of malate in starch-containing guard cells of *Vicia faba* is catalysed by PEP-carboxylase (PEPCase) via CO_2 fixation from PEP (Outlaw & Kennedy, 1979; Donkin & Martin, 1980; Schnabl, 1981). The level of malate, the major organic anion that accumulates during stomatal opening (Allaway, 1973; Raschke & Schnabl, 1978; Outlaw & Kennedy, 1979), was shown to be enhanced by a K^+-induced increase of PEPCase activity in swelling guard cell protoplasts (Schnabl *et al.*, 1982; Schnabl & Kottmeier, 1984*a*).

The guard cells of *Vicia faba* contain chloroplasts which are filled with starch, but in some species, e.g. *Allium cepa*, the guard cells possess only poorly developed plastids which do not contain starch. The guard cell chloroplasts in these species therefore appear to lack the substrate for the production of the quantities of malate required for osmotic purposes. Schnabl & Raschke (1980) showed that these stomata satisfy their anion requirement solely through import of Cl^-. Thus, in addition to its role in balancing the positive charge difference which results from proton extrusion, Cl^- import provides a second mechanism for maintaining electroneutrality during guard cell opening.

On the basis of these data K^+ has been proposed as one of the positive effectors of stomatal opening. In order to obtain a more detailed picture of the cause–effect relationship between K^+, PEPCase and the accumulation of malate during the swelling of guard cells in the dark and the light, we measured the levels of malate in different organelles during stomatal opening.

The use of isolated and purified guard cell protoplasts (GCP) provides a means of examining these cells without contamination by surrounding cells and without any influence of the cell wall, which has been removed. The distribution of substrates and the shuttle of substrates between different organelles was determined with the use of swelling GCP. However, the removal of the cell wall is not entirely advantageous. Therefore, it is necessary to compare the values obtained with protoplasts with those of stomata. The aim of our experiments with GCP was to determine how the distribution and pool sizes of malate and adenine nucleotides changed within the plastids, mitochondria and cytosolic fractions during the K^+-induced swelling of isolated protoplasts. This information was then used, together with kinetic data on the regulation of PEPCase activity, to formulate a model relating all of these changes in metabolism to the processes involved in protoplast swelling.

K^+ fluxes and swelling of GCP

Changes in K^+ concentrations of opened and closed guard cells have been measured in epidermal strips floated on K^+ solutions (100 mM). The potassium content calculated in opened stomata was c. 300–450 mM, equivalent to a K^+ level of 2.8–4.5 pmol per guard cell pair (Outlaw, 1983).

These values are in accordance with those obtained when guard cell protoplasts were incubated in a K^+ solution (10 mM) in darkness for 15 min. The kinetics of K^+ accumulation are shown in Fig. 1, demonstrating that the K^+ level is saturated at 2300 fmol GCP^{-1} after 4 min.

During the process of K^+ accumulation an increase in the volume of GCP is observed which is correlated with stomatal opening (Schnabl *et al.*, 1978). The regulation of stomatal turgor is therefore believed to be directly related to the changes in volume of the isolated protoplasts. The swelling response to K^+ is a specific property of guard cells, and can be applied as a sensitive indicator for checking the viability of guard cell protoplasts.

When treated with K^+ in the dark or the light the GCPs show a difference in their diameters of 30%. They are smaller after a dark than after a light treatment (14.8 μm in the dark; 16.2 μm in the light). In Fig. 2, both light- and dark-pre-treated GCPs are shown to swell in a solution containing K^+ (10 mM) until they have reached almost identical diameters (17.5 μm) after 10 min. The K^+-induced volume increase of both types of protoplasts which is levelled out at a plateau between 3 and 5 min is continued after 5 min. Thus, the effect of K^+ on the volume of dark- and light-treated GCPs was shown to be biphasic.

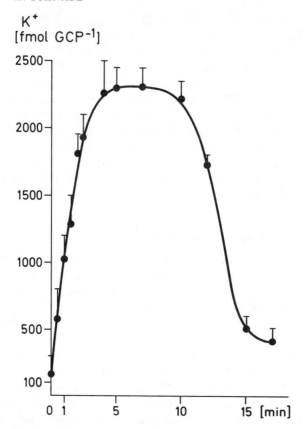

Fig. 1. K$^+$ kinetics of GCP incubated in 10 mM K$^+$ solution in darkness for 15 min.

Regulation of PEPCase activity

PEPCase, together with malate dehydrogenase, is responsible for the synthesis of malic acid, which accumulates during stomatal opening in starch-containing guard cells (Outlaw & Kennedy, 1979) and during the swelling of GCP (Schnabl et al., 1982; Schnabl, 1983). The strong increase in PEPCase activity during the K$^+$-induced swelling of GCP strengthens the hypothesis that PEPCase is involved as an important trigger for the volume increase in GCP. A correlation between swelling of protoplasts and the K$^+$-induced decrease in the K_m of PEPCase and an increase in its V_{max} indicates a triggering function (Schnabl & Kottmeier, 1984a; Kottmeier & Schnabl, 1986).

Experiments were undertaken in order to obtain a more detailed pic-

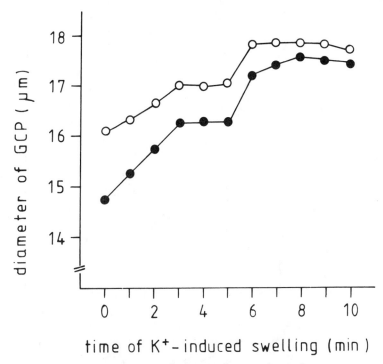

Fig. 2. K$^+$-induced swelling of dark (●) and light (○) treated GCP. The protoplasts were preincubated for 30 min at 20 °C in the dark, or in the light (300 μmol photons m^{-2} s^{-1}, 30 min, 20 °C). K$^+$ iminodiacetate (10 mM) was added at time zero for stimulating the swelling (10 min).

ture of the K$^+$ effects on PEPCase during the swelling of GCP. The activity of PEPCase in dark- and light-treated GCP was modulated by K$^+$ rhythmically in time periods of 2.5 min by a factor of 2–5 (1.7–5 pmol GCP^{-1} h^{-1}; Fig. 3). In contrast, the volume of GCP and PEPCase activity were constant in the absence of K$^+$ (Michalke & Schnabl, 1990).

At the moment, we may only speculate about the mechanism involved in the K$^+$-induced rapid activation and deactivation mechanism of PEP-Case. One possible explanation is that there may be different aggregation or disaggregation forms of PEPCase. The appearance of inactive or less active monomers and dimers, and highly active tetramers (as found in C4 plants by Huber *et al.*, 1986), was recently confirmed by experiments undertaken with stomatal epidermals in this laboratory (Denecke *et al.*, 1992). It was found that, under certain conditions, the tetramer of stomatal PEPCase (molecular weight 465 000) is transformed into the

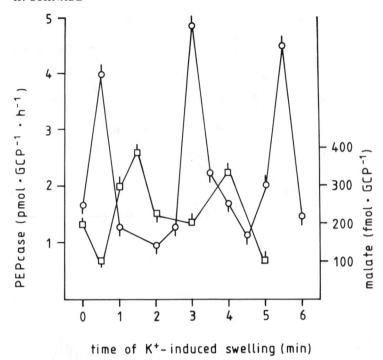

time of K⁺- induced swelling (min)

Fig. 3. PEPcase activity (○) and malate pools of mitochondria (□) during K⁺-induced swelling of GCP treated for 30 min in the light before being transferred for 5 min to a buffer solution with K⁺ iminodiacetate in the light at time zero ($n = 15$).

monomer with a molecular weight of 110 000. It has been suggested that the state of (dis-)aggregation of the enzyme is controlled by the malate/PEP ratio (studied in CAM by Wedding & Black, 1986). Thus, data on the regulation of PEPCase in both C4 and CAM plants appear to be consistent with our observations from stomatal material.

Another hypothesis which may explain the rapid modulations of active/inactive forms of PEPCase is presented by the different states of (de-)phosphorylation of the enzyme (studied in CAM by Nimmo et al., 1987; Kluge et al., 1988). Recent experiments in this laboratory have found that PEPCase phosphorylation can be observed both in vivo and in vitro by feeding ³²P to either the concentrated enzyme (precipitated by fractionation with ammonium sulphate between 0 and 70% saturation) or to isolated guard cell protoplasts. Subsequent analysis by SDS-PAGE provides evidence that the two polypeptides (112 and 110 kDa bands) which are labelled in vitro represent isoforms or subunits of stomatal PEPCase.

The *in vivo* labelling experiments indicated that, in contrast to the situation in C4 or CAM plants, there is no detectable change in the phosphorylation state of PEPCase during light/dark transitions in guard cell protoplasts (H. Schnabl, M. Denecke and M. Schulz, unpublished data). Hormonal interactions with other events in the dual signal pathway have been centred on phosphatidylinositol turnover and protein phosphorylation. Auxin and cytokinin have been found to stimulate the degradation of phosphatidylinositol. We therefore propose that a similar direct, or indirect, mechanism operates whereby an external stimulus, such as the addition of K^+, stimulates protein phosphorylation.

At present, precise data on the mechanism of PEPCase activation and deactivation are still lacking. Therefore, investigations are needed involving the immunological and physiological analysis of PEPCase in GCPs in order to elucidate the mechanism behind the oscillations in activity of this enzyme.

Pool sizes of malate

Malate levels have been shown to increase during stomatal opening (Allaway, 1973; Travis & Mansfield, 1977) and this is paralleled by a decrease in starch content (Outlaw & Manchester, 1979). The reverse situation occurs during stomatal closure. By measuring malate in 'rolled' epidermal peelings of *Vicia faba*, Allaway (1973) reported a malate accumulation of 1000 fmol per stoma, which was confirmed by Ogawa *et al.* (1979) and Raschke & Schnabl (1978). By using a quantitative histochemical technique, Outlaw & Kennedy (1979) calculated malate values to be 500 fmol per stoma. We have extended these studies to incorporate a cell fractionation technique which allows the subcellular compartmentation of metabolites to be determined in different organelles. In this way, information is obtained regarding the mechanisms which serve to supply and degrade these substrates (Michalke & Schnabl, 1987). In dark-treated GCP, malate (total amount 272 fmol GCP^{-1}) is distributed between the mitochondria, plastids and the supernatant fraction (vacuole and cytoplasm) in the respective proportions of 35, 6 and 60%. After light treatment malate (565 fmol GCP^{-1}) is distributed between these organelles, respectively at 42, 2 and 79%. In the absence of K^+ ions, these distributions remain relatively constant during the storage of GCP in either the light or the dark. It can be seen that the data showing the malate levels in GCP are in agreement with those obtained with stomata (Michalke & Schnabl, 1987).

In the presence of K^+, malate pool sizes show time-dependent oscillations in the different subcellular compartments (Michalke & Schnabl,

1990). As shown in Fig. 3, the observed peaks in malate concentration in the mitochondria (after 1.5 and 4 min) are preceded by an increase of PEPCase activity after 0.5 and 3 min, a process which is coupled with a higher malate synthesis. Malate transfer from the cytosol into the mitochondria is indicated by the dramatic increase in mitochondrial malate from 100 to almost 400 fmol GCP^{-1} after 1.5 min of swelling. The mitochondria, which are depleted of malate by 50% after 0.5 min of K^+-induced swelling are rapidly supplied with malate again by synthesis via PEPCase and malate dehydrogenase in the cytosol. The cytosol-associated malate is partially transported into the vacuole and partially into the mitochondria. The time shift of about 1 min between the stimulation of PEPCase and its reflection as mitochondrial malate indicates a malate shuttle across the mitochondrial membrane.

Sources of ATP

The energy supply for K^+ transport into the guard cells is provided by respiration and/or photosynthesis. In order to distinguish between the two pathways which may supply energy for stomatal opening, Schwartz & Zeiger (1984) tested the effects of a photosynthetic inhibitor, DCMU, and a respiratory inhibitor, KCN, on light- and dark-treated epidermals. In the dark, DCMU had no effect on stomatal aperture, whereas KCN was inhibitory. Schwarz & Zeiger (1984) discussed the possibility that, whereas oxidative phorphorylation may be significant in the dark, this was probably less important in the light when photophosphorylation was active. The role of stomatal respiration in the light remained open to question, although the authors proposed that oxidative phosphorylation was probably inhibited in the light (Schwartz & Zeiger, 1984).

In order to obtain information on the distribution of respiratory substrates and their role in the swelling of GCP, the microgradient technique (Michalke & Schnabl, 1987) was used to determine pool sizes of adenine nucleotides in plastids, mitochondria and the supernatant fraction during the K^+-induced volume increase of light- and dark-treated GCP (Schnabl & Michalke, 1988). Figure 4 shows that after 3 min of K^+ incubation the ATP content of the whole GCP and of the mitochondrion has been depleted by 50%. This period of ATP deficiency is reflected by the inhibition of GCP swelling which occurs between 3 and 5 min (Fig. 2). However, after the third minute of K^+-induced swelling the mitochondrial ATP as well as that of the whole extract is replenished – both attain the concentrations measured immediately before adding K^+. Indeed, ATP concentrations continue to increase by a further 20–30% in the 5 min of swelling. Hence, GCP are able to continue their volume increase in the

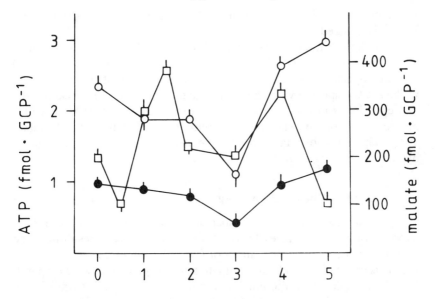

time of K$^+$-induced swelling (min)

Fig. 4. Pools of mitochondrial malate (□) and ATP of mitochondria (●) and whole extract (○) during K$^+$-induced swelling of GCP treated for 30 min in the light before being transferred for 5 min to a buffer solution with K$^+$ iminodiacetate in the light at time zero ($n = 9$).

light and the dark (Fig. 2). The decrease of the mitochondrial malate from 380 to 200 fmol GCP^{-1} between 1.5 and 3 min of K$^+$-induced swelling (Figs 3 and 4) is accompanied by an ATP increase of 0.7 fmol GCP^{-1} in the mitochondria and of 1.8 fmol GCP^{-1} in the whole extract during the third and the fifth minute. Thus, mitochondrial malate is assumed to serve as the substrate for ATP synthesis in these organelles.

The stomatal mitochondria which are obviously ATP-depleted in the third minute of swelling are apparently able to make up this deficiency of ATP. Malate, synthesised by PEPCase and malate dehydrogenase activity in the cytosol, is transported partly into the vacuole and partly into the mitochondria, where it appears to be respired via the tricarboxylic acid (TCA) cycle, thereby producing ATP. The disappearance of mitochondrial malate is coupled with a strong respiratory activity of GCP (Shimazaki *et al.*, 1983). Thus, malate is concluded to function as a mitochondrial substrate for ATP supply necessary for K$^+$-induced swelling of GCP. The availability of sufficient ATP is an important prerequisite for GCP swelling and for the membrane-bound ATPase required for

K+ import and PEPCase activation. If ATP supply is restricted, K+ import is blocked.

In the dark, mitochondria are concluded to deliver ATP via oxidative phosphorylation for the ATP-consuming process of K+ import associated in the cytoplasm. The cytoplasm itself acts as a source of ADP and AMP which are transported via a rapid adenine nucleotide shuttle back into the mitochondria. In the light, the plastids are able to contribute almost 30% of the ATP content via photophosphorylation (Schnabl & Michalke, 1988). The mitochondrial respiration seems to be inhibited in the light as hypothesised by Schwartz & Zeiger (1984).

A model of GCP swelling

The potassium ions imported via specific channels into the guard cells are hypothesised to activate PEPCase directly or indirectly, resulting in an increase in the cytosolic malate pool (Fig. 5). The increase in cytosolic malate is coupled with a higher mitochondrial malate level which is paralleled by an increase in the mitochondrial ATP pool. ATP is transported into the cytosol where it may be hydrolysed by plasmalemma-bound ATPase to ADP/AMP which are transferred back into the mitochondria (Schnabl & Michalke, 1988). A rapid shuttle for ATP and ADP/AMP between mitochondria and cytoplasm is proposed as a source of energy for the volume increase of GCP (Fig. 5). Adenine nucleotide shuttle systems are widely known in mitochondrial membranes. Pfaff & Klingenberg (1968) and Pfanner & Neupert (1985) have found an adenine nucleotide shuttle with a high substrate and direction specificity. These systems have been characterised in the mitochondria (Klingenberg, 1985; Adrian et al., 1986) yet remain to be identified in guard cells.

Conclusions

The pools of adenine nucleotides of dark- and light-treated guard cell protoplasts of Vicia faba have been determined in plastids, mitochondria and the cytosolic fraction (cytoplasm and vacuole) during the K+-induced volume increase (0–5 min). In darkness, the oxidative phosphorylation in guard cell mitochondria delivers energy for protoplast swelling. In the light, the plastids contribute almost 30% via photophosphorylation.

Malate was measured to function as a mitochondrial substrate for the ATP supply. If the mitochondria became depleted of ATP, the protoplasts were unable to swell. The mitochondrial malate level is replenished by an increase of cytosolic malate synthesised by PEPCase, an enzyme which was shown to be modulated by K+. On the basis of our

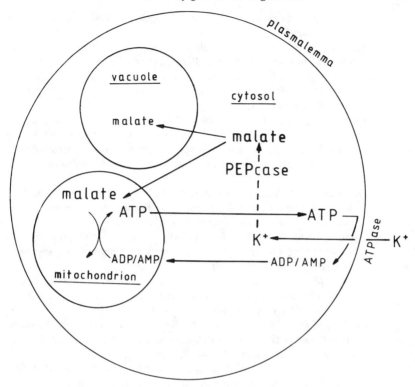

Fig. 5. A simplified model showing the relationship between K^+, PEP-case activity, malate levels in the cytosol, the vacuole, and the mitochondria, and ATP–ADP/AMP levels for GCP.

data a model is proposed to explain the cause–effect relationship between K^+ ions, PEPCase, the cytosolic and mitochondrial malate levels and adenine nucleotide pool sizes during the K^+-induced increase of protoplast volume.

Acknowledgements

This work was supported by a grant from Deutsche Forschungs-gemeinschaft.

References

Adrian, G.S., McCammon, M.T., Montgomery, D.L. & Douglas, M.G. (1986). Sequences required for delivery and localization of the ADP/

ATP translocator to the mitochondrial inner membrane. *Molecular and Cellular Biology* **6**, 626–34.

Allaway, W.G. (1973). Accumulation of malate in guard cells of *Vicia faba* during stomatal opening. *Planta* **110**, 63–70.

Assmann, S.M., Simoncini, L. & Schroeder, J.I. (1985). Blue light activates electrogenic ion pumping in guard cell protoplasts of *Vicia faba*. *Nature* **318**, 285–7.

Blum, W., Key, G. & Weiler, W. (1988). ATPase activity in plasmalemma-rich vesicles isolated by aqueous two-phase partitioning from *Vicia faba* mesophyll and epidermis: Characterization and influence of abscisic acid and fusiococcin. *Physiologia Plantarum* **72**, 279–87.

Denecke, M., Schulz, M., Fischer, Ch. & Schnabl, H. (1992). Purification and characterization of stomatal phosphoenolpyruvate carboxylase from *Vicia faba* L. *Planta* (in press).

Donkin, M.E. & Martin, E.S. (1980). Studies on the properties of carboxylating enzymes in the epidermis of *Commelina communis*. *Journal of Experimental Botany* **31**, 357–63.

Fujino, M. (1967). Role of adenosine triphosphate and adenosine triphosphatase in stomatal movement. *Science Bulletin, Faculty of Education (Nagasaki University)* **18**, 1–47.

Fricker, M.D. & Willmer, C.M. (1987). Vanadate sensitive ATPase and phosphatase activity in guard cell protoplasts of *Commelina*. *Journal of Experimental Botany* **38**, 642–8.

Gotow, K., Sakaki, T., Kondo, N., Kobayashi, K. & Syono, K. (1985). Light-induced alkalinisation of the suspending medium of guard cell protoplasts from *Vicia faba* L. *Plant Physiology* **79**, 825–8.

Huber, S.C., Sugiyama, T. & Akazawa, T. (1986). Light modulation of maize leaf phosphoenolpyruvate carboxylase. *Plant Physiology* **82**, 550–4.

Kasamo, K. (1979). Characterization of membrane-bound Mg^{++}-activated ATPase isolated from the lower epidermis of tobacco leaves. *Plant and Cell Physiology* **20**, 281–92.

Kluge, M., Maier, P., Brulfert, J., Faist, K. & Wollny, E. (1988). Regulation of phosphoenolpyruvate carboxylase in Crassulacean acid metabolism: *in vitro* phosphorylation of the enzyme. *Journal of Plant Physiology* **133**, 252–6.

Klingenberg, M. (1985). Principles of carrier catalysis elucidated by comparing two similar membrane translocators from mitochondria, the ADP/ATP carrier and the uncoupling protein. *Annals of the New York Academy of Science* **465**, 279–88.

Kottmeier, C. & Schnabl, H. (1986). The K_m-value of phosphoenolpyruvate carboxylase as an indicator of swelling state of guard cell protoplasts. *Plant Science* **43**, 213–17.

Lurie, S. & Hendrix, D.L. (1979). Differential ion stimulation of plasmalemma adenosine triphosphatase from leaf epidermis and mesophyll of *Nicotiana rustica* L. *Plant Physiology* **63**, 936–9.

Michalke, B. & Schnabl, H. (1987). The status of adenine nucleotides and malate in chloroplasts, mitochondria and supernatant of guard protoplasts from *Vicia faba*. *Journal of Plant Physiology* **130**, 243–53.

Michalke, B. & Schnabl, H. (1990). Modulation of the activity of phosphoenolpyruvate carboxylase during potassium-induced swelling of guard-cell protoplasts of *Vicia faba* L. after light and dark treatments. *Planta 180*, 188–93.

Nejidat, A., Itai, Ch. & Roth-Bejerano, N. (1989). Regulation of Mg, K-ATPase activity in guard cells of *Commelina communis* by phytochrome and ABA. *Plant and Cell Physiology* **30**, 945–9.

Nejidat, A., Roth-Bejerano, N. & Itai, Ch. (1986). K, Mg-ATPase activity in guard cells of *Commelina communis*. *Physiologia Plantarum* **68**, 315–19.

Nimmo, G.A., Wilkins, M.B., Fewson, C.A. & Nimmo, H.G. (1987). Persistent circadian rhythms in phosphorylation state of phosphoenolpyruvate carboxylase from *Bryophyllum fedtschenkoi* leaves and its sensitivity to inhibition by malate. *Planta* **170**, 408–15.

Ogawa, T., Ishikawa, H., Shimida, K. & Shibata, K. (1979). Synergistic action of red and blue light and action spectra of malate formation in guard cells of *Vicia faba* L. *Planta* **142**, 61–5.

Outlaw, W.H. (1983). Current concepts on the role of potassium in stomatal movements. *Physiologia Plantarum* **59**, 302–11.

Outlaw, W.H. & Kennedy, J. (1979). Enzymic and substrate basis for the anaplerotic step in guard cells. *Plant Physiology* **62**, 648–52.

Outlaw, W.H. & Manchester, J. (1979). Guard cell starch concentration quantitatively related to stomatal aperture. *Plant Physiology* **64**, 79–82.

Pfaff, E. & Klingenberg, M. (1968). Adenine nucleotide translocation of mitochondria. 1. Specificity and control. *European Journal of Biochemistry* **6**, 66–79.

Pfanner, N. & Neupert, W. (1985). Transport of proteins into mitochondria: A potassium diffusion potential is able to drive the import of ADP/ATP carrier. *EMBO Journal* **4**, 2819–25.

Raghavendra, A.S. (1981). Energy supply for stomatal opening in epidermal strips of *Commelina benghalensis*. *Plant Physiology* **67**, 385–7.

Raghavendra, A.S., Rao, I.M. & Das, V.S.R. (1976). Adenosine triphosphatase in epidermal tissue of *Commelina benghalensis*: Possible involvement of isoenzymes. *Plant Science Letters* **7**, 391–6.

Raschke, K. & Fellows, M.P. (1971). Stomatal movement in *Zea mays*: Shuttle of potassium and chloride between guard cells and subsidiary cells. *Planta* **101**, 296–316.

Raschke, K. & Humble, G.D. (1973). No uptake of anions required by opening stomata of *Vicia faba*: Guard cells release hydrogen ions. *Planta* **115**, 47–57.

Raschke, K. & Schnabl, H. (1978). Availability of chloride affects the balance between potassium chloride and potassium malate in guard cell of *Vicia faba* L. *Plant Physiology* **62**, 84–7.

Schnabl, H. (1981). The compartmentation of carboxylating and decarboxylating enzymes in guard cell protoplasts. *Planta* **152**, 307–13.

Schnabl, H. (1983). The key role of phosphoenolpyruvate carboxylase during the volume changes of guard cell protoplasts. *Physiologie Végétale* **21**, 955–62.

Schnabl, H., Bornman, Ch. & Ziegler, H. (1978). Studies in isolated starch-containing and starch-deficient guard cell protoplasts. *Planta* **143**, 33–9.

Schnabl, H., Denecke, M. & Schulz, M. (1992). *In vitro* and *in vivo* phosphorylation of stomatal phosphoenolpyruvate carboxylase from *Vicia faba* L. *Botanica Acta* (in press).

Schnabl, H., Elbert, Ch. & Krämer, G. (1982). The regulation of starch-malate balances during volume changes of guard cell protoplasts. *Journal of Experimental Botany* **33**, 996–1003.

Schnabl, H. & Kottmeier, C. (1984a). Properties of phosphoenolpyruvate carboxylase in desalted extracts from isolated guard cell protoplasts. *Planta* **162**, 220–5.

Schnabl, H. & Kottmeier, C. (1984b). Determination of malate levels during the swelling of vacuoles isolated from guard cell protoplasts. *Planta* **161**, 27–31.

Schnabl, H. & Michalke, B. (1988). The role of oxidative phosphorylation in stomatal opening. Adenylate pool sizes in light- and dark-treated swelling guard cell protoplasts. *Plant Physiology* (Life Sci. Adv.) **7**, 203–7.

Schnabl, H. & Raschke, K. (1980). Potassium chloride as stomatal osmoticum in *Allium cepa* L., a species devoid of starch in guard cells. *Plant Physiology* **65**, 88–93.

Schroeder, J.I. (1988). K^+ transport properties of K^+ channels in the plasma membrane of *Vicia faba* guard cells. *Journal of General Physiology* **92**, 667–83.

Schroeder, J.I., Hedrich, R. & Fernandez, J.M. (1984). Potassium-selective single channels in guard cell protoplasts of *Vicia faba*. *Nature* **312**, 361–2.

Schroeder, J.I., Raschke, K. & Neher, E. (1987). Voltage dependent K^+ channels in guard cell protoplasts. *Proceedings of the National Academy of Sciences, USA* **84**, 4108–12.

Schwartz, A. & Zeiger, E. (1984). Metabolic energy for stomatal opening. Roles of photophosphorylation and oxidative phosphorylation. *Planta* **161**, 129–36.

Serrano, R. (1985). Purification and reconstitution of the proton-pumping ATPase of fungal and plant plasma membranes. *Archives of Biochemistry and Biophysics* **227**, 1–8.

Shimakazi, K., Gotow, K., Sakaki, T. & Kondo, N. (1983). High respiratory activity of guard cell protoplasts from *Vicia faba* L. *Plant and Cell Physiology* **24**, 1049–56.

Shimazaki, K. & Kondo, N. (1987). Plasma membrane H^+-ATPase in guard-cell protoplasts from *Vicia faba* L. *Plant and Cell Physiology* **28**, 893–900.

Travis, A.J. & Mansfield, T.A. (1977). Studies of malate formation in 'isolated' guard cells. *New Phytologist* **78**, 541–6.

Wedding, R.T. & Black, K. (1986). Malate inhibition of phospho-enolpyruvate carboxylase from *Crassula*. *Plant Physiology* **82**, 985–90.

JULIE W. MEADOWS, JAMIE B.
SHACKLETON, DIANE C. BASSHAM, RUTH
M. MOULD, ANDREW HULFORD and COLIN
ROBINSON

Transport of proteins into chloroplasts

All cells transport proteins across membranes, but the complexity of protein traffic in plant cells is especially striking because of the variety of organelle types involved. Many proteins are inserted, during translation, into the lumen of the endoplasmic reticulum, after which they are transported via the endomembrane system to the Golgi apparatus, vacuole, protein bodies or plasma membrane. Other proteins are transported post-translationally into glyoxysomes, mitochondria and plastids. In each case, the protein is synthesised with an appropriate signal which ensures targeting to the correct organelle, and a number of studies have attempted to define the characteristics of these targeting signals (reviewed by Bennett & Osteryoung, 1991; Robinson, 1991).

In terms of protein transport events, the biogenesis of the chloroplast is particularly complex, primarily owing to the architecture of the organelle. The chloroplast is bounded by a double-membrane envelope, between whose membranes is a soluble phase, the functions of which are presently obscure. Within the organelle is the soluble stromal phase (site of CO_2 fixation, amino acid synthesis and many other key reactions) and the extensive internal thylakoid membrane. The thylakoid network also encloses a further soluble phase, usually termed the thylakoid lumen. Thus, the chloroplast comprises in total three distinct membranes and three discrete soluble phases. Most of the proteins located in each of these organellar compartments are encoded by nuclear genes, synthesised in the cytosol, and transported into the organelle. Clearly, therefore, chloroplast biogenesis requires both the specific, efficient targeting of a large number proteins into the organelle, and the operation of intraorganellar 'sorting' mechanisms to distribute imported proteins to their correct destinations. Many of the underlying mechanisms are at present poorly understood, but substantial advances have been made in recent years in studies on the biogenesis of thylakoid proteins. In this chapter we will consider three key aspects of the import and sorting of thylakoid lumen proteins: the targeting signals involved, the mechanism

Society for Experimental Biology Seminar Series 50: *Plant organelles*, ed. A. K. Tobin.
© Cambridge University Press 1992, pp. 281–92.

of removal of these signals, and the energetics of protein translocation across the thylakoid membrane.

Biogenesis of thylakoid lumen proteins

Most of the abundant thylakoid proteins are components of the four protein complexes that account for the bulk of thylakoid protein: photosystem I, photosystem II, the cytochrome *b/f* complex, and the ATP synthetase complex. Many of these proteins are synthesised in the cytosol, and these proteins must therefore cross the envelope membranes and the stromal phase to reach the thylakoid membrane. Of particular interest is the biogenesis of hydrophilic thylakoid lumen proteins, since these must cross all three chloroplast membranes to reach their sites of function. The best-studied lumenal proteins are plastocyanin, a small copper-containing electron carrier, and three extrinsic photosystem II proteins of 33, 23 and 16 kDa which are components of the oxygen-evolving complex (Murato & Miyao, 1985; Andersson, 1986). The locations of these three proteins (33K, 23K and 16K) are shown diagrammatically in Fig. 1.

The import pathway taken by these proteins can be divided into two phases (Fig. 2). Initially, the proteins are synthesised in the cytosol as larger precursors containing amino-terminal precursors, transported into the stroma, and cleaved to intermediate-sized forms by a stromal processing peptidase (SPP). Thereafter, the intermediates are transported

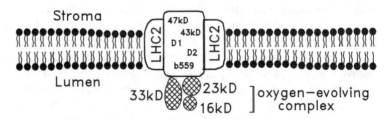

Fig. 1. Simplified diagram of the Photosystem II complex. The PSII complex contains a reaction centre core complex, which consists of the D1 and D2 proteins, cytochrome *b*559, and polypeptides of 43 and 47 kDa; a light-harvesting protein-pigment network of which the major protein component is the light-harvesting chlorophyll-binding protein (LHC-2); and a peripheral oxygen-evolving complex. The major components of the oxygen-evolving complex are three extrinsic polypeptides of 33, 23 and 16 kDa; these polypeptides and LHC-2 are nuclear-encoded and imported from the cytosol whereas the core complex polypeptides are chloroplast-encoded.

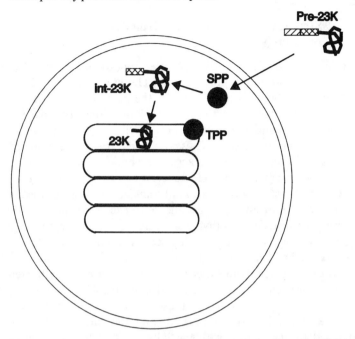

Fig. 2. Two-step model for the import of thylakoid lumen proteins. Lumenal proteins such as the 23 kDa protein of the oxygen-evolving complex (23K) are synthesised in the cytosol with a bipartite pre-sequence. After synthesis, pre-23K is imported into the stroma and cleaved to an intermediate form (int-23K) by a stromal processing peptidase (SPP). Int-23K is subsequently transported across the thylakoid membrane and processed to the mature size by a thylakoidal processing peptidase, TPP.

across the thylakoid membrane and processed to the mature sizes by a second, thylakoidal processing peptidase TPP (Hageman *et al.*, 1986; Smeekens *et al.*, 1986; James *et al.*, 1989). In keeping with this import mechanism, the pre-sequences of lumenal proteins contain two distinct domains which have differing characteristics. The amino-terminal 'envelope transfer' domains are structurally and functionally equivalent to the pre-sequences of imported stromal proteins, being rich in positively charged and hydroxylated residues. The second, 'thylakoid transfer' sequences have markedly different features which are discussed in the following section.

Structure of the thylakoid transfer signals of lumenal proteins

A comparison of the carboxy-terminal regions of the pre-sequences of several thylakoid lumen proteins (von Heijne *et al.*, 1989; Halpin *et al.*, 1989) revealed two common features: the presence of short-chain amino acids (usually alanine) at the -3 and -1 positions, relative to the TPP processing site, and the presence of a hydrophobic stretch of residues upstream from this motif. These features are also shared by 'signal' peptides which direct proteins across the endoplasmic reticulum and the bacterial plasma membrane. However, it was not possible to establish any other characteristics of the thylakoid transfer domains because, since SPP does not cleave at any recognisable consensus sequence, the lengths of the envelope transfer and thylakoid transfer domains within a given pre-sequence cannot be deduced using sequence data alone. In order to be able to define the characteristics of several thylakoidal transfer sequences, we carried out experiments to determine the SPP cleavage site within the pre-sequences of wheat and spinach 23K, wheat 33K and *Silene pratensis* plastocyanin. Each of these precursors was synthesised in the presence of a labelled amino acid, and the intermediate forms were generated by incubation with partially purified SPP as previously described (Hageman *et al.*, 1986; James *et al.*, 1989). The intermediates were then subjected to automated Edman degradation, and the SPP cleavage sites were deduced from the cycle numbers at which the radiolabelled amino acids were released. Figure 3 shows some of the features of the thylakoid transfer signals, and compares these features with those of typical signal sequences, after analysis of the thylakoid transfer sequences within the total pre-sequence (Smeekens *et al.*, 1985; Meadows *et al.*, 1991).

One of the most surprising findings to emerge from this study concerned the length of these targeting signals. Signal peptides are typically about 20 residues in length, but the 23K and 33K thylakoid transfer domains are at least twice as long as this. The corresponding sequence from plastocyanin is, however, more similar to signal peptides in terms of length, but this sequence is probably atypically short since the pre-sequence as a whole is unusually small.

The 23K and 33K transfer sequences are also markedly different in terms of charge distribution from typical signal sequences. An important feature in both prokaryotic and eukaryotic signal peptides is the presence of one or two positive charges between the amino terminus and the hydrophobic section (von Heijne, 1986). In contrast, this sequence in the 23K and 33K thylakoid transfer domains is much longer and contains

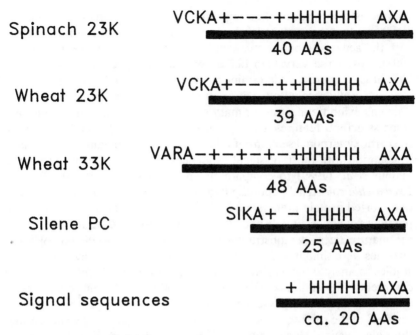

Fig. 3. Structural features of thylakoid transfer sequences. The carboxy-terminal regions of thylakoid transfer domains resemble signal sequences in two respects: the presence of a stretch of hydrophobic residues (H) and the presence of short-chain residues, usually alanine, at the -3 and -1 positions. The two types of peptide can have markedly different amino-terminal sections, especially in the cases of the 33K and 23K sequences, where the transfer sequences are much longer than signal sequences, and contain numerous negatively and positively charged residues. Signal sequences usually contain one or two positive charges in this region.

numerous positive *and* negative charges in equal numbers. The functional significance of this charged region is presently unclear, but the presence of this feature means that these thylakoid transfer domains are, in overall terms, similar to signal sequences in some respects but also dissimilar in others.

Mechanism of the thylakoidal processing peptidase

During or shortly after translocation of import intermediates across the thylakoid membrane, the thylakoid transfer sequence is removed by TPP to release the mature-size protein. Studies on TPP have shown that this enzyme is highly specific for imported lumenal proteins (Kirwin et al., 1987) and that the enzyme is a hydrophobic thylakoid membrane protein with the active site on the lumenal face of the membrane (Kirwin et al., 1988). Given the very high degree of reaction specificity exhibited by TPP, it is of interest to determine the features within the substrate which are specifically recognised. Different cleavage sites of thylakoid transfer domains exhibit almost no primary sequence homology, suggesting that some structural features are recognised instead. As pointed out above, the carboxy-terminal sections of thylakoid transfer domains have features in common with those of signal sequences. These similarities prompted Halpin et al. (1989) to compare the reaction specificities of TPP and Escherichia coli signal peptidase (which is responsible for the maturation of exported protein precursors). It was found that purified E. coli signal peptidase could accurately and efficiently process pre-23K and pre-33K to the mature sizes, demonstrating that the reaction specificities of these enzymes were similar, if not identical. More recently, we have carried out studies to analyse the TPP reaction mechanism in more detail, based on comprehensive studies on E. coli signal peptidase. This enzyme has been shown to require the presence of short-chain residues at the -3 and -1 residues of the pre-sequence. At the -3 position, alanine, glycine, serine, threonine, valine, leucine and isoleucine are tolerated. The -1 position is more restrictive, and only alanine, serine, glycine and cysteine are tolerated (von Heijne, 1986; Folz et al., 1988; Fikes et al., 1990).

To test the requirements for cleavage by TPP, we used site-specific mutagenesis to substitute the -3 and -1 residues of wheat pre-33K (both of which are alanine) by a variety of amino acids. The mutant precursors were then expressed by in vitro transcription–translation and imported into intact chloroplasts to monitor the effects of the mutations on membrane translocation and maturation. Figure 4 shows the results of import assays using wild-type pre-33K and two of the mutants, -1 Ser and -1 Lys. Wild-type pre-33K is imported into the thylakoids where it is resistant to added protease, showing that the protein is located in the lumen. Both of the mutants are also transported into the lumen, and studies (not shown) on the remaining mutants give a similar result: none of the -3 or -1 mutations affect transport across either the envelope or the thylakoid membranes. However, in almost every case, cleavage by TPP is dramatically affected. In the case of -1 Ser, very little of the imported protein is processed to the mature size, and the majority is instead processed only to a 36 kDa polypeptide (36K). This polypeptide is slightly smaller than the stromal intermediate generated by SPP, and our interpretation is that TPP, being unable to cleave efficiently at the correct processing site as a result of the mutation, is recognising a cryptic processing site elsewhere in

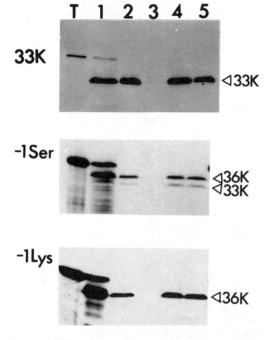

Fig. 4. Import of mutated pre-33Ks into isolated chloroplasts. The pre-33K cDNA was subjected to site-specific mutagenesis in order to substitute a variety of amino acids for the -3 and -1 alanine residues of the pre-sequence. The autoradiogram shows the import and localisation of wild-type pre-33K and two of the mutants: -1 Ser and -1 Lys. Lanes T, translation product; lanes 1, total polypeptides after import; lanes 2, after protease treatment of the chloroplasts; lanes 3 and 4, after fractionation into stroma and thylakoids, respectively; lanes 5, after protease treatment of the thylakoids. 33K: mobility of mature 33K; 36K: 36kDa cleavage product.

the thylakoid transfer sequence (Shackleton & Robinson, 1991). In the case of imported -1 Lys mutant, only the 36K polypeptide is detected in the thylakoid fraction indicating that this mutation completely blocks cleavage by TPP at the correct site.

Time course analyses of the maturation of imported protein were carried out for all of the mutants, and the results are summarised in Table 1. Of the 10 mutants tested, only one (-3 Val) is efficiently processed by TPP at the correct site. The other -3 mutations (-3 Leu, Glu and Lys) all drastically inhibit cleavage, leading to the generation of mostly 36K polypeptide. With the -1 mutations, four of the substitutions completely

Table 1. *Maturation of pre-33K mutants by TPP*

Construct	Rate of processing by TPP (% of wild-type rate)
Wild-type pre-33K	100
-3 Val	100
-3 Leu	5>
-3 Lys	5>
-3 Glu	5>
-1 Ser	5>
-1 Gly	5>
-1 Thr	0
-1 Glu	0
-1 Lys	0
-1 Leu	0

Time-course chloroplast import assays were carried out and the levels of mature-size 33K quantitated by laser densitometry. In all cases, inhibition of cleavage at the correct site by TPP resulted in accumulation of a 36 kDa intermediate, as described in Fig. 4.

block cleavage by TPP at the correct site (-1 Thr, Leu, Glu and Lys). Of particular interest, however, the -1 Gly and -1 Ser substitutions also drastically inhibit cleavage, even though these residues are very similar in overall structure to the -1 Ala in the wild-type precursor.

The conclusion from this study is that, although TPP appears to be similar in terms of overall reaction mechanism to signal peptidases, TPP has far more stringent requirements at the -3 and -1 residues of the substrate than either bacterial or ER signal peptidases. Whereas the latter enzymes can tolerate leucine at the -3 position, the presence of this residue almost completely blocks cleavage by TPP. It appears likely that the valine side-chain represents the maximum permitted length at this position. At the -1 position, TPP has very strict requirements in that alanine appears to be essential for efficient cleavage to take place. Even glycine and serine, which are commonly found at the -1 position in bacterial and eukaryotic signal peptides, are not tolerated.

These results raise interesting possibilities concerning the evolution of

TPP. It is possible that TPP evolved from bacterial signal peptidase, for example by gene duplication in an ancestral cyanobacterium; if this is the case, however, TPP has for some reason acquired much more stringent requirements regarding the -3 and -1 residues of the substrates. Alternatively, TPP may have evolved quite separately, in which case it is intriguing that the reaction mechanism should be so similar to those of signal peptidases. Whatever the reason, it will be of interest to obtain information of the sequence of the TPP gene, since this type of information may well resolve some of the key outstanding questions.

Energetics of protein translocation across the thylakoid membrane

Until recently, very little was known about the mechanism of protein transport across the thylakoid membrane. This was for largely technical reasons: using the standard intact chloroplast import assay, it is in practise very difficult to analyse in detail protein translocation across the internal membrane network. However, Kirwin *et al.* (1989) demonstrated that isolated thylakoid vesicles are capable of importing 33K in the presence of stromal extract and ATP. More recently, we have attempted to optimise the import assay further in order to analyse the mechanism of the transport system. We found that a critical requirement for efficient import of both 23K and 33K is the presence of light (Mould & Robinson, 1991). It seemed likely that this reflected a role, either directly or indirectly, of thylakoidal electron transport in promoting protein translocation, and experiments were carried out to test this possibility. Import assays were carried out in the presence of electron transport inhibitors (a combination of dichlorophenyl-dimethylurea (DCMU) and methyl viologen) to block formation of the thylakoidal protonmotive force (Δp, also known as PMF). Other assays were carried out in the presence of valinomycin (which selectively dissipates the electrical potential component, $\Delta\psi$, of the Δp) and nigericin, which collapses the proton gradient component, ΔpH. Figure 5 shows the result of a thylakoid import assay using pre-23K as a substrate. In the presence of stroma and light, pre-23K is converted to the intermediate form (by SPP in the stromal extract) and also to the mature size (lane 1). When this mixture is protease-treated after incubation (lane 5) the mature-size 23K is protected from digestion, showing that these molecules have been imported into the lumen. In the presence of DCMU/methyl viologen, pre-23K is efficiently processed to the intermediate form but no mature-size polypeptide appears, showing that import has been completely inhibited (lanes 2 and 6). The presence of valinomycin (lanes 3 and 7) leads to a very slight inhibition of import,

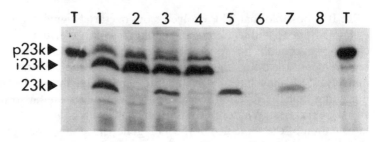

Fig. 5. Light-dependent import of 23K by isolated thylakoids. Pre-23K was incubated with stroma and thylakoids in the light (lane 1), and in the presence of DCMU/methy viologen (lane 2), valinomycin (lane 3) or nigericin (lane 4). Lanes 5–8, as in lanes 1–4 except that the thyalkoids were protease-treated after incubation. Lane T, translation product; p23K, i23K, precursor and intermediate forms of 23K.

but nigericin again blocks import completely (lanes 4 and 8). We conclude from this experiment that protein transport across the thylakoid membranes requires an energised membrane, and that the dominant component driving the translocation process is the proton gradient, and not the electrical potential, of the total protonmotive force. It remains to be determined precisely how Δp drives protein transport across the thylakoid membrane.

References

Andersson, B. (1986). Proteins participating in photosynthetic water oxidation. In *Photosynthesis III. Encyclopedia of Plant Physiology*, Vol. 19, ed. L.A. Staehelim & C.J. Arntzen, pp. 447–56. Berlin: Springer-Verlag.

Bennett, A.B. & Osteryoung, K.W. (1991). Protein transport and targeting within the endomembrane system of plants. In *Plant Genetic Engineering*, ed. D. Grierson, pp. 199–230. Glasgow: Blackie.

Fikes, J.D., Barkocy-Gallaher, G.A., Klapper, D.G. & Bassford, P.J. (1990). Maturation of *Escherichia coli* maltose-binding protein by signal peptidase I. Sequence requirements for efficient processing and demonstration of an alternate cleavage site. *Journal of Biological Chemistry* **265**, 3417–23.

Folz, R.J., Notwehr, S.F. & Gordon, J.I. (1988). Substrate specificity of eukaryotic signal peptidase. Site-saturation mutagenesis at position -1 regulates cleavage between multiple sites in human preproapolipoprotein A-II. *Journal of Biological Chemistry* **263**, 2070–8.

Hageman, J., Robinson, C., Smeekens, S. & Weisbeck, P. (1986). A thylakoid processing peptidase is required for complete maturation of the lumen protein plastocyanin. *Nature* **324**, 567–9.

Halpin, C., Elderfield, P.E., James, H.E., Zimmermann, R., Dunbar, B. & Robinson, C. (1989). The reaction specificities of the thylakoidal processing peptidase and *Escherichia coli* leader peptidase are identical. *EMBO Journal* **8**, 3917–22.

James, H.E., Bartling, D., Musgrove, J.E., Kirwin, P.M., Herrmann, R.G. & Robinson, C. (1989). Transport of proteins into chloroplasts. Import and maturation of precursors to the 33, 23 and 16 kDa proteins of the oxygen-evolving complex. *Journal of Biological Chemistry* **264**, 19573–6.

Jansen, T., Rother, C., Steppuhn, J., Reinke, J., Bayreuther, K., Jansson, C., Andersson, B. & Herrmann, R.G. (1987). Nucleotide sequence of cDNA clones encoding the complete 23 kDa and 16 kDa precursor proteins associated with the photosynthetic oxygen-evolving complex from spinach. *FEBS Letters* **216**, 224–40.

Kirwin, P.M., Elderfield, P.D. & Robinson, C. (1987). Transport of proteins into chloroplasts. Partial purification of a thylakoidal processing peptidase involved in plastocyanin biogenesis. *Journal of Biological Chemistry* **262**, 16386–90.

Kirwin, P.M., Elderfield, P.D., Williams, R.S. & Robinson, C. (1988). Transport of proteins into chloroplasts. Organisation, orientation and lateral distribution of the plastocyanin processing peptidase in the thylakoid network. *Journal of Biological Chemistry* **263**, 18128–32.

Kirwin, P.M., Meadows, J.W., Shackleton, J.B., Musgrove, J.E., Elderfield, P.D., Hay, N.A. & Robinson, C. (1989). ATP-dependent import of a lumenal protein by isolated thylakoid vesicles. *EMBO Journal* **8**, 3917–21.

Meadows, J.W., Hulford, A., Raines, C.A. & Robinson, C. (1991). Nucleotide sequence of a cDNA clone encoding the precursor of the 33KDa protein of the oxygen-evolving complex from wheat. *Plant Molecular Biology* **16**, 1085–7.

Mould, R.M. & Robinson, C. (1991). A proton gradient is required for the transport of two lumenal oxygen-evolving proteins across the thylakoid membrane. *Journal of Biological Chemistry* **266**, 12189–93.

Murata, N. & Miyao, M. (1985). Extrinsic membrane proteins in the photosynthetic oxygen-evolving complex. *Trends in Biochemical Science* **10**, 122–4.

Robinson, C. (1991). Targeting of proteins to chloroplasts and mitochondria. In *Plant Genetic Engineering*, ed. D. Grierson, pp. 179–98. Glasgow: Blackie.

Shackleton, J.B. & Robinson, C. (1991). Transport of proteins into chloroplasts. The thylakoidal processing peptidase is a signal-type peptidase with stringent substrate requirements at the -3 and -1 positions. *Journal of Biological Chemistry* **266**, 12152–6.

Smeekens, S., de Groot, M., van Binsbergen, J. & Weisbeck, P. (1985).

Sequence of the precursor of the chloroplast thylakoid lumen protein plastocyanin. *Nature* **317**, 456–8.

Smeekens, S., Kauelle, C., Hageman, J., Keegstra, K. & Weisbeck, P. (1986). The role of the transit peptide in the routing of precursors toward different chloroplast compartments. *Cell* **46**, 365–75.

von Heijne, G. (1986). A new method for predicting signal sequence cleavage sites. *Nucleic Acids Research* **14**, 4683–90.

von Heijne, G., Steppuhn, J. & Herrmann, R.G. (1989). Domain structure of mitochondrial and chloroplast targeting peptides. *European Journal of Biochemistry* **180**, 535–41.

ALYSON K. TOBIN and W. JOHN ROGERS

Metabolic interactions of organelles during leaf development[1]

In any discussion of plant cell metabolism it is essential to recognise the heterogeneity that exists within an organ such as the leaf. This encompasses the variety of cell types, such as the mesophyll, vascular, epidermal, etc., and also the range of cell ages within a leaf. The metabolic role of the organelles may therefore be quite different depending on the cell in which they are localised and also on the developmental stage of that cell. In this chapter we discuss the changes which occur in the size, frequency and metabolic activity of organelles during leaf cell development and differentiation.

Patterns of leaf cell development

Leaf cell development is most readily studied in grasses, such as wheat (*Triticum aestivum* L.) and barley (*Hordeum vulgare* L.). This is because all of the cells of the leaf originate from a single meristematic region at the leaf base (intercalary meristem). Thus a developmental gradient is generated whereby the youngest cells are always at the leaf base and there is a measurable range of increasing cell age towards the tip of the leaf. Because of its simplicity, the developing light-grown wheat leaf has been a popular tool with which to study chloroplast and photosynthetic development (see, for example, reviews by Leech (1985) and Baker (1985) in this series). In contrast, dicotyledon leaf development is far more complex as there are several growing regions across the lamina (Maksymowych, 1973). The resulting heterogeneity makes it difficult to separate out areas where cell division, expansion and development are occurring within a dicotyledonous leaf. Studies of dicotyledon leaf development, therefore, generally employ de-etiolation or 'greening' techniques whereby a dark-grown plant is exposed to light and the change in enzyme levels is monitored progressively with organelle 'biogenesis'.

[1] This chapter is dedicated to the memory of Mina Patel, who worked with us in the early stages of our research into wheat leaf development.

Society for Experimental Biology Seminar Series 50: *Plant organelles*, ed. A. K. Tobin.
© Cambridge University Press 1992, pp. 293–323.

Such conditions are rarely, if ever, encountered by a plant growing in its natural environment and, although useful information has been derived from this type of study, it is apparent that there are numerous differences in organelle development between 'greening' and light-grown leaves (Leech, 1984). An alternative means of analysing leaf development in dicotyledons is to sample whole leaves from different parts of the plant to obtain a range of ontogenetic ages, or to take leaves from different ages of plant. This method enables the plants to be grown under normal light/dark cycles but the samples are, nevertheless, heterogeneous because of the cell and age range within the whole leaves.

The advantages of using monocotyledons in studies of leaf development are therefore quite clear and, although we will refer to some studies with dicotyledons, much of the work described here has been carried out using light-grown wheat (*T. aestivum* cv. Maris Huntsman) primary leaves. The leaves are sampled from 8-day-old plants, when the growth rate of the primary leaf is at its maximum (Tobin *et al.*, 1985), thus ensuring that the developmental gradient between leaf base and tip is also maximal. In most cases, measurements are expressed on a 'per cell' basis, as other parameters such as weight, chlorophyll and protein concentration change considerably during leaf development.

Changes in cell size and subcellular compartmentation during leaf development

The sequestration of metabolites within organelles is a means of altering the localised concentration of a substrate. In any compartmentalised pathway that involves fluxes of metabolites between organelles it is therefore important to know the relative size of the compartments. In this section we present evidence of changes in the relative size and number of organelles within wheat leaf mesophyll cells during leaf development.

Mesophyll cell size

In wheat primary leaves, the cells expand and elongate within a zone of cell elongation which extends from immediately above the intercalary meristem to a region approximately 3.0 cm distal to this, where maximum cell size is attained (Fig. 1; Boffey *et al.*, 1980; Ellis *et al.*, 1983; Tobin *et al.*, 1985). The precise cellular volume is extremely difficult to quantify because of the irregular shape of the mesophyll cells which become increasingly lobed as they mature. From three-dimensional reconstructions of electron micrographs we have estimated an approximately 30% increase in mesophyll cell volume between 1.0 and 15 mm from the leaf

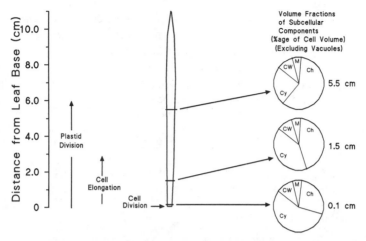

Fig. 1. Growth and subcellular compartmentation of a wheat primary leaf. Plants were grown under a 16 h photoperiod and the primary leaf was harvested at 8 days (Tobin *et al.*, 1985). Pie charts show V_v (% fractional volume of the mesophyll cell) for each subcellular compartment at 1.0, 15.0 and 55.0 mm from the basal meristem. The vacuole V_v has been omitted for reasons of clarity: it was constant at 80% in all sections. M, mitochondria; Ch, chloroplasts; CW, cell wall; Cy, cytosol. Actual V_vs are: M, 0.6% in all sections; CW, 3.0% in all sections; Ch, 5.7% (at 1.0 mm), 10.3% (at 15.0 mm), 13.0% (at 55.0 mm); Cy, 10.7% (at 1.0 mm), 8.3% (at 15.0 mm), 6% (at 55.0 mm) (W. J. Rogers, J. R. Thorpe and A. K. Tobin, unpublished data).

base (Table 1). This contrasts with an estimated doubling of mesophyll cell length over a similar region of the wheat primary leaf (Ellis *et al.*, 1983), although, in this study, the cell volume was not determined.

Developmental changes in size and number of organelles in wheat mesophyll cells

Within this framework of mesophyll cell development there are progressive changes in the structure and size of organelles. Most notable are the chloroplasts. At the base of the leaf they are small and relatively undeveloped 'proplastids' with few granal regions. Division of proplastids within the meristem keeps pace with cell division and maintains a constant number of plastids per cell (Leech, 1985). In cells above the meristem, however, the chloroplasts continue to divide and this leads to an increase in their number per cell (Table 1; Boffey *et al.*, 1979; Tobin *et al.*, 1985). This increase in number, together with an increase in

Table 1. *Changes in mesophyll cell size, chloroplast and mitochondrial number and mitochondrial size during development of light-grown wheat primary leaves*

Distance above basal meristem (cm)	Number of plastids per cell	Number of mitochondria per cell	Volume of mitochondrion (μm^3)	Volume of cell (μm^3)
0.1		1252	0.13	27 066
1.0				
1.5		1863	0.12	35 427
2.0	59			
2.5			0.21	
3.0	84			
3.5			0.25	
4.0	109			
4.5			0.34	
5.0	126			
5.5		625	0.32	35 147
6.0	137			
7.0	146			
8.0	150			

The volume of mitochondrion refers to the mean volume of individual mitochondria. All cell measurements refer to mesophyll cells.
Plastid data from Tobin *et al.*, 1985. Other data are from W.J. Rogers, J.R. Thorpe and A.K. Tobin (unpublished).

chloroplast size, results in an increasing proportion of the mesophyll cell becoming occupied by the chloroplasts as the leaf develops (Fig. 1; Ellis & Leech, 1985).

With the exception of the chloroplasts and cytosol, there is no change in the volume fraction (V_v, i.e. the fractional volume (%) of the cell occupied by a component (Weibel, 1973)) of any subcellular compartment during mesophyll cell development (Fig. 1). The vacuole occupies a constant 80%, the mitochondria 0.6% and the cell wall approximately 3% of the mesophyll cell volume throughout its development (Fig. 1). The chloroplast V_v, however, increases from 5.7%, at 1.0 mm from the base, to 13.0% at 55.0 mm (Fig. 1; W. J. Rogers, J. R. Thorpe and A. K. Tobin, unpublished data) and finally to 17% in mature mesophyll cells (Ellis & Leech, 1985). The V_v of the cytosol decreases from 10.7% in the youngest cells to 6.0% in the oldest (Fig. 1). This means that as the wheat mesophyll cells develop, an increasing proportion of their volume consists of chloroplasts and this leaves less free cytosol surrounding the organelles. This may have important consequences regarding the interactions between organelles as these become more closely juxtaposed with leaf cell maturity.

Although the mitochondrial V_v remains constant there are changes in the composition of the mitochondrial population with mesophyll cell development (Table 1). In the basal region of the leaf, where the cells are expanding, there is an increase in the number of mitochondria per cell while their mean size remains constant. The increase in mitochondrial number keeps pace with cell expansion so that the mitochondrial V_v remains constant. From this it may be deduced that the mitochondria are dividing, although we have no direct evidence of this. There have, however, been reports of 'dumb-bell shaped' mitochondria, which may represent dividing mitochondria, in meristematic tissues and callus (Bagsaw *et al.*, 1969), hence this is not without precedent in plant cells. Once cell elongation is complete and the cells have attained their maximum volume there is an increase in the individual size of the mitochondria and an apparent decrease in their number per cell (Table 1). These changes in mitochondrial size and number are reciprocal so that the mitochondrial V_v still remains unchanged after cells have reached their maximum size.

The decrease in number of mitochondria in maturing cells could be due either to autolysis, i.e. degeneration of mitochondria, or to mitochondrial fusion. Although we have no direct evidence in support of either possibility we tend to favour the latter because of the change in the size distribution of the mitochondrial populations. If autolysis were responsible then there would be a reduction in the number of mitochondria throughout the population, regardless of mitochondrial size, so that the range of mitochondrial sizes would always be the same at each stage of development. If, however, the reduction in numbers was brought about by mitochondrial fusion then there would be an increasing trend towards larger mitochondria and a decrease in the number of smaller mitochondria as the cells develop. This is the type of pattern that we observed from a plot of the mitochondrial size distributions at six different stages of mesophyll cell development (Fig. 2). From 25 mm and upwards along the leaf the distributions become increasingly skewed towards larger mitochondrial profile diameters and this, we believe, is preliminary evidence for the fusion of mitochondria in maturing mesophyll cells. It is possible that fusion occurs as the chloroplasts take up an increasing proportion of the cytoplasm and effectively confine the mitochondria into a declining volume of free cytosol. This situation is somewhat analogous to

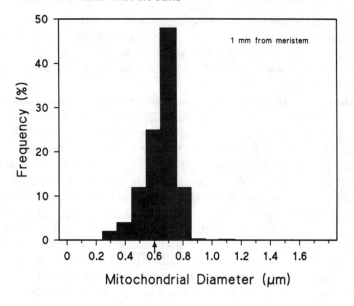

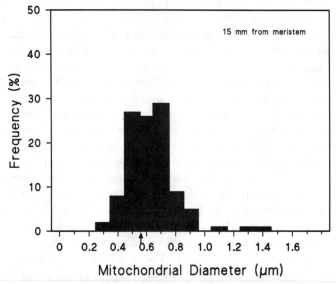

Fig. 2. Frequency distribution of mitochondrial profile diameters in wheat mesophyll cells at six different regions above the basal meristem of the primary leaf. Arrows indicate the population mean. Diameters were determined by measurement of electron micrographs. A minimum of 130 mitochondrial profiles was measured at each region (W. J. Rogers, J. R. Thorpe and A. K. Tobin, unpublished data).

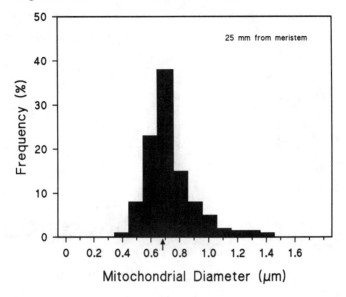

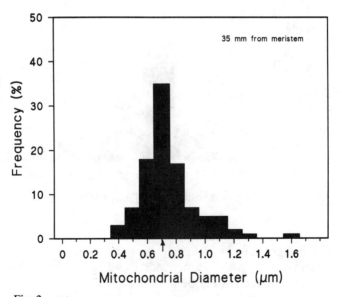

Fig. 2–cont.

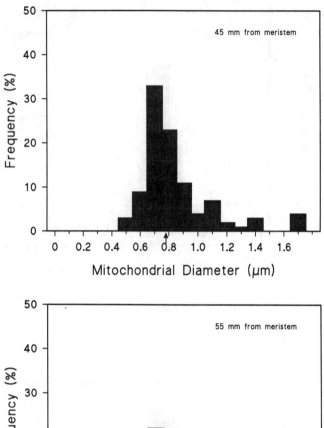

Fig. 2–cont.

that of the flight muscle of ageing blowflies, where close packing of muscle fibres confines the mitochondria and results in their fusion to produce larger organelles (Tribe & Ashurst, 1972).

It is apparent from the above that the size and subcellular compartmentation of the mesophyll cell changes with its development. The chloroplasts increase in size and number and take up an increasing proportion of the cell volume. The mitochondria appear to occupy a constant proportion of the cell, although there are more of them in young cells at the leaf base and they increase in size and decrease in number once maximum cell size is reached. All of these changes may relate to the differing metabolic activity and subcellular interactions of the developing mesophyll cell, as discussed below.

Development of photosynthetic activity

Immature leaf tissue is heterotrophic and has to import carbon from mature leaf tissue or from the seed (Dale, 1985). With increasing maturity of the chloroplasts comes an increasing ability to photosynthesise and consequently a predominantly autotrophic metabolism. Thus, between the base and tip of a leaf there is a gradation of leaf cell metabolism from heterotrophic, non-photosynthetic cells through to autotrophic, fully photosynthetic cells at the leaf tip. In barley, for example, the rate of $^{14}CO_2$ uptake (per unit leaf area) is much greater at the tip than at the base of the leaf (Dale, 1972). Measurements of CO_2-dependent O_2 evolution along the length of a wheat leaf further illustrate this transition. At the base of the wheat leaf there is no detectable rate of photosynthesis. Photosynthetic activity then increases with leaf cell development (Fig. 3).

The absence of detectable rates of photosynthesis at the leaf base may be attributable to the inability of proplastids to synthesise ATP at very early stages of leaf development (Webber *et al.*, 1984). There are, however, many possible limitations to the rate of photosynthesis and these are likely to change with leaf development. It is therefore simplistic to refer to a single 'rate-limiting' parameter, as control of a pathway is often shared between several steps in the pathway (see Kacser & Burns, 1973, 1979; Kacser & Porteus, 1987; Kacser, 1987, for discussion). Levels of Calvin cycle enzymes, for example, are also relatively low in immature leaf cells. Rubisco protein is detectable in the basal cells of wheat (Dean & Leech, 1982) and barley (Viro & Kloppstech, 1980) leaves but the amount per cell increases *c.* 20-fold towards the tip (Dean & Leech, 1982). Several authors have concluded that Rubisco activity limits net photosynthetic rate (P_N) during leaf ontogeny (Besford *et al.*, 1985; Ostareck & Lieckfeldt, 1989) although other Calvin cycle enzymes show a similar correlation. In the third leaf of wheat, for example, there is a strong correlation between Rubisco activity and P_N at all stages of development but this is also true of phosphoglycerate kinase (PGA

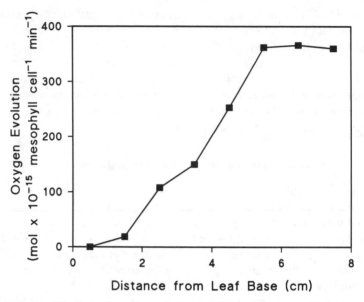

Fig. 3. CO_2-dependent O_2 evolution in different regions of light-grown wheat primary leaves. Measurements were made using a leaf disc electrode (Hansatech, UK) at saturating concentrations of CO_2 (from Tobin *et al.*, 1988).

kinase) and fructose 1,6-bisphosphatase (FBPase) activities (Suzuki *et al.*, 1987). In the dicotyledon *Dianthus chinensis*, Rubisco and NADP-dependent glyceraldehyde 3-phosphate (NADP-GAP) dehydrogenase activities are minimal in the leaf primordia and increase along the blade as the leaf expands out of the bud and becomes exposed to light (Croxdale & Vanderveer, 1986; Croxdale & Pappas, 1987). NADP-GAP dehydrogenase and FBPase activities have also been found to increase along the length of light-grown barley leaves (Sibley & Anderson, 1989). Thus, even within a single leaf there are considerable changes in the rate of photosynthesis and in the activities of the photosynthetic enzymes. Limitations to photosynthesis are likely to change during leaf development (see, for example, Catsky *et al.*, 1976; Zima & Sestak, 1979; Catsky & Ticha, 1980; Sestak, 1985) but the trend is always consistent: photosynthetic acivity increases with leaf cell maturity.

Changes in photorespiratory rate during leaf development

Photorespiration (as discussed by Wallsgrove *et al.*, this volume) arises from the oxygenase reaction of Rubisco in the chloroplast. The pathway then involves interaction between the chloroplasts, cytosol, peroxisomes and mitochondria in a series of reactions whose prime role appears to be to return the carbon lost at the oxygenase step back into a form which can be utilised in the Calvin cycle. The development of this pathway thus involves a degree of coordination of the biogenesis of these organelles. Studies of the development of photorespiration therefore provide an approach to the problem of how this coordination is regulated.

We have already discussed the observation that photosynthesis increases with leaf and chloroplast development and there are few, if any, reports to the contrary. The relationship between photorespiration and leaf development has, however, been a subject of conflicting reports owing primarily to the difficulty of measuring the rate of photorespiration in leaves (see Sharkey, 1988). There are two issues to consider: first, is there a change in the ratio of photosynthesis to photorespiration and, secondly, is there a change in the absolute rate of photorespiration during leaf development?

The partitioning of carbon between the photosynthetic and photorespiratory pathways is determined at the level of Rubisco, i.e. the ratio of carboxylase to oxygenase activity (see Wallsgrove *et al.*, this volume). This ratio of activities may alter as a result of changes in the relative CO_2 and O_2 concentrations at the active site of the enzyme (Keys, 1986) or by changes in the properties of Rubisco which result in a change in the relative carboxylase and oxygenase activities. The latter is rather unlikely because although the carboxylase/oxygenase ratio of purified Rubisco may vary between species (Keys, 1986) there is no evidence of any such change in relation to leaf development within a single species. The increase in Rubisco protein during normal leaf development, as discussed above, has been reported by a number of independent groups (Viro & Kloppstech, 1980; Dean & Leech, 1982; Nivison & Stocking, 1983; Zielinski *et al.*, 1989). As there is also a coordinated development of Rubisco protein and of Rubisco activase (Zielinski *et al.*, 1989), then there is likely to be an increase in Rubisco activity with leaf development. The relationship between the amount of Rubisco protein and its activity is not, however, a necessarily direct one because the catalytic activity of the enzyme (V_{max}) is altered substantially by the concentrations of many substances, e.g. CO_2, Mg^{2+}, orthophosphate (P_i) and NADPH (Badger & Lorimer, 1981; Parry *et al.*, 1985). It also follows that there is not necessarily a direct relationship between the amount of Rubisco and the absolute rates of photosynthesis or photorespiration. Nevertheless, the amount of Rubisco protein per cell increases to such an extent (20-fold, according to Dean & Leech, 1982), that there is likely to be an increase in the rate of photorespiration with the development of wheat leaves.

Unless there is also a change in the CO_2 or O_2 concentration at the active site of Rubisco then one would anticipate that the development of photorespiration would parallel that of photosynthesis.

One method of measuring the rate of photorespiration is to treat the leaves with the glutamine synthetase (GS) inhibitor, L-methionine sulphoximine (MSO), thus blocking the assimilation of ammonia. The rate at which ammonia accumulates in the leaf then gives an estimate of the flux through the photorespiratory ammonia cycle, providing that certain controls are also performed. One consideration, for example, is that there are alternative sources of ammonia in leaves. These include nitrite reduction, asparaginase, arginase, and ureide catabolism (Joy, 1988). The contribution from these other sources may be estimated by determining the effects of photorespiratory inhibitors on the accumulation of ammonia. This method was used to determine the extent to which the rate of photorespiration changes along the length of wheat primary leaves. Ammonia accumulation, in the presence of MSO, increases with wheat

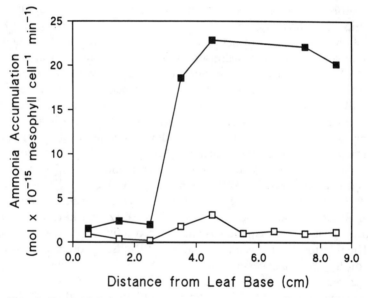

Fig. 4. Rate of light-dependent ammonia accumulation in different regions of light-grown wheat primary leaves. Leaf slices (5 mm transverse sections) were incubated with: □—□, water; ■—■, 2.0 mM MSO; O—O, 10.0 mM $KHCO_3$; ●—●, 2.0 mM MSO + 10.0 mM $KHCO_3$; ▲—▲, 10.0 mM glycine; △—△, 2.0 mM MSO + 10.0 mM glycine; ▽—▽, 2.0 mM MSO + 10.0 mM HPMS; ▼—▼, 2.0 mM MSO + 10.0 mM HPMS + 10.0 mM glycine (from Tobin et al., 1988).

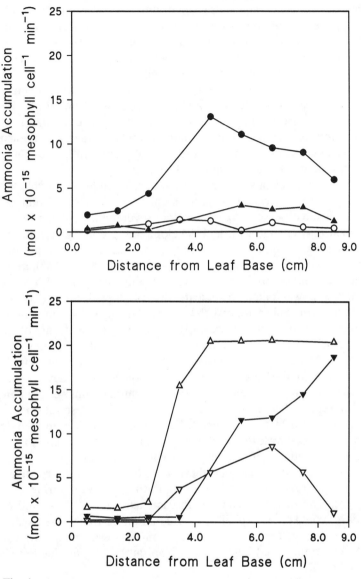

Fig. 4.–cont.

leaf development (Fig. 4; Tobin *et al.*, 1988). The accumulation is light-dependent at all stages of development. In mature leaf sections, most of the ammonia that accumulates is of photorespiratory origin as it is inhibited by pyrid-2yl-hydroxymethane sulphonic acid (HPMS–glycolate oxidase inhibitor) and is reduced in the presence of bicarbonate (which

306   A.K. TOBIN AND W.J. ROGERS

increases the CO_2 concentration around the leaf, stimulates Rubisco carboxylase and inhibits Rubisco oxygenase). In younger cells at the leaf base the ammonia accumulation is still inhibited by HPMS but it is not affected by increased CO_2 concentrations (Fig. 4). This suggests that the ammonia generated in basal leaf cells is not entirely attributable to photorespiration. The presence of low levels of nitrate reductase activity in these cells indicates one additional source (Tobin et al., 1988). The pattern of increase in photorespiratory ammonia accumulation resembles that for the development of photosynthetic oxygen evolution (Fig. 3). This indicates that, in wheat, the ratio of photosynthesis to photorespiration remains constant with leaf development.

Other studies of photorespiration in relation to leaf development have relied on gas exchange measurements. These must be interpreted with care as the observed flux is a result of several different factors, any one of which may change with leaf development. Kennedy & Johnson (1981), for example, found that the CO_2 compensation point was very high in young apple leaves and then decreased with leaf age. As there was also a decrease in the oxygen sensitivity of photosynthesis with leaf development this was taken to indicate that photorespiration decreased while photosynthesis increased with leaf age. The high CO_2 compensation point of young leaves may, however, be caused by the high rate of respiration (see below) and the low photosynthetic capacity of young leaves. The decline in the CO_2 compensation point is more likely attributable to the decrease in the rate of dark respiration and the development of an increased photosynthetic activity with leaf age. Salin & Homann (1971) also reported a decrease in the rate of photorespiration during leaf development in Citrus species. The photosynthetic rate of older leaves was increased by lowering the oxygen concentration from 21% to 2%, but this was without effect on photosynthesis in younger leaves. Homann (1975) later concluded that there was no change in the relative activities of the photosynthetic and photorespiratory pathways during leaf development. The observed changes in gas exchange of young and old Citrus leaves could all be explained by the limitations to net photosynthetic CO_2 fixation in young leaves by high stomatal resistances, low chlorophyll concentrations and high rates of respiratory CO_2 production (Homann, 1975).

Despite these problems of interpretation, there is general support for the view that photorespiration increases with leaf development and parallels the development of photosynthesis. In addition to our studies on wheat primary leaves, described above, this has been found in wheat primary leaves sampled from plants of different ages (Ostareck & Lieckfeldt, 1989); in leaves of white lupin, where there was an increase in photorespiratory ammonia production with time after leaf initiation (Atkins *et al.*, 1983); and in poplar trees (*Populus* × *Euramericana*) where CO_2 exchange was measured for attached leaves of different ages (Dickman *et al.*, 1975).

The photosynthetic and photorespiratory pathways are intrinsically linked in that they both originate from the activity of Rubisco. Experiments with photorespiratory mutants (described by Wallsgrove *et al.*, this volume) clearly show the serious effects on the plant which result from blocking the photorespiratory pathway. The effects are two-fold. First, they result in the accumulation of toxic photorespiratory intermediates, such as ammonia, and secondly, they inhibit the recycling of carbon and consequently deplete the Calvin cycle of intermediates and this inhibits photosynthesis. From this, it follows that the only way in which the ratio of photosynthesis to photorespiration may change during leaf development is via a change in the ratio of Rubisco oxygenase to carboxylase activity, as discussed earlier. Developmental changes in the properties of Rubisco have yet to be reported, nor is there any evidence of a change in the relative flux through the oxygenase and carboxylase reactions (for example, as a result of changes in internal CO_2 concentrations) at different stages of leaf development.

Changes in photorespiratory enzyme activities during leaf development

The activity of photorespiratory enzymes increases with leaf development and parallels that of the photosynthetic enzymes described above (Fig. 5; Tobin *et al.*, 1985, 1988, 1989; Rogers *et al.*, 1991). In all cases, there is an initially slow rate of increase in enzyme activity followed by a rapid increase which then declines in later stages of leaf development. The most rapid increase in enzyme activity occurs during the final stages of chloroplast division, at 4.0 cm from the leaf base. These increases in activity (per cell) are not simply related to the fact that the cells have expanded and increased in size, because the magnitude of increase in enzyme activity far exceeds the increase in cell volume. GS2 activity, for example, increases more than 50-fold per cell between the base and tip of a wheat primary leaf (Tobin *et al.*, 1985). Not only is there an increase in the cellular concentration of these enzymes (i.e. activity per unit cell volume): there is also an increase in the activity within the organelles. The activity of GS2 within the chloroplast, for example, increases more than five-fold (Tobin *et al.*, 1988). The concentration of glycine decarboxylase complex (GDC) within the mitochondria also increases during leaf development (see below).

There appears to be a coordinated development of the photorespiratory enzymes during leaf growth, regardless of their subcellular localisation. Glycolate oxidase (peroxisomal), GS2 (chloroplastic) and GDC (mitochondrial) all show the same pattern of development (Fig. 5). The precise control of this process is unknown, although several factors may be implicated. There is evidence, for example, that the synthesis of a number of photorespiratory enzymes is regulated by light. GS2 activity increases during the greening of etiolated seedlings (Guiz et al., 1979; Mann et al., 1979; Nishimura et al., 1982; Canovas et al., 1986) and this may be under the control of phytochrome (Kansara et al., 1989). Ferredoxin-dependent glutamate synthase (Fd-GOGAT) activity increases during greening (Nicklisch et al., 1977; Suzuki et al., 1982; Wallsgrove et al., 1982; Hecht et al., 1988). Again, phytochrome has been implicated in this process but in this example a direct involvement of photosynthetic development was ruled out (Hecht et al., 1988). The peroxisomal enzymes glycolate oxidase, serine:glyoxylate aminotransferase and hydroxypyruvate reductase all increase in a light-dependent manner in developing cotyledons (Hondred et al., 1987; Betsche & Eising, 1989). The glycine decarboxylase complex (GDC) is the

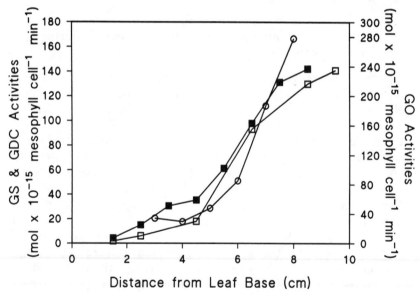

Fig. 5. Changes in photorespiratory enzyme activities during development of wheat primary leaves. □–□, GS2; ■–■, glycolate oxidase; O–O, glycine decarboxylase activity (actual values ×10 to aid comparison) (from Tobin et al., 1988, 1989).

only mitochondrial enzyme, to date, for which there is evidence of an involvement of light in the regulation of its synthesis (Arron & Edwards, 1980; Walker & Oliver, 1986*a*). This will be discussed in more detail in the following section.

Changes in respiration during leaf development

In general, rates of 'dark' respiration decrease with leaf development (Kidd *et al.*, 1921; Smillie, 1962; Azcón-Bieto *et al.*, 1983). Rates of oxygen uptake are particularly high in protoplasts isolated from the meristematic region at the base of barley primary leaves (Owen *et al.*, 1986). As with photosynthesis, there are numerous factors contributing to the overall rate of respiration in leaves and it is likely that the relative influence of factors such as substrate availability, adenylate control, metabolite transport, enzyme levels and mitochondrial compartment size will change with leaf development.

There is some evidence of a decline in the rate of glycolysis with leaf development. Rates of sucrose oxidation by bean leaf slices in the presence of uncoupler decrease with leaf age (Azcón-Bieto *et al.*, 1983). In *Dianthus chinensis*, the activity of the glycolytic enzyme phosphofructo-kinase (PFK) is high in very young leaf primordia and decreases in older leaves (Croxdale, 1983). It has been suggested that glycolysis is the main pathway for carbohydrate degradation in very young leaf cells but that the oxidative pentose phosphate pathway (OPPP) becomes more important at later stages of development (Croxdale & Pappas, 1987). The activities of the OPPP enzymes glucose 6-phosphate (G6P) dehydrogenase and 6-phosphogluconate (6PG) dehydrogenase, for example, are highest in the leaf primordia in which PFK activity has already begun to decrease (Croxdale & Outlaw, 1983). In *Phaseolus vulgaris* leaves and in young wheat seedlings the glycolytic enzyme NAD-GAP dehydrogenase and the OPPP enzymes G6P dehydrogenase and 6PG dehydrogenase are all present at early stages of leaf development, reach maximum activities in 5-day-old leaves and then decrease along with a decrease in the rate of respiration (Hoffmann & Schwarz, 1975). All of the enzymes involved in carbohydrate oxidation decrease in activity at the stage where photosynthetic enzymes, such as Rubisco and NADP-GAP dehydrogenase, begin to increase in activity, thus marking a transition from heterotrophy to autotrophy (Croxdale & Vanderveer, 1986; Croxdale & Pappas, 1987; Hoffmann & Schwarz, 1975).

Mitochondrial activity in developing leaves

The presence of an alternative, non-phosphorylating, cyanide-insensitive oxidase in plant mitochondria has been discussed in previous chapters of this volume (Wiskich & Meidan; Moore *et al.*). A number of early studies suggested that there is a difference in the cyanide-sensitivity of respiration in young and old leaves. Respiration in young leaves was severely inhibited by cyanide, whereas in older leaves the addition of the cytochrome oxidase inhibitor had very little effect on their respiration (Marsh & Goddard, 1939; MacDonald & DeKok, 1958; Ducet & Rosenberg, 1962). This was taken to indicate that cytochrome oxidase activity was absent from mature leaves (Marsh & Goddard, 1939). In the presence of cyanide, however, electrons are diverted from the cytochrome pathway to the alternative pathway (as discussed by Moore *et al.*, this volume). The lack of sensitivity to cyanide cannot therefore be taken to indicate a loss of cytochrome oxidase activity from mature leaves. These early studies, nevertheless, provided evidence of a change in the balance between cytochrome oxidase and alternative oxidase activities during leaf development. Azcón-Bieto *et al.* (1983) later found that the decline in the rate of dark respiratory oxygen uptake (per unit leaf area) during development of bean leaves was attributable mainly to a decrease in the activity and capacity of the cytochrome pathway. The capacity of the alternative pathway, however, remained relatively constant with leaf age and hence there was an increase in the percentage of cyanide resistance with development. These changes in bean leaf respiration were also observed in the respiration of mitochondria isolated from pea leaves of different ages (Azcón-Bieto *et al.*, 1983). The rate of oxygen uptake (per unit mitochondrial protein) was higher for the mitochondria isolated from younger leaves. The alternative pathway capacity of the isolated mitochondria remained the same irrespective of the age of the leaf. Thus, at least some of the observed developmental changes in leaf respiration may be as a result of changes in the activity of the mitochondria.

There are distinct patterns of development of mitochondrial enzymes along the length of a wheat primary leaf. Cytochrome oxidase (membrane marker) and glutamate dehydrogenase (GDH; matrix marker) activities (per cell), for example, are relatively high at the leaf base and show a gradual increase in activity to reach a maximum at around 5 cm from the base (Fig. 6). The relative abundance of the α-subunit of the mitochondrial ATPase protein per cell remains relatively constant between the base and 4.0 cm along a 7-day-old wheat primary leaf (Topping & Leaver, 1990). GDC activity develops in a different pattern and,

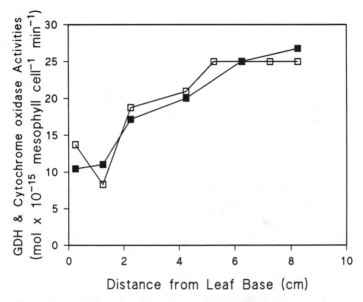

Fig. 6. Changes in cytochrome oxidase (■−■) and glutamate dehydrogenase (□−□) activities during development of wheat primary leaves (from Tobin *et al.*, 1988).

as discussed above, parallels the development of other photorespiratory and photosynthetic enzymes (Fig. 5).

GDC is a large, multi-enzyme complex localised in the mitochondrial matrix. It consists of four subunits – P, H, T and L proteins – each of which has a separate catalytic activity (Walker & Oliver, 1986*b*). The increase in GDC activity which occurs during wheat leaf development is primarily caused by an increase in the amount of GDC protein (Fig. 7). The P, H and T protein subunits of GDC all increase together in a highly coordinated manner (Fig. 7; Rogers *et al.*, 1991). The L protein, however, deviates from this pattern and is present at relatively high concentrations at the leaf base. This, perhaps, reflects an additional role for the L protein. Of the four GDC proteins, the L protein (lipoamide dehydrogenase) is the only one believed to be involved in other enzyme reactions. Lipoamide dehydrogenase activity is associated with the pyruvate dehydrogenase and 2-oxoglutarate dehydrogenase complexes (Denton *et al.*, 1975; Furuta *et al.*, 1977; McManus & Cohn, 1975). Only one form of lipoamide dehydrogenase has been detected in plant mitochondria. Furthermore, a monoclonal antibody which inhibits GDC

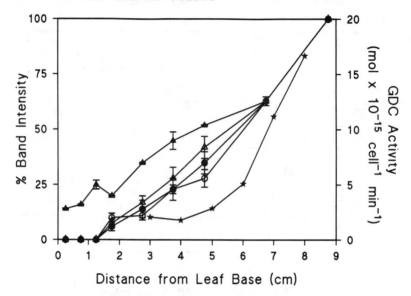

Fig. 7. Development of the four GDC subunit proteins, P (O—O), H (●—●), T (△—△) and L (▲—▲), and GDC activity (★—★) during development of wheat primary leaves. The graph represents the combined results of laser densitometer scans of Western blots of the four GDC subunits, with bands of the 8–8.5 cm leaf section set at 100% for each subunit. P, H and T proteins were undetectable below 1.5 cm from the basal meristem (from Rogers *et al.*, 1991).

activity by reacting with the L protein, equally inhibits pyruvate dehydrogenase from pea leaves (Walker & Oliver, 1986*b*). This suggests that there may only be one lipoamide dehydrogenase and that the L protein of GDC is shared between the other keto acid dehydrogenases. The pattern of development of the L protein may therefore be a reflection of its involvement in TCA cycle oxidations as well as in the GDC reaction.

The discrete developmental pattern of the L protein indicates that there is a distinct regulatory mechanism for this protein. We have already referred to the involvement of light in the regulation of expression of photorespiratory enzymes. Mitochondria isolated from green leaves oxidise glycine at a much faster rate than those isolated from etiolated leaves (Day *et al.*, 1985). The activities of the GDC subunits (Walker & Oliver, 1986*a*) and the amount of GDC protein (Rogers *et al.*, 1991) also increase during greening of etiolated tissue. Again, the L protein shows a quite different response from that of the other GDC proteins. Compared with the P, H and T proteins, much higher levels of L protein (up to 70%

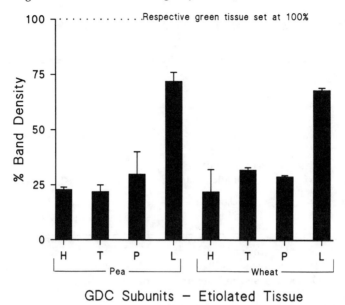

GDC Subunits — Etiolated Tissue

Fig. 8. Comparative levels of the four GDC subunits in etiolated, compared with light-grown, leaf tissues from wheat and pea. The results, obtained by laser densitometer scanning of Western blots, are based on the loading of SDS-PAGE wells to represent equal amount of fumarase activity. Levels in light-grown leaves are set at 100% (from Rogers *et al.*, 1991).

of that of mature light-grown leaves) are maintained in etiolated leaves (Fig. 8). In contrast, P, H and T proteins are present at only 25–30% of the light-grown level. The maintenance of relatively high concentrations of GDC subunits in etiolated tissue suggests that light is not the only factor involved in regulating GDC expression.

There is evidence of cell-specific localisation of GDC in photosynthetic tissue, which indicates a spatial control of gene expression. Mitochondria isolated from different parts of a plant show considerable variation in the rate at which they oxidise glycine. Leaf and stem mitochondria oxidise glycine at much higher rates than do the mitochondria from roots or leaf veins (Gardeström *et al.*, 1980). Immunogold localisation studies in our laboratory indicate a high degree of spatial control. The mitochondria within the photosynthetic mesophyll cells of wheat and pea leaves have a much higher concentration of GDC P protein than do the mitochondria of epidermal or vascular parenchyma cells (Table 2; Tobin *et al.*, 1989). This discrimination between cell types is accentuated during leaf develop-

Table 2. *Immunogold labelling of GDC Protein in pea and wheat leaves*

Species	Cell type	GDC labelling density (Gold dots/μm^2 mitochondrion)
Pea	Mesophyll	116.4 (5.9)
	Vascular	18.4 (2.9)
	Epidermal	15.0 (3.9)
	Guard cell	40.6
	Bundle sheath	49.3
Wheat	Mesophyll	608.7 (31.0)
	Vascular	48.2 (6.8)

Values presented are the means (standard errors in parentheses) of the number of gold particles per unit area (μm^2) of mitochondria. Values have been corrected for background labelling and null serum controls. Vascular cells in all cases were parenchyma (xylem and phloem).
From Tobin *et al.*, 1989.

ment, in a temporal manner. The concentration of P protein within the mesophyll cell mitochondria increases with wheat leaf development, whereas that of the vascular cell mitochondria decreases (Fig. 9). Thus, there are both spatial and temporal controls over GDC expression in photosynthetic tissue.

The observed changes in mitochondrial size and activity, when combined together, serve to illustrate the way in which a single type of organelle changes during leaf development. The activity of cytochrome oxidase and of GDC, and the amount of each of the GDC subunits all increase, per mitochondrion, between 1.5 and 5.5 cm from the wheat leaf base (Table 3). If, however, we take into account the change in mitochondrial size over this region of the leaf, a different picture emerges. Cytochrome oxidase activity remains relatively constant while GDC activity and all four of its subunits increase per unit mitochondrial volume, i.e. the concentration of GDC in the mitochondria increases with leaf development. The relationship between GDC protein concentration and GDC activity has been shown, *in vitro*, to be parabolic (Bourgignon *et al.*, 1988). There appears to be a critical concentration of GDC subunits below which there is negligible activity and this may be one explanation for the observed lag between the development of the subunits and the appearance of GDC activity in wheat leaves (Fig. 7; Rogers *et al.*, 1991).

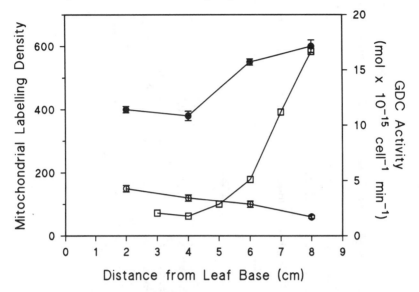

Fig. 9. Developmental changes in the immunogold labelling of GDC in the mitochondria of mesophyll (●—●) and vascular (○—○) parenchyma cells of wheat primary leaves. GDC activity (□—□), in extracts from wheat mesophyll protoplasts, is presented for comparison. Mitochondrial labelling density refers to the intensity of immunogold labelling, i.e. the number of gold particles per μm^2 mitochondrial profile (from Tobin *et al.*, 1989).

The difference in the mitochondrial concentrations of cytochrome oxidase and GDC cannot be attributed simply to the fact that the former is a membrane protein while the latter is localised in the matrix. Glutamate dehydrogenase is a matrix enzyme, yet it develops in the same way as cytochrome oxidase (Fig. 6).

It is interesting to speculate that these changes in mitochondrial size and in enzyme concentration may be indicative of a changing role for the mitochondria during leaf cell development. At early stages of mesophyll cell development, as discussed above, the chloroplasts are immature and the cells are heterotrophic (Baker, 1985). The rate of 'dark respiration' is high and the mitochondria will thus be the main site for cellular ATP synthesis. At later stages of development, with the increasing maturity of the chloroplast comes an increasing autotrophy and, perhaps, a diminishing requirement for mitochondrial ATP synthesis. In return, with the development of photorespiration, there is an increasing demand for mitochondrial glycine oxidation. It remains to be seen whether larger

Table 3. *Changes in the mitochondrial levels of cytochrome oxidase and GDC during development of the wheat primary leaf*

Distance from leaf base (cm)	Cytochrome oxidase activity		GDC protein (subunits)			
	Per mitochondrion[a]	Per mitochondrial volume $(\mu m^{-3})^b$	Per mitochondrion[c]		Per mitochondrial volume $(\mu m^{-3})^d$	
			P, H, T	L	P, H, T	L
0.1	6.58	50.80	0	0.011	0	0.09
1.5	6.08	49.10	0.01	0.012	0.08	0.10
5.5	24.19	76.10	0.08	0.095	0.25	0.30

Data were obtained by dividing the enzyme measurements on whole tissue extracts (as in Figs 6 and 7) by the mitochondrial volumes and numbers estimated as in Table 1. Cytochrome oxidase activity is given in mol $\times 10^{-16}$ cytochrome c oxidised per minute per mitochondrion[a] or per μm^{-3} mitochondrial volume[b]. GDC concentrations are relative, i.e. the relative percentage band intensity (from Fig. 7) is divided by the mitochondrial number[c] or mitochondrial volume[d].
From Tobin *et al.*, 1988; Rogers *et al.*, 1991.

mitochondria are less efficient at synthesising ATP. Based on relative estimates of flux, it would appear that the overriding demand for mitochondrial activity in a mature mesophyll cell is the requirement for glycine oxidation (see Wiskich & Meidan, this volume). Perhaps the mitochondria achieve this at the expense of efficient coupling to ATP synthesis, i.e. the demand for substrate oxidation may be such that the system has to be 'uncoupled' to some extent from adenylate control. This could be achieved by increasing the extent to which the non-phos-phorylating bypasses, such as the rotenone-resistant NADH dehydro-genase and the alternative oxidase, are engaged during leaf development. Alternatively, substrate shuttles (as described by Wiskich & Meidan, Chapter 1, and Heldt & Flügge, this volume) may be increasingly employed to oxidise mitochondrially generated reducing equivalents as the mesophyll cells mature. There is evidence of a change in the capacity for mitochondrial ATP synthesis during greening. The rate of oxidative phosphorylation by mitochondria isolated from etiolated *Avena* leaves increased during the first 2 h of illumination and then decreased with further exposure of the leaves to light (Hampp & Wellburn, 1980). It remains to be seen whether these changes occur during the normal development of mitochondria within light-grown leaves.

Conclusions

There are many changes in the subcellular compartmentation and the activity of organelles during the natural development of a leaf. The bio-energetics of developing leaf cells is, at present, poorly understood. What is clear, however, is that, within a single leaf, there is a transition from heterotrophic to autotrophic metabolism and an associated change in the activities of the chloroplasts and mitochondria. Plant cell organelles are heterogeneous entities which have specific functions within different types of cell and, within a single cell type, at different stages in its development. The chloroplast may well dominate the metabolic activity of a mature mesophyll cell but this is by no means the typical situation, even within a leaf. Only 50% of leaf cells are mesophyll, the remainder are predominantly vascular (40%) and epidermal (10%) (Jellings & Leech, 1982). The mesophyll cells will be at different stages of develop-ment, even with a fully expanded leaf, and not all of them will be photosynthetically active. The metabolic activity of an organelle and its interaction with other organelles is clearly a dynamic situation which is intrinsically linked with the development and differentiation of the cell.

318   A.K. TOBIN AND W.J. ROGERS

Acknowledgements

We thank The Royal Society and the Agricultural and Food Research Council for financial support.

References

Arron, G.P. & Edwards, G.E. (1980). Light-induced development of glycine oxidation by mitochondria from sunflower cotyledons. *Plant Science Letters* **18**, 229–35.

Atkins, C.A., Pate, J.S., Peoples, M.B. & Joy, K.W. (1983). Amino acid transport and metabolism in relation to the nitrogen economy of a legume leaf. *Plant Physiology* **71**, 841–8.

Azcón-Bieto, J., Lambers, H. & Day, D.A. (1983). Respiratory properties of developing bean and pea leaves. *Australian Journal of Plant Physiology* **10**, 237–45.

Badger, M.R. & Lorimer, G.H. (1981). Interaction of sugar phosphate with the catalytic site of ribulose-1,5-bisphosphate carboxylase. *Biochemistry* **20**, 2219–25.

Bagsaw, V., Brown, R. & Yeoman, M.M. (1969). Changes in the mitochondrial complex accompanying callus growth. *Annals of Botany* **33**, 35–44.

Baker, N.R. (1985). Energy transduction during leaf growth. In *Control of Leaf Growth*. Society for Experimental Biology Seminar Series No. 27, ed. N.R. Baker, W.J. Davies & C.K. Ong, pp. 115–34. Cambridge: Cambridge University Press.

Besford, R.T., Withers, A.C. & Ludwig, L.J. (1985). Ribulose bisphosphate carboxylase activity and photosynthesis during leaf development in the tomato. *Journal of Experimental Botany* **36**, 1530–41.

Betsche, T. & Eising, R. (1989). CO_2-fixation, glycolate formation, and enzyme activities of photorespiration and photosynthesis during greening and senescence of sunflower cotyledons. *Journal of Experimental Botany* **40**, 1037–43.

Boffey, S.A., Ellis, J.A., Sellden, G. & Leech, R.M. (1979). Chloroplast division and DNA synthesis in light-grown wheat leaves. *Plant Physiology* **64**, 502–5.

Boffey, S.A., Sellden, G. & Leech, R.M. (1980). Influence of cell age on chlorophyll formation in light-grown and etiolated wheat seedlings. *Plant Physiology* **65**, 680–4.

Bourgignon, J., Neuburger, M. & Douce, R. (1988). Resolution and characterisation of the glycine cleavage reaction in pea leaf mitochondria. Properties of the forward reaction catalysed by glycine decarboxylase and serine hydroxymethyl aminotransferase. *Biochemical Journal* **255**, 169–78.

Canovas, F.M., Avilla, C., Botella, J.R., Valpuesta, A. & Nunez de

Castro, I. (1986). Effect of light-dark transition on glutamine synthetase activity in tomato leaves. *Physiologia Plantarum* 66, 648–52.

Catsky, J. & Ticha, I. (1980). Ontogenetic changes in the internal limitations to bean-leaf photosynthesis. 5. Photosynthetic and photorespiration rates and conductances for CO_2 transfer as affected by irradiance. *Photosynthetica* 14, 392–400.

Catsky, J., Ticha, I. & Solorova, J. (1976). Ontogenetic changes in the internal limitations to bean-leaf photosynthesis. 1. Carbon dioxide exchange and conductances for carbon dioxide transfer. *Photosynthetica* 10, 394–402.

Croxdale, J.G. (1983). Quantitative measurements of hexokinase activity in the shoot apical meristem, leaf primordia, and leaf tissues of *Dianthus chinensis* L. *Plant Physiology* 73, 66–70.

Croxdale, J.G. & Outlaw, W.H. (1983). Glucose-6-phosphate dehydrogenase activity in the shoot apical meristem, leaf primordia, and leaf tissues of *Dianthus chinensis* L. *Planta* 157, 289–97.

Croxdale, J.G. & Pappas, T. (1987). Activity of glyceraldehyde-3-phosphate dehydrogenase-NADP in developing leaves of light-grown *Dianthus chinensis* L. *Plant Physiology* 84, 1427–30.

Croxdale, J.G. & Vanderveer, P.J. (1986). RuBPCase activity in relation to leaf development. *Plant Physiology* 80, S-701.

Dale, J.E. (1972). Growth and photosynthesis in the first leaf of barley. The effect of time and application of nitrogen. *Annals of Botany* 36, 967–79.

Dale, J.E. (1985). The carbon relations of the developing leaf. In *Control of Leaf Growth*. Society for Experimental Biology Seminar Series, No. 27, ed. N.R. Baker, W.J. Davies & C.K. Ong, pp. 135–53. Cambridge: Cambridge University Press.

Day, D.A., Neuburger, M. & Douce, R. (1985). Biochemical characterization of chlorophyll-free mitochondria from pea leaves. *Australian Journal of Plant Physiology* 12, 219–28.

Dean, C. & Leech, R.M. (1982) Genome expression during normal leaf development. I. Cellular and chloroplast numbers and DNA, RNA and protein levels in tissues of different ages within a seven-day old wheat leaf. *Plant Physiology* 69, 904–10.

Denton, R.M., Randle, P.J., Bridges, B.J., Cooper, R.H., Kerbey, A.L., Pask, H.T., Severson, D.L., Stansbie, D. & Whitehouse, S. (1975). Regulation of mammalian pyruvate dehydrogenase. *Molecular and Cellular Biochemistry* 9, 27–53.

Dickmann, D.I., Gjerstad, D.H. & Gordon, J.C. (1975). Developmental patterns of CO_2 exchange, diffusion resistance and protein synthesis in leaves of *Populus* × *Euramericana*. In *Environmental and Biological Control of Photosynthesis*, ed. R. Marcelle, pp. 171–81. The Hague: Dr W. Junk.

Ducet, G. & Rosenberg, A.J. (1962). Leaf respiration. *Annual Review of Plant Physiology* 13, 171–200.

Ellis, J.R., Jellings, A.J. & Leech, R.M. (1983). Nuclear DNA content and the control of chloroplast replication in light-grown wheat leaves. *Planta* 157, 376–80.

Ellis, J.R. & Leech, R.M. (1985). Cell size and chloroplast size in relation to chloroplast replication in light-grown wheat leaves. *Planta* 165, 120–5.

Furuta, S., Shindo, Y. & Hashimoto, T. (1977). Purification and properties of pigeon breast muscle alpha-keto acid dehydrogenase complexes. *Journal of Biochemistry* 81, 1839–47.

Gardeström, P., Bergman, A. & Ericson, I. (1980). Oxidation of glycine via the respiratory chain in mitochondria prepared from different parts of spinach. *Plant Physiology* 65, 389–91.

Guiz, C.D., Hirel, B., Shedlovsky, G. & Gadal, P. (1979). Occurrence and influence of light on the relative proportions of two glutamine synthetases in rice leaves. *Plant Science Letters* 15, 271–7.

Hampp, R. & Wellburn, A.R. (1980). Translocation and phosphorylation of adenine nucleotides by mitochondria and plastids during greening. *Zeitschrift für Pflanzenphysiologie* 98, 289–303.

Hecht, U., Oelmuller, R., Schmidt, S. & Mohr, H. (1988). Action of light, nitrate and ammonium on the levels of NADH- and ferredoxin-dependent glutamate synthases in the cotyledons of mustard seedlings. *Planta* 175, 130–8.

Hoffmann, P. & Schwarz, Zs. (1975). Characterisation of regulative interactions between the autotrophic and heterotrophic system in *Phaseolus vulgaris* and *Triticum aestivum* seedlings. In *Environmental and Biological Control of Photosynthesis*, ed. R. Marcelle, pp. 191–200. The Hague: Dr W. Junk.

Homann, P.H. (1975). Carbon dioxide exchange of young tobacco leaves in light and darkness. In *Environmental and Biological Control of Photosynthesis*, ed. R. Marcelle, pp. 183–90. The Hague: Dr W. Junk.

Hondred, D., Wadle, D.-M., Titus, D.E. & Becker, W.M. (1987). Light-stimulated accumulation of the peroxisomal enzymes hydroxypyruvate reductase and serine-glyoxylate aminotransferase and their translatable mRNAs in cotyledons of cucumber seedlings. *Plant Molecular Biology* 9, 259–75.

Jellings, A.J. & Leech, R.M. (1982). The importance of quantitative anatomy on the interpretation of whole leaf biochemistry in species of *Triticum*, *Hordeum* and *Avena*. *New Phytologist* 92, 39–48.

Joy, K.W. (1988). Ammonia, glutamine and asparagine: a carbon–nitrogen interface. *Canadian Journal of Botany* 66, 2103–9.

Kacser, H. (1987). Control of metabolism. In *The Biochemistry of Plants*, Vol. 11, ed. D.D. Davies, pp. 39–67. Orlando, FL: Academic Press.

Kacser, H. & Burns, J.A. (1973). The control of flux. In *Rate Control of*

Biological Processes. Symposia of the Society for Experimental Biology No. 27, pp. 65–107. Cambridge: Cambridge University Press.

Kacser, H. & Burns, J.A. (1979). Molecular democracy: who shares the controls? *Biochemical Society Transactions* **7**, 1149–60.

Kacser, H. & Porteus, J.W. (1987). Control of metabolism: what do we have to measure? *Trends in Biochemical Science* **12**, 5–14.

Kansara, M.S., Bharsali, J.P. & Srivastava, S.K. (1989). Regulation of glutamine synthetase by phytochrome in pea terminal buds. *Journal of Plant Physiology* **134**, 294–7.

Kennedy, R.A. & Johnson, D. (1981). Changes in photosynthetic characteristics during leaf development in apple. *Photosynthesis Research* **2**, 213–23.

Keys, A.J. (1986). Rubisco: its role in photorespiration. *Philosophical Transactions of the Royal Society of London B* **313**, 325–36.

Kidd, F., Briggs, G.E. & West, C. (1921). A quantitative analysis of the growth of *Helianthus annuua*. I. The respiration of the plant and its parts throughout the life cycle. *Proceedings of the Royal Society B* **92**, 368–84.

Leech, R.M. (1984). Chloroplast development in Angiosperms: current knowledge and future prospects. In *Topics in Photosynthesis*, Vol. 5, Chloroplast Biogenesis, ed. N.R. Baker & J. Barber, pp. 1–21. Amsterdam: Elsevier.

Leech, R.M. (1985). The synthesis of cellular components in leaves. In *Control of Leaf Growth*. Society for Experimental Biology Seminar Series, No. 27, ed. N.R. Baker, W.J. Davies & C.K. Ong, pp. 93–114. Cambridge: Cambridge University Press.

MacDonald, I.R. & De Kok, P.C. (1958). The stimulation of leaf respiration by respiratory inhibitors. *Physiologia Plantarum* **11**, 464–77.

McManus, I.R. & Cohn, M.L. (1975). Properties of multiple molecular forms of lipoamide dehydrogenase. In *Isoenzymes*, Vol. I. Molecular Structure, ed. C.L. Markert, pp. 621–36. New York: Academic Press.

Maksymowych, R. (1973). *Analysis of Leaf Development*. Cambridge: Cambridge University Press.

Mann, A.F., Fentem, P.A. & Stewart, G.R. (1979). Identification of two forms of glutamine synthetase in barley (*Hordeum vulgare*). *Biochemical and Biophysical Research Communications* **888**, 515–21.

Marsh, P.B. & Goddard, D.R. (1939). Respiration and fermentation in the carrot, *Daucus carota*. I. Respiration. *American Journal of Botany* **26**, 724–8.

Nicklisch, A., Tsenova, E.N. & Hofmann, P. (1977). Über die Bedeutung von Glutamasynthase und Glutamatdehydrogenase für die N-Assimilation im Weizenprimarblatt. *Biochemie und Physiologie der Pflanzen* **171**, 375–84.

Nishimura, M., Bhusawang, P., Strzalka, K. & Akazawa, T. (1982).

Developmental formation of glutamine synthetase in greening pumpkin cotyledons and its subcellular localization. *Plant Physiology* **70**, 353–6.

Nivison, H.T. & Stocking, C.R. (1983). Ribulose bisphosphate carboxylase synthesis in barley leaves. A developmental approach to the question of coordinated subunit synthesis. *Plant Physiology* **73**, 906–11.

Ostareck, D. & Lieckfeldt, E. (1989). Ribulose-1,5 bisphosphate carboxylase/oxygenase and gas-exchange characteristics during the development of primary wheat leaves (*Triticum aestivum*). *Photosynthetica* **23**, 486–93.

Owen, J.H., Laybourn-Parry, J.E.M. & Wellburn, A.R. (1986). Leaf respiration during early plastidogenesis in light-grown barley seedlings. *Physiologia Plantarum* **68**, 100–6.

Parry, M.A.J., Schmidt, C.N.G., Cornelius, M.J., Keys, A.J., Millard, B.N. & Gutteridge, S. (1985). Stimulation of ribulose bisphosphate carboxylase activity by inorganic orthophosphate without an increase in bound activating CO_2: co-operativity between subunits of the enzyme. *Journal of Experimental Botany* **36**, 1396–404.

Rogers, W.J., Jordan, B.R., Rawsthorne, S. & Tobin, A.K. (1991). Changes to the stoichiometry of glycine decarboxylase subunits during wheat (*Triticum aestivum* L.) and pea (*Pisum sativum* L.) leaf development. *Plant Physiology* **96**, 952–6.

Salin, M.L. & Homann, P.H. (1971). Changes of photorespiratory activity with leaf age. *Plant Physiology* **48**, 193–6.

Sestak, Z. (ed.) (1985). *Photosynthesis during Leaf Development*. Dordrecht: Dr W. Junk.

Sharkey, T.D. (1988). Estimating the rate of photorespiration in leaves. *Physiologia Plantarum* **73**, 147–52.

Sibley, M.H. & Anderson, L.E. (1989). Light/dark modulation of enzyme activity in developing barley leaves. *Plant Physiology* **91**, 1620–4.

Smillie, R.M. (1962). Photosynthetic and respiratory activities of growing pea leaves. *Plant Physiology* **37**, 716–21.

Suzuki, A., Vidal, J. & Gadal, P. (1982). Glutamate synthase isoforms in rice. Immunological studies of enzymes in green leaf, etiolated leaf and root tissues. *Plant Physiology* **70**, 827–32.

Suzuki, S., Nakamoto, H., Ku, M.S.B. & Edwards, G.E. (1987). Influence of leaf age on photosynthesis, enzyme activity, and metabolite levels in wheat. *Plant Physiology* **84**, 1244–8.

Tobin, A.K., Ridley, S.M. & Stewart, G.R. (1985). Changes in the activities of chloroplast and cytosolic isoenzymes of glutamine synthetase during normal leaf growth and plastid development in wheat. *Planta* **163**, 544–8.

Tobin, A.K., Sumar, N., Patel, M., Moore, A.L. & Stewart, G.R.

(1988). Development of photorespiration during chloroplast biogenesis in wheat leaves. *Journal of Experimental Botany* **39**, 833–43.

Tobin, A.K., Thorpe, J.R., Hylton, C.M. & Rawsthorne, S. (1989). Spatial and temporal influences on the cell-specific distribution of glycine decarboxylase in leaves of wheat (*Triticum aestivum* L.) and pea (*Pisum sativum* L.). *Plant Physiology* **91**, 1219–25.

Topping, J.F. & Leaver, C.J. (1990). Mitochondrial gene expression during wheat leaf development. *Planta* **182**, 399–407.

Tribe, M.A. & Ashurst, E.A. (1972). Biochemical and structural variations in the flight muscle mitochondria of ageing blowflies, *Calliphora erythrocephala*. *Cell Science* **10**, 443–69.

Viro, M. & Kloppstech, K. (1980). Differential expression of the genes for ribulose-1,5-bisphosphate carboxylase and light-harvesting chlorophyll a/b protein in the developing barley leaf. *Planta* **150**, 41–5.

Walker, J.L. & Oliver, D.J. (1986a). Light-induced increases in the glycine decarboxylase multienzyme complex from pea leaf mitochondria. *Archives of Biochemistry and Biophysics* **248**, 626–38.

Walker, J.L. & Oliver, D.J. (1986b). Glycine decarboxylase multienzyme complex. Purification and partial characterisation from pea leaf mitochondria. *Journal of Biological Chemistry* **261**, 2214–21.

Wallsgrove, R.M., Lea, P.J. & Miflin, B.J. (1982). The development of NAD(P)H-dependent and ferredoxin-dependent glutamate synthase in greening barley and pea leaves. *Planta* **154**, 473–6.

Webber, A.N., Baker, N.R., Platt-Aloia, K.A. & Thomson, W.W. (1984). Appearance of a state 1-state 2 transition during chloroplast development in the wheat leaf: energetic and structural considerations. *Physiologia Plantarum* **60**, 171–9.

Weibel, E.R. (1973). Stereological techniques for electron microscopic morphometry. In *Principles and Techniques of Electron Microscopy*, Vol. 3, ed. M.A. Hayat, pp. 237–96. New York: Van Nostrand Reinhold.

Zielinski, R.E., Werneke, J.M. & Jenkins, M.E. (1989). Coordinate expression of Rubisco activase and Rubisco during barley leaf cell development. *Plant Physiology* **90**, 516–21.

Zima, J. & Sestak, Z. (1979). Photosynthetic characteristics during ontogenesis of leaves. 4. Carbon fixation pathways, their enzymes and products. *Photosynthetica* **13**, 83–106.

Index

328 INDEX

glutamine synthetase (*cont.*)
  expression of 88
  in C3–C4 plants 129, 130
  inhibition of 304
  mutant 83, 85
glycine
  intracellular diffusion of, in C3–C4
    plants 129
  oxidation in C3–C4 plants 133
  oxidation in mitochondria 7, 9, 11, 12,
    15, 37, 204
  transport in mitochondria 37
glycine decarboxylase complex (GDC) 9,
    11, 12, 80, 87, 88
  activity during leaf development 308
  and mitochondrial size 115, 117, 128
  concentration in mitochondria 307,
    313–14
  differential distribution of, in C3 117,
    313–14
  differential distribution of, In C3–C4
    88, 117, 128
  immunogold localisation of 11, 117,
    313–14
  in C4 plants 88, 102
  light-regulation of expression 309,
    312–13
  subunit levels during leaf
    development 310–12
glycolate oxidase 87
  activity in C3 plants 116
  activity in C3–C4 plants 116
  activity during leaf development 308
glycolysis 6, 7, 64
  in CAM plants 114, 146, 147
  during leaf development 309
glyoxylate aminotransferase
  activity in C3 plants 116
  activity in C3–C4 plants 116, 130
guard cells
  ABA effect on 265
  K+ content of 267
  K+ currents in 266
  K+ fluxes in 267
  K+ -induced increase in PEP
    carboxylase of 266
  K+-induced swelling of 267, 268
  model of swelling of 274–5
  oxidative phosphorylation in 265
  PEP carboxylase in 266, 272
  photophosphorylation in 265
  plasmalemma ATPase in 265
  pool sizes of malate in 271, 272
  protoplasts of 266
  source of ATP in 265, 272–274
  synthesis of PEP in 266

synthesis of malate in 266
uptake of K+ in 265

H+-ATPase
  in tonoplast of CAM plants 150, 151
  in tonoplast of C3 plants 170
  properties of 170
H+ conductance 14
  of the inner mitochondrial
    membrane 189, 192
hydroxypyruvate
  dehydrogenase in C3 plants 116
  dehydrogenase in C3–C4 plants 116
  reductase 87, 308
  reductase-deficient mutant 83, 84
  reduction in peroxisomes 7, 30, 33, 84,
    133, 160

inositol (1,4,5) trisphosphate (InsP3) 171
  -gated Ca2+ channel 172, 179, 181
  receptor in tonoplast 180–1
invertase 27
  in vacuoles of CAM plants 151
isocitrate dehydrogenase (IDH) 86
  NAD-dependent 86
  NADP-dependent 86

leaf development
  changes in mesophyll cell size
    during 294–5
  in C3–C4 plants 115
  in dicotyledons 293–4
  in grasses 293
  organelle interactions during 293–317
lipid(s) (see also fatty acids)
  metabolism 211–28
lipoamide dehydrogenase 311

malate
  accumulation in CAM plants 144
  /aspartate shuttle 3
  content in CAM vacuoles 150, 153,
    154
  decarboxylation in CAM plants 154
  efflux from vacuole in CAM plants 144
  oxidation in mitochondria 9, 10, 11,
    148
  pool sizes in guard cell organelles
    271–4
  stimulation of glycine oxidation by 11
  synthesis in guard cells 266, 267
  transport in CAM mitochondria 160
  transport in CAM vacuoles 145, 146
malate dehydrogenase 6, 7, 9, 11, 12
  in C3–C4 plants 120
  in CAM plants 145